Microorganisms

Edited by Miroslav Blumenberg, Mona Shaaban, Abdelaziz Elgaml

Published in London, United Kingdom

IntechOpen

Supporting open minds since 2005

Microorganisms
http://dx.doi.org/10.5772/intechopen.82960
Edited by Miroslav Blumenberg, Mona Shaaban, Abdelaziz Elgaml

Contributors
Daniel Acosta-Avalos, Fernanda Abreu, Ivan Kushkevych, Tamaki Mizuno, Anusuya Debnath, Shin-Ichi Miyoshi, Hassaan Bin-Asif, Syed Abid Ali, Dele Ogunremi, Ruimin Gao, Linru Wang, Mona Shaaban, Jose Maria Da Luz, Prajakta P. Kamble, Suresh Suryawanshi, Maheshkumar Kore, Nahid Irani, Jyoti Jadhav, Yasmin Attar, Marina Nisnevitch, Faina Nakonechny, Michelle J. Rivera-Rivera, Elvira Cuevas, Qomarudin Helmy, Edwan Kardena, Sri Gustiani, Barachel Odaro-Junior Umukoro, Arturo Cano-Flores, Francisco Javier Díaz-García, Saúl Flores-Medina, Mohammed El-Mowafy, Abdelaziz Elgaml, Diana Mercedes Soriano-Becerril, Marliane de Cássia Soares da Silva, Leonardo Ferreira dos Santos, Maria Catarina Megumi Kasuya, Javier Gómez, Iker S. Escalona-Torres, Benjamín Velasco-Bejarano

First published in London, United Kingdom, 2020 by IntechOpen
IntechOpen is the global imprint of INTECHOPEN LIMITED, registered in England and Wales, registration number: 11086078, 7th floor, 10 Lower Thames Street, London, EC3R 6AF, United Kingdom
Printed in Croatia

British Library Cataloguing-in-Publication Data
A catalogue record for this book is available from the British Library

Additional hard and PDF copies can be obtained from orders@intechopen.com

Microorganisms
Edited by Miroslav Blumenberg, Mona Shaaban, Abdelaziz Elgaml
p. cm.
Print ISBN 978-1-83880-187-8
Online ISBN 978-1-83880-188-5
eBook (PDF) ISBN 978-1-83880-363-6

For EU product safety concerns:
IN TECH d.o.o., Prolaz Marije Krucifikse Kozulić 3, 51000 Rijeka, Croatia, info@intechopen.com or visit our website at intechopen.com.

Meet the editors

Miroslav Blumenberg, PhD, was born in Subotica and received his BSc in Belgrade, Yugoslavia. He completed his PhD at MIT in Organic Chemistry; he followed up his PhD with two postdoctoral study periods at Stanford University. Since 1983, he has been a faculty member of the RO Perelman Department of Dermatology, NYU School of Medicine, where he is a codirector of a training grant in cutaneous biology. Dr. Blumenberg's research is focused on the epidermis, expression of keratin genes, transcription profiling, keratinocyte differentiation, inflammatory diseases and cancers, and most recently the effects of the microbiome on skin. He has published more than 100 peer-reviewed research articles and graduated numerous PhD and postdoctoral students. Dr. Blumenberg lives in New York, USA, with his wife and two children.

Dr. Mona Shaaban received her PhD at Mansoura University/Egypt through a joint scholarship with the University of Wisconsin Madison, USA. Currently, she is working as an associate professor in the Microbiology and Immunology Department, Mansoura University, Egypt. She has more than 40 publications in international journals. She is interested in the study of bacterial virulence and resistance. She is also working on the inhibition of quorum sensing as an antipathogenic approach to counter bacterial resistance and virulence. She is a reviewer of *PLOS One*, scientific reports, *Frontiers in Microbiology*, *Microbial Pathogenesis*, *Microbiology Research International*, and other impacted journals. She is also an editor of perpetual journals.

Abdelaziz Ahmed Elgaml, PhD, MSc, BSc, is an associate professor of microbiology and immunology at the Microbiology and Immunology Department, Faculty of Pharmacy, Mansoura University, Mansoura, Egypt. He received his PhD degree at the Environmental and Applied Microbiology Laboratory at the Graduate School of Medicine, Dentistry and Pharmaceutical Sciences, Okayama University, Japan. Moreover, he received his MSc and BSc degrees at the Microbiology and Immunology Department, Faculty of Pharmacy, Mansoura University, Mansoura, Egypt. His main research interests are bacteriology, Gram-negative bacteria, Gram-positive bacteria, bacterial toxins, vibrios, vibrosis, bacterial resistance, bacterial virulence factors, regulation systems of virulence factor production, quorum sensing, microbial biodiversity, phylogeny, virology, hepatitis B virus, hepatitis C virus, and complement systems.

Contents

Preface

The recent breakthrough in microbial studies has applied next-generation sequencing (NGS), a massive omics analysis, to the composition and structure of microbial communities. NGS can identify microbes without the need for their cultivation. Their mere presence can be ascertained and often quantitated, and even their metabolic capabilities of microbial constituents predicted. This breakthrough led to an explosive growth in research on microbes.

Many important advances have been made in human health-related studies. Indeed, gut microbial communities have been extensively analyzed and differences between healthy and diseased microbiomes have been determined. Studies of the effects of changes of diet, of antibiotic treatments, and of probiotics on the gut microbiome have been published. It was found that the human gut microbiome varies significantly from individual to individual, and longitudinal studies have shown the major effects of antibiotics as well as the time course of return to a "normal," healthy microbiome.

Ethnic and geographical influences have been analyzed. While most studies focused on individuals in the Western world and their corresponding diet, comparisons with less accessible communities have been very informative. In particular, analyses of indigenous communities, never exposed to antibiotics, have illuminated the effects of their wide use. In other studies it was found, for example, that the use of antibiotics in early childhood can have life-long consequences, perhaps predisposing to allergic and atopic sequelae.

Similar microbiome studies have been performed on human skin, oral cavities, and genitals. Different sites of the human body carry different niche-specific microbiota. Dry, moist, sebaceous, UV-exposed, and protected regions of skin all have different microbiota. The immense interpersonal variation precludes describing a "healthy" skin microbiome; however, one's skin microbiomes tend to be stable over time. Interestingly, the effects of hygiene products, soaps and deodorants, are apparently underwhelming and transient. Human skin and the oral cavity contain very rich and varied microbiomes with myriad species. In contrast, vaginal microbiomes in the posterior fornix tend to have a limited number of species, nearly monoclonal lactobacilli environment.

Specific research has been devoted to human pathogens, their mechanisms of causing disease, and the potentials for their management and alleviation. For example, eczema, atopic dermatitis, is associated with overgrowth of *Staphylococcus aureus*. In contrast, psoriasis, another inflammatory skin disease, is not specifically associated with a particular microbiome.

Interestingly, the richness and interpersonal differences in human microbiomes are equalized by the metabolic capabilities of such microbiomes. In other words, while the compositions of the microbiomes vary greatly, the overall metabolic processes, synthesis, and degradation of nutrients and other natural products do not vary much; different assemblies of microorganism can perform similar tasks.

The increased understanding of microbial communities led to attempts to use microbes as antagonists of pathogens, i.e. as treatments. For example, fecal transplants are used in the treatment of *Clostridium difficile* infections, a serious nosocomial challenge. Identification of gut microbial influence on mood and brain function may yield corresponding neurological therapies. The gut microbiome is known to influence the metabolism of various drugs and chemicals, e.g. cytostatic medications in cancer treatment, making some patients less sensitive to chemotherapy.

The microbiome studies of natural habitats, terrestrial and aquatic, have benefited from NGS methodology. Such studies revealed that large-scale chemical transformations can occur in oceans and seas due to microorganisms. Similarly, nutrient cycles in microbiota in various tropical habitats have been described.

The remarkable growth of microbial DNA sequence databases has provided a background for novel uses of microbes in industrial processes, either for the synthesis of important compounds or for the degradation and handling of waste. Microbial production of specific enzymes for industrial uses is an interesting option.

In this volume, chapters dealing with the cutting-edge research in all these fields are presented.

Miroslav Blumenberg
New York University Langone Medical Center,
USA

Mona Shaaban and Abdelaziz Elgaml
Mansoura University,
Egypt

Section 1

Bacterial Physiology

Chapter 1

Biology and Physics of Magnetotactic Bacteria

Fernanda Abreu and Daniel Acosta-Avalos

Abstract

Magnetotactic bacteria are able to align their swimming direction to the geomagnetic field lines because they possess a magnetic moment. These bacteria biomineralize magnetic nanoparticles, magnetite or greigite, inside a membrane. The membrane + nanoparticle set is known as magnetosome and intracellular magnetosomes are disposed in a linear chain. Cytoskeleton-like filaments are resposible for the mechanical stability of this chain. The genes responsible for the magnetosome membrane and for the cytoskeleton proteins have been largely studied: the mam genes. The magnetosome chain also confers to the bacterial body a magnetic moment that can be measured through different physical techniques. Because of their response to magnetic field inversions, magnetotactic bacteria are good models to study bacterial motion. Theoretical and experimental studies show that magnetotactic bacteria swim following a trajectory similar to cylindrical helix. Magnetotactic microorganisms have been observed avoiding regions with UV or violet-blue light of high intensity. If the intensity is lower, magnetotactic microorganims show photokinesis, increasing their velocity in the presence of red light and decreasing their velocity in the presence of green light, both relative to the velocity with blue light.

Keywords: magnetotactic bacteria, magnetosomes, magnetic moment, mam genes, magnetotaxis

1. Introduction

Bacteria are one of the simplest organisms found in nature. They are distinguished from eukaryotes and superior organisms because their genetic material is not contained in a nucleus but is free in the cytoplasm. However, despite their relative simplicity, bacteria inhabit Earth for longer than many other organisms and constitute the most abundant type of cell on our planet [1]. Magnetotactic bacteria (MTB) are microorganisms that biomineralize magnetic nanoparticles inside their cytoplasm. These magnetic nanoparticles are involved by a lipidic membrane, and each "membrane + magnetic nanoparticle" set is known as magnetosome [2]. Magnetosomes are arranged in linear chains in the cytoplasm, conferring a magnetic moment to the bacterial body, being able to interact with the geomagnetic field to orient its navigational direction to the geomagnetic field lines. This response is known as magnetotaxis, resulting from the magnetic torque among the geomagnetic field and the magnetic moment of the magnetosome chain, and for that reason, MTB are described as "living compasses." MTB were discovered

 IntechOpen

independently by Salvatore Bellini in 1963 and Richard Blakemore in 1975 [3]. The first to observe MTB was Salvatore Bellini, an Italian physician from Pavia, Italy. In 1958, physicians from Pavia were asked to analyze the quality of water for human consume. Bellini was part of the team that studied the water quality, and he observed in water samples some bacteria that consistently accumulate in one side of water drops. After trying different stimuli, he discovered that they were affected by magnets. He called that bacteria as magnetosensitive. The first published observation of MTB was done in 1975 by Richard Blakemore in Massachusetts, USA [4]. His discovery was accidental, because his goal was to isolate *Spirochaeta plicatilis* from marine marsh muds. During his observations, he noted microorganisms migrating to one end of the drop of the mud on the microscope slide, and discover that the presence of magnets altered their swimming direction. He called magnetotaxis as the tactic response to magnetic fields.

2. Biology of magnetotactic bacteria

2.1 Cell biology

MTB comprise a diverse group of prokaryotes that share the ability to synthesize magnetosomes, which are composed by a magnetic nanocrystal, magnetite (Fe_3O_4), or greigite (Fe_3S_4), enclosed by a lipid bilayer (**Figure 1**; [2]). Thus, MTB has no taxonomic meaning regarding phylogeny, morphology, and physiology. The morphology of cultured and uncultured MTB described until now are cocci, spirilla, rods, ovoids, vibrios, and multicellular spherical/ellipsoidal forms [5]. In all morphotypes, magnetosomes are arranged in one or multiple chains along the major axis of the cell, imparting the cell a magnetic moment, as mentioned previously [2]. However, in some uncultured MTB, apparently disorganized chains have already been observed [6, 7]. Besides the magnetosomes, other common features are the Gram-negative cell wall structure, motility by flagella and lipid, polyphosphate and sulfur inclusions [2].

MTB present microaerophilic or anaerobic metabolism and all inhabit aquatic environments characterized by vertical chemical stratification [2]. When observing a drop of environmental sample containing MTB on a light microscope with a magnet next to the slide, MTB tend to accumulate in one of the borders of the drop, which correspond to a magnetic pole, North or South. In the Southern Hemisphere, when MTB are being observed they usually accumulate at the border of the drop of sample corresponding to the magnet's North magnetic pole, swimming to Geomagnetic South pole indicated by a compass. Therefore, these bacteria are

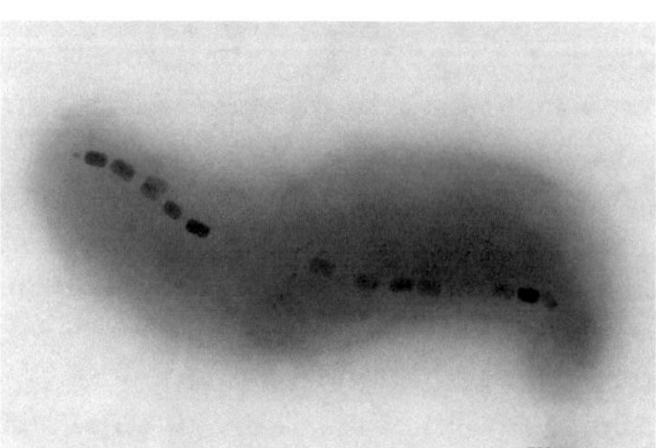

Figure 1.
Transmission electron microscopy of the magnetotactic bacterium Magnetovibrio blakemorei *strain MV-1 showing a single chain of prismatic magnetite magnetosomes.*

called South-seeking. In the Northern Hemisphere, the opposite situation occurs and MTB presenting this behavior are called North-seeking.

The study of MTB movement in an oxygen gradient showed that MTB change flagellar rotation and, consequently, the direction of movement depending on the oxygen concentration, migrating to the optimal conditions [8]. These observations were used to infer MTB dislocation along the vertical chemical gradient in the environment, which is based on the vertical component of the magnetic field and depends on the cellular state regarding oxygen. In the upper layers of the chemically stratified environment, where oxygen is abundant and higher than the optimal for MTB growth, the bacterium is on the oxidized state and rotate flagella to migrate downward, where the environment is more reduced. While in reducing conditions, in which oxygen is not abundant and concentrations are lower than that required for MTB growth, bacteria rotate the flagella in the opposite direction to reach upper layers of the gradient with optimal oxygen concentrations. Therefore, the presence of specific cell structures in MTB, as the magnetosome chain, flagella, and storage inclusions, represent adaptive advantages for dislocation across the chemical gradient to explore resources in the environment. For example, during the day, the oxygen gradient changes among the stratified layers of the environment and microorganisms dislocate across these layers to reach regions with suitable conditions for survival and growth. Because MTB are microaerophilic and/or capable of anaerobic respiration, which means that they are sensitive to high oxygen concentrations, an efficient mechanism to orient and migrate in the environment guarantees species survival.

Only a few species of MTB have been isolated in axenic cultures [9], and fewer type strains are available in cell line repositories. Many uncultured MTB species have been characterized from environmental samples, because it is possible to purify these cells based on their response to an applied magnetic field using a magnet [10]. **Figure 2** shows examples of cultured and uncultured species of MTB and their characteristics according to their phylogenetic affiliation. Note that MTB are spread among different phyla in Bacteria domain and that greigite magnetosomes are only synthesized by MTB belonging to Deltaproteobacteria class.

Magnetospirillum species, which include *M. gryphiswaldense* strain MSR-1, *M. magneticum* strain AMB-1, and *M. magnetotacticum* strain MS-1, among other strains, are spirilla with flagella at each pole of the cell and represent the most characterized MTB. Information about the biomineralization process and magnetosome organization within the cell is mainly based on species belonging to this genus [11]. Cryoelectron tomography studies using *M. magneticum* strain AMB-1 have shown that the magnetosome vesicle is a result of the cytoplasmic membrane invagination, which occurs before the synthesis of the magnetite nanocrystal [11, 12]; forming the magnetosome membrane vesicle in which proteins related to the biomineralization will be anchored. These proteins that are involved in all steps of the magnetosome formation are anchored to this invaginated portion of the membrane and will participate in the process by recruiting other proteins that integrate the process, for example, iron transport, crystal nucleation and growth, size and shape control, and organization of the magnetosomes [11, 13].

According to studies performed in *M. gryphiswaldense* strain MSR-1, mature magnetite magnetosomes are found already organized in a chain within the cytoplasm of the cell 15 min after formation has started [14]. All *Magnetospirillum* species produce a single chain of cuboctahedral magnetite magnetosomes that are 40–45 nm in size [10]. Other MTB species are capable of synthesizing magnetite magnetosomes with cuboctahedral, prismatic, or anisotropic shapes [2] and/or greigite magnetosomes, which are usually classified as irregular. Usually, a magnetotactic bacterium species is capable of producing magnetosomes with one

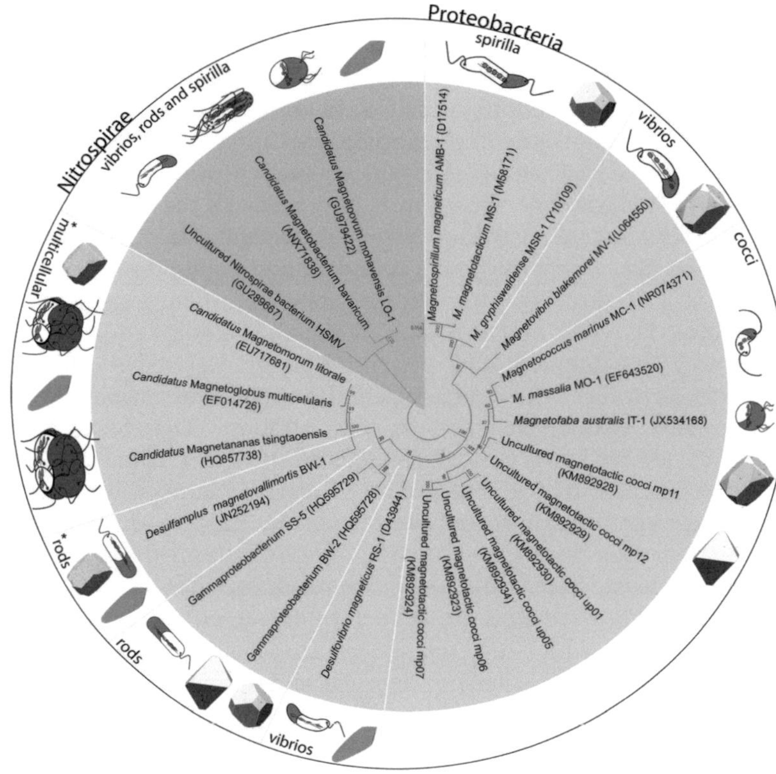

Figure 2.
Phylogenetic tree based on the 16S rRNA gene showing the distribution of some cultured and uncultured MTB in Nitrospirae and Proteobacteria phylum of Bacteria domain. White lines separate morphotypes of MTB in Proteobacteria phylum, showing the distribution of spirilla, vibrios, cocci, including ovoid and fava-*like cells, rods, and multicellular microorganisms. The shape of magnetosomes found in each morphotypes also displayed next to the cells. However, it is important to remember that each MTB species is usually capable of producing only one type of magnetosome (composition and shape). Morphotypes labeled with an asterisk (*) are the only capable of producing greigite magnetosomes; some of them, for example,* Desulfamplus magnetovallimortis *strain BW-1 produces both types of magnetosomes, irregular greigite and bullet-shaped magnetite magnetosomes.*

mineral composition and regular size and shape [15]. Rarely, a magnetotactic bacterium is capable of producing both magnetite and greigite magnetosomes; when it occurs, these magnetosomes with different composition and shape are arranged in the same chain(s) [16, 17]. Although differences have been observed in the formation of magnetosomes in MTB species [12, 18], the process described for bacteria belonging to *Magnetospirillum* genus is considered the model of magnetite biomineralization in MTB.

In the environment, magnetotactic cocci are the most abundant morphotype of MTB and present high phylogenetic diversity and variety of size, shape, and organization of magnetosomes [10, 19]. For example, magnetotactic cocci have been found in marine sediments from Antarctica, suggesting the existence of psychrophilic MTB [7]. Interestingly, these samples presented at least three types of magnetotactic cocci based on the magnetosomes crystal size, shape, and organization [7]. Ultrastructure characterization of cultured magnetotactic cocci showed that these cells present two bundles of flagella and can achieve speed of approximately 300 μm/s, which is extremely high if we consider that the bacterium has nearly 1 μm. Each flagellar apparatus of *Magnetococcus massalia* strain MO-1 is formed by seven flagellar filaments surrounded by a sheath that might interact with the bundle of flagella to decrease the friction of the high-speed rotation of individual flagella and promote efficient swimming at high speed [20]. One of the most peculiar morphotype of MTB are the multicellular forms, named magnetotactic multicellular prokaryotes (MMP), which can be divided into spherical and ellipsoidal. MMP are formed by Gram-negative magnetotactic cells organized in a sphere that swims as a unit [21]. Cellular organization in MMP is not random and represents the best configuration to optimize the magnetic response of the microorganism [22]. These microorganisms present an exclusive multicellular life cycle in which cells of the microorganisms grow, divide, and rearrange before splitting into two identical multicellular microorganisms [23]. Individual cells of this type of MTB have never been observed and viability assays suggest that when a cell disaggregates from the multicellular structures, it does not remain viable [24]. MMPs are capable of synthesizing irregular greigite or bullet-shaped magnetosomes [21]. MMPs with both types of magnetosomes have already been reported [25].

2.2 Genetics

The origin of magnetotaxis and its distribution among the different phyla of the Bacteria domain are not well understood. Despite the great phylogenetic diversity, MTB have unique genes related to biomineralization, which are located in a generally unstable region in the genome [26]. The genomic and ultrastructure characterizations of nonmagnetotactic spontaneous mutants of *M. gryphiswaldense* strain MSR-1 showed the absence of 130 kb genomic region and complete lack of magnetosomes within these cells [27]. Genomic comparison among MTB affiliated to different phyla showed that genes in this region are conserved within MTB group, even when magnetite- and greigite-producing MTB were analyzed [11]. This region containing the genes responsible for the synthesis of magnetosomes was named magnetosome island and the genes are referred to as *mam* (*magnetosome membrane*), *mms* (*magnetosome membrane specific*), and *mtx* (*magnetotaxis*) genes because proteins encoded by these genes are localized on the magnetosome membrane or participate on the magnetotaxis motility behavior.

The genes for biomineralization are grouped into four operons in *M. gryphiswaldense* strain MSR-1 and other freshwater spirilla, called *mamAB*, *mamGFDC*, *mms6,* and *mamXY* in the magnetosome island [28]. Although species of MTB have different sizes in the region that encompasses the genes involved in biomineralization, some genes are conserved in all species. The content and organization of genes on the magnetosome island vary between magnetotactic species, and often, some genes are deleted or inserted without any change in the formation of the magnetosomes [27]. In general, proteins encoded by *mam* genes are involved in four major steps of magnetosome formation. These steps include: (1) formation of the magnetosome membrane (MamI, MamL, and MamAB); (2) formation of magnetite crystal (MamE, Mms6, MamB, and MamM); (3) maturation of the magnetite crystal (MamE, MmsF, MamGFDC, and Mam P, S, T); and (4) alignment of the chain magnetosome (MamJ and MamK) [29]. The mechanism by which these genes were and can be transferred between species of bacteria is unknown till date. In the past years, hypotheses were elaborated to explain the evolution of magnetotaxis. One of them was based on the observation that the evolution and divergence of the conserved Mam proteins and the 16S rRNA gene among MTB are congruent and support the monophyletic origin, in which all MTB would present a single common ancestor [9]. The other hypothesis states that the present diversity of MTB and magnetotaxis-related gene distribution is a result of multiple events of horizontal gene transfer, possibly with a common ancestral, gene modification and/or loss in different cell lines [26]. Functional analysis of the magnetosome island based on deletion of genes in *Ms. gryphiswaldense* strain MSR-1 indicated that genes in the *mamAB* operon are sufficient for magnetosome biomineralization [30]. Examples of genes found in this operon and their functions are listed in **Table 1**. Interestingly, the transference of genes from the *mamAB* operon from *M. gryphiswaldense* strain MSR-1 to *Rhodospirillum rubrum* resulted in magnetite magnetosome production within the photosynthetic cell [39]. Because of the magnetic properties of magnetosomes, which will be discussed on the following topic, these nanometric magnetic structures have great importance for the development of new applications and processes in Biotechnology. However, one of the limitations of their use in biotechnological applications is the fastidiousness of MTB, which makes the production of magnetosomes in bioreactors expensive and with low yield. The transference of the ability to synthesize magnetosomes from MTB to

Protein	Function	Effect on cell upon deletion	Reference/ strain used on the study
MamA	Activation of biomineralization	Decrease in the number of magnetosomes	[31] Strain AMB-1
MamB	Transport of iron into the magnetosome vesicle; interacts with PDZ domains (MamE)	Magnetosomes are not produced; no magnetic response	[29] Strain AMB-1; [32] strain MSR-1
MamE	Crystal maturation	Synthesis of magnetosomes smaller than <20 nm; cells without magnetic response	[29, 33] Strain AMB-1
MamH	Balance in the redox state of iron on the magnetosome membrane	Decrease in magnetosome number and size; less efficient magnetic response	[29] Strain AMB-1; [34] strain MSR-1
MamI	Synthesis of the magnetosome membrane	Magnetosomes are not produced; no magnetic response	[29] Strain AMB-1
MamJ	Alignment of the magnetosome chain: it interacts with the surface of the magnetosome and with a structure similar to the cytoskeleton (MamK filament)	Magnetosomes are not aligned in chains; reduced magnetic orientation	[35] Strain MSR-1
MamK	Controls the assembly, segregation, and positioning of the magnetosome chain inside the cell	Alignment of the magnetosomes in short chains and decrease in the number of magnetosomes	[36] Strain MSR-1
MamM	Transport of iron into the magnetosome: magnetite nucleation, crystal growth, and stabilization of MamB	Increase in the formation of polycrystalline particles of magnetite; formation of crystals of hematite; no magnetic response	[29] Strain AMB-1; [32] strain MSR-1
MamO	Nucleation of the magnetite crystal	Magnetosomes are not produced; cells without magnetic response	[29] Strain AMB-1; [37] strain MSR-1
MamP	Magnetite crystal size control	Decrease in the size of magnetosomes; less efficient magnetic response	[29, 38] Strain AMB-1
MamR MamT	Magnetite crystal size control	Decrease in the size of magnetosomes; less efficient magnetic response	[29, 38] Strain AMB-1
MamU	Possibly lipid metabolism	No changes were observed	[29] Strain AMB-1

Table 1.
Proteins encoded by genes within mamAB *operon in* Magnetospirillum *and their function.*

other cells represents a new frontier in Microbiology and greatly expands the use of magnetosomes in nanotechnological and biomedical applications [39].

3. Physical characteristics of magnetosomes

Two different processes of mineral formation by living beings have been recognized. One process of mineral formation is known as biologically induced mineralization (BIM), and is characterized by bulk extracellular and/or intercellular mineral formation, without the elaboration of organic matrices. It produces minerals having crystal habits similar to those produced by precipitation from inorganic solutions. BIM processes are less controlled than organic matrix-mediated mineralization, and looks like a primitive stage in the evolution of biogenic mineral formation. The other process is known as biologically controlled mineralization (BCM). In general, the organism constructs an organic mold into which the appropriate ions are actively introduced to crystallize a mineral. The mineral type, orientation of crystallographic axes, and microarchitectures are under genetic control [40].

Magnetotactic bacteria distinguish from other bacteria because they biomineralize, through BCM, magnetic nanoparticles of magnetite, or greigite. Magnetite is a very interesting iron oxide because it is magnetic and a good conductor. Its free charge density is similar to that of some metals [41]. It is also a hard mineral, being used by chitons for tooth hardening [42]. Several studies show that greigite has similar electrical [43] and hardening use [44] as magnetite. The magnetic properties of nanoparticles have a strong dependence on the size: very small particles have a magnetic moment nonstable in the body, changing randomly its orientation and producing a null average magnetic moment. Those particles are known as superparamagnetic. If the size increases, the anisotropic energy also increases and creates an energy barrier that maintains the magnetic moment in a fixed direction. In that case, the nanoparticles behave as stable magnets and are known as single domains [45]. Magnetosomes are in the size range of magnetic single domains. The magnetic flux lines created by the magnetosome in the chain can be observed using the magnetic electron holography technique [46], showing the flux lines entirely aligned to the chain as corresponds with a dipolar field created by a single magnetic moment. So, it is appropriate to say that the magnetosome chain behaves as a compass needle. The linear arrange of magnetosomes is not energetically stable, because after some number of magnetosomes the best configuration is a ring. To maintain the linear configuration, magnetosomes are attached to the cytoskeletal filaments [47].

The first analysis done in magnetosomes was energy-dispersive X-ray microanalysis (EDS or EDX), showing that they are composed mainly by iron and oxygen [4]. To show that they are the iron oxide magnetite, the Mossbauer technique was used [48], showing that the Mossbauer spectra behave as a mixture of magnetite and maghemite. Also, electron diffraction shows the diffraction patterns corresponding to magnetite [49]. Several studies with EDS show that this magnetite is highly pure. However, in some cases, some metallic ions can be absorbed in the magnetosome structure, depending on the ambient pollution [50]. Studies done with high-resolution electron microscopy show that magnetosomes are produced in specific geometric morphologies [51]. Those morphologies are truncated cuboctahedron, elongated cuboctahedron, and hexagonal prisms. In the case of greigite, the crystalline morphologies are truncated cuboctahedrons and elongated rectangular prisms [51]. This iron sulfide was discovered in magnetosomes of multicellular magnetotactic prokaryotes, and identified through EDX spectroscopy and electron diffraction [52].

4. Determination of bacterial magnetic moment

As the magnetosome chain determines a magnetic moment to MTB, let us talk about the different techniques used to estimate that magnetic moment. The first theoretical estimate for the magnetic moment was done counting the contribution of several nanoparticles arranged in a chain [53]. For a magnetosome chain composed of 22 particles of magnetite with every nanoparticle having 1.25×10^{-16} cm^3 of volume, it is possible to calculate the total magnetic moment as $M = nV_{ind}M_V$, where n is the number of particles in the chain, V_{ind} is the volume of each particle (assuming that all are equal), and M_V is the magnetic moment per unit volume of the magnetic material. For magnetite, $M_V = 480 \times 10^{-3}$ Am^2/cm^3. In this way, a magnetic moment of 1.3×10^{-15} Am^2 is obtained. This magnetic moment value means a magnetic to thermal energy rate of about 16 (assuming a temperature of 300 K). This method can be used whenever is possible to observe and count the number of magnetosomes in the chain. This method is not applied in the case of live bacteria and for bigger microorganisms with lots of magnetosomes, as is the case for "*Candidatus* Magnetobacterium bavaricum" and "*Candidatus* Magnetoglobus multicellularis."

A statistical analysis of the swimming orientation of magnetotactic bacteria, assuming that they behave as paramagnetic particles, produces the orientation to be equivalent to the average of cosθ ($\langle \cos\theta \rangle$), being θ the angle among the bacterial velocity and the magnetic field. Kalmijn showed that $\langle \cos\theta \rangle$ is function of the magnetic to thermal energy ratio [54]: $\langle \cos\theta \rangle = L\,(MH/kT) = \coth(MH/kT) - kT/MH$, where M is the bacterium magnetic moment, H is the magnetic field intensity, k is the Boltzmann constant, T is the absolute temperature, and L(x) is the Langevin function: coth(x)—1/x. For $MH/kT \approx 10$, the Langevin function is about 0.9, which means that the bacterial trajectory is well oriented to the magnetic field direction. The analysis of the velocity as function of the magnetic field [54] or of the orientation as function of the magnetic field [55] permits the estimative of the bacterial magnetic moment. Kalmijn stressed the fact that this kind of study is valid only for the orientation of a single bacterium and not for the average orientation of several bacteria [54]. Using this method, it has been shown that "*Candidatus* Magnetoglobus multicellularis" shows values of L(x) lower than 0.9 in the presence of the geomagnetic field. A measuring method for the magnetic moment of individual MTB was developed in [56] and consists in the analysis of the U-turn trajectory, which is the form of the trajectory followed by an MTB when the sense of the external magnetic field vector is inverted. The theoretical analysis assumes that the bacteria and the magnetosome chain forms a rigid body, the bacteria following the movement of the magnetic moment. In the low Reynolds number regime and ignoring the flagellar forces, the sum of the magnetic torque and the viscous torque is equal to zero. From that equation, mathematical expressions are obtained for the time τ and diameter L of the reversal trajectory: $L = 8\pi\eta R^3 v/(MH)$ and $\tau = [8\pi\eta R^3/(MH)]\ln[2MH/(kT)]$, where M is the bacterium magnetic moment, H is the magnetic field intensity, k is the Boltzmann constant, T is the absolute temperature, R is the bacterium radius (assuming it is a coccus), v is the velocity, η is the viscosity, and ln is the natural logarithm function. The measurement of those parameters for the U-turn trajectory makes possible to calculate the value of the magnetic moment of magnetotactic bacteria. The experimental measurement of the magnetic moment of bacteria with different sizes and shapes, done by Esquivel and Lins de Barros [56], showed that the magnetic moment can have values from 0.3×10^{-15} to 54×10^{-15} Am^2, generating magnetic to thermal energy ratios from 3 to 326. Those results challenge the idea that the magnetosome contains the sufficient magnetic nanoparticles, arranged in the proper configuration, to efficiently orient the bacteria in the geomagnetic field direction. A problem with this method is that the U-turn trajectory must be in the focal plane for a good measurement of L, but that is not the case in the general. An alternative is to use only the U-turn time τ because it can be determined well for any U-turn trajectory [57]. The U-turn analysis done by Esquivel and Lins de Barros [56] also assumes that bacteria have spherical geometry, that it is not the general case. When the bacterium is enlarged, as a small cylinder, another approximation must be done. So, assuming that this small cylinder behaves as a set of attached spheres, the contribution to the total torque can be calculated. Doing the experimental analysis in that way, Bahaj et al. [58] calculated a value of 6.1×10^{-16} Am^2 for the magnetotactic spirillum *Magnetospirillum magnetotacticum*. They also calculated the variation of magnetic moment with the growth time, and observed that it grows from 2.8×10^{-16} Am^2 at 35 h to 6.5×10^{-16} Am^2 at 240 h [59]. Another technique widely used to determine the magnetic moment of magnetotactic bacteria is the analysis of the movement in a rotating magnetic field [60]. In that method, a set of four coils (two crossed pairs) is adapted to an optical microscope stage to generate a rotating magnetic field with frequency f. That experimental setup is known as bacteriodrome. The resultant trajectory is a circle, observed clearly by dark-field images. Again, ignoring the flagellar movement and in the low Reynolds number regime, the magnetic torque must be equal to the viscous torque. The magnetic torque depends on the angle among the bacterial magnetic moment and the external magnetic field. That angle increases when the frequency f increases, and its upper limit is 90° meaning that there is a critical value of f_c. For values of f higher than f_c, the trajectory is not more a circle. The determination of f_c permits to calculate the magnetic moment as: $M \approx c\eta 2\pi f_c\, l^3/H$, where M is the bacterium magnetic moment, H is the magnetic field intensity, c is a shape factor, η is the viscosity, and l is the bacterium length. It is difficult to determine the shape factor, and Petersen et al. [60] proposed an approximated value of 8π. Using this technique, Petersen et al. [60] determined the magnetic moment of magnetotactic bacteria, of natural samples from Southern Germany, of about 4×10^{-15} Am^2, and Pan et al. [61] calculated a value of about 1.8×10^{-15} Am^2 for MYC-1, an uncultivated magnetotactic coccus from China.

Other techniques have been used for measuring the magnetic moment of magnetotactic bacteria. Using a SQUID magnetometer, an average magnetic moment of $1.8 \pm 0.4 \times 10^{-12}$ emu for bacteria from natural sediments had been determined [62]. This method is interesting because it is a direct measurement and does not need to assume unknown values for parameters from the studied cell. There are two interesting physical techniques involving light for measuring the magnetic moment of magnetic bacteria. One is the analysis of the birefringence arising in a pull of magnetotactic bacteria when in presence of an external magnetic field [63]. The birefringence transforms an input linear polarized light beam in an output elliptically polarized light beam, with a phase shift between the fast and slow components. This phase shift is measured and it depends on the intensity of the external magnetic field and on the magnetic moment. Experiments were done with live and dead bacteria, killed with drops of formalin. The measured values, at normal concentration conditions, for live bacteria were about 1.21×10^{-13} emu and for dead bacteria about 1.33×10^{-13} emu. Apparently, for dead bacteria, the measured values are higher than for live bacteria. It was assumed that this difference could be an effect of motile behavior in live bacteria and the concept of "effective temperature" T_{eff} was introduced, meaning that live bacteria feels a disorienting thermal energy kT_{eff} higher than the ambient thermal energy in 10–20%. The other technique is the analysis of the light scattered by a pull of magnetic bacteria [64], based on the fact that the presence of an external magnetic field determines an angular distribution in the orientation of bacteria. This angular distribution affects the structure factor in the scattered light intensity. With this method, the average length and average magnetic moment can be determined. For two different cultures of *Aquaspirillum magnetotacticum* were determined values of $(2.2 \pm 0.2) \times 10^{-13}$ emu and $(4.3 \pm 0.5) \times 10^{-13}$ emu, which are in good agreement with the value obtained by electron microscopy, or about 4.4×10^{-13} emu. Using a similar experimental approach in [65] was determined the magnetic moment of a wild-type *Magnetospirillum gryphiswaldense* strain and obtained a value of about 25.3 $(\pm 1.6) \times 10^{-13}$ emu. Other methods found in literature are based basically in the analysis of the bacterial body rotation caused by the magnetic torque and in the analysis of the equation magnetic torque = viscous torque. For example, in Ref. [66], it was measured the magnetic moment of single *Magnetospirillum gryphiswaldense* cells using magnetic tweezers, observing and analyzing the rotation of the bacterial body after a magnetic field reversion. They observed that the measured magnetic moment has a dependence on the magnetic field intensity, as occurs in magnetization measurements of magnetic materials, starting from a remanence magnetization at zero magnetic field and progressively increasing until the magnetization saturates at higher magnetic fields. They measured for $6\ mT < H < 23\ mT$ a magnetic moment of $2.4\ (\pm 1.1) \times 10^{-13}$ emu and for $90\ mT < H < 130\ mT$ a magnetic moment of $7.7\ (\pm 3.4) \times 10^{-13}$ emu. **Table 2** resumes the magnetic moment measured with the different techniques, remembering that $1\ emu = 10^{-3}\ Am^2$. It can be observed that the magnetic moment obtained by the direct measurement from the magnetosome chain is always bigger than that obtained from indirect physical methods. In the study by Zahn et al. [66], this fact is explained identifying the direct measurement in the magnetosome chain as the saturation magnetization, that is only observed for higher magnetic fields.

Technique	Organism	Magnetic moment (Am^2)	References
Electron microscopy	Magnetotactic spirillum MS1	1.3×10^{-15}	[53]
U-turn analysis	Several microorganisms from fresh to marine water	0.3×10^{-15} to 54×10^{-15}	[56]
U-turn modified	*Magnetospirillum magnetotacticum*	6.1×10^{-16}	[58]
SQUID	Fresh water uncultured bacteria	1.8×10^{-15}	[62]
Light scattering	*Aquaspirillum magnetotacticum*	(live) 2.2×10^{-16} (dead) 4.3×10^{-16}	[64]
Light scattering	*Magnetospirillum gryphiswaldense*	$25.3 \pm 1.6 \times 10^{-16}$	[65]
Birefringence	*Aquaspirillum magnetotacticum*	1.21×10^{-16}	[63]
Rotating magnetic field	Natural samples	4×10^{-15}	[60]
Rotating magnetic field	Uncultivated coccus MYC-1	1.8×10^{-15}	[61]
Magnetic tweezers	*Magnetospirillum gryphiswaldense*	(low H) $2.4 \pm 1.1 \times 10^{-16}$ (high H) $7.7 \pm 3.4 \times 10^{-16}$	[66]

Table 2.
Magnetic moment value for MTB using different physical techniques.

5. The movement of magnetotactic bacteria

Several experimental observations show that magnetotaxis functions together with aerotaxis, determining the so-called magneto-aerotaxis [8, 67]. Basically, two different behaviors have been identified in magneto-aerotaxis: polar magnetotaxis, that consists in the North-seeking or South-seeking behaviors in the search for the better oxygen concentrations; and axial magnetotaxis, in that case, MTB move in the magnetic field direction but without preferential sense. MTB from natural samples always present polar magnetotaxis. Axial magnetotaxis has been observed only in cultured MTB.

MTB are easily identified because of their response to the inversion of the local magnetic field direction: after the inversion bacteria swim following the new magnetic field direction. It can be stated that magnetic field inversions stimulate MTB to swim, making them a model for the study of microorganism swimming. Bacteria swim in the low Reynolds number regime, where viscous forces and torques act to null the resultant force and torque [68]. In that regime, microorganisms swim following an helical trajectory [69] whose parameterization in Cartesian coordinates (x, y, z), considering the helix axis as the z axis, can be written as (Rcos(ωt), Rsin (ωt), Vt), where R is the helix radius, V is the axial velocity, and $\omega = 2\pi f$ being f the helix frequency. In the case of magnetotactic microorganisms, the helical trajectory of the multicellular magnetotactic prokaryote "*Candidatus* Magnetoglobus multicellularis" has been studied for two different applied magnetic fields (3.9 and 20 Oe) [70] and for magnetic fields from 0.9 to 32 Oe [55]. Those studies show that for spherical multicellular magnetotactic prokaryotes, the axial velocity V is about 90 μm/s, the radius R is about 8 μm for lower magnetic fields, and the helix frequency f is about 1.1 Hz. For uncultured magnetotactic coccus, the helical movement has been studied recently (data not published), in the presence of magnetic fields of about 0.7 Oe, and the helical parameters measured were: axial velocity of about 90 μm/s, radius of about 2.5 μm, and helix frequency of about 1 Hz. For other magnetotactic microorganisms, it has been observed that the 2D trajectory is similar to the projection of a 3D helix in the microscope focal plane (for example, see [19, 68, 71].

For the theoretical study of microorganisms, motion in the low Reynolds number regime is necessary to know all the forces and torques acting on the microorganism. Nogueira and Lins de Barros [68] developed a model in that regime, considering a spherical MTB with a single flagellum and a magnetosome chain aligned to the flagellum line. The equations to be considered are $\mathbf{F}_{flagella} + \mathbf{F}_{viscous} = 0$ and $\boldsymbol{\tau}_{flagella} + \boldsymbol{\tau}_{viscous} + \boldsymbol{\tau}_{magnetic} + \boldsymbol{\tau}_{body} = 0$. Using the appropriate expressions for the forces and torques in that model, they were able to calculate numerically the temporal evolution of the center of mass coordinates (x, y, z) and of the Euler's angles for the rigid body (θ, ϕ, ψ), being the trajectory similar to a cylindrical helix. In the other hand, Refs. [72, 73] studied the motion of nonspherical MTB, to include the effect of the bacterial body geometry on the viscous forces. Also, Yang et al. [73] studied MTB with two flagellar bundles. To do that, they calculated numerically the motion using the second Newton's law, considering all the forces and torques and calculating the appropriate inertial terms for the geometrical body form. They also studied the effect of the relative inclination λ between the magnetosome chain and the flagella. Those studies showed that when $\lambda \neq 0$, the velocity decreases when the magnetic field increases, effect also observed experimentally in the work by Pan et al. [74] when studying the circular movement of the MYC-1 strain. In that case, it was measured the velocity in the circular trajectory obtained in a bacteriodrome as function of the applied magnetic field, in the hope to obtain a growing Langevin curve as predicted by Kalmijn [54]. But they observed that the velocity decreases as the magnetic field increases, in the contrary of a Langevin curve. To explain this, they assumed that the magnetosome chain has an inclination relative to the flagellar bundle. Interestingly, it has been observed that some MTB strains have the magnetosome chain with different inclinations relative to the flagellar bundle, in some cases being almost perpendicular to it, not orienting the magnetic moment to the magnetic field direction during their swimming [73].

The movement of magnetotactic microorganisms also depends on the presence of light, and the response depends on the wavelength and the intensity. This behavior has been studied mainly in multicellular magnetotactic prokaryotes. Negative photo response has been observed when they are illuminated with high-intensity UV light (365 nm), violet-blue light (395–440 nm filter) of about 80 W m^{-2} of intensity, and blue light (450–490 nm filter) of about 200 W m^{-2} of intensity. For longer wavelengths, no photo response was observed, and apparently long exposure to green light is lethal [70, 75]. That negative photo response is not observed when very low intensities are used. Photokinesis has been observed in multicellular magnetotactic prokaryotes, decreasing the velocity when illuminated with green light (517 nm, 0.46 W m^{-2}) and increasing the velocity when illuminated with red light (628 nm, 0.16 W m^{-2}), both respectively to the velocity observed when illuminated with blue light (469 nm, 0.8 W m^{-2}) [76, 77]. Recently, De Melo and Acosta-Avalos [78] showed that the photokinesis in multicellular magnetotactic prokaryotes is related to the combined presence of monochromatic light and a constant magnetic field, and that it can be canceled in the presence of radio-frequency electromagnetic fields oscillating at the Zeeman resonance frequency associated to the constant magnetic field, showing the involvement of a radical pair mechanism, a very well-known magnetoreception mechanism used by migratory birds. Interestingly, magnetotactic microorganisms have the proper physical tools to sense the geomagnetic field with light. Perhaps, magnetotaxis and the radical pair mechanism are involved in a more elaborate magnetic sensing in MTB.

6. Conclusions

Since the discovery of MTB, they attracted the attention of the scientific community for several reasons. Firstly, they show clearly that living beings can interact with the geomagnetic field through magnetic nanoparticles and became a model that has been extensively used in magnetoreception research. The study of the magnetosome synthesis process within MTB is being used to develop new strategies to produce magnetic nanoparticles with potential use in Biotechnology. For example, genes responsible for magnetosome synthesis could be transferred and expressed in a host cell to increase the yield of magnetosomes production in bioreactors. If high amounts of magnetosomes were achieved at low cost, magnetosome use as biotechnological tools would be possible. For physicists, MTB are interesting to apply models of magnetism, used in solid-state theory, in living beings behavior. The different techniques developed to measure the MTB magnetic moment have shown that considering MTB as paramagnetic particles is as insufficient model, defying previous models about MTB magnetotaxis. The study of motion is also giving support to new understandings about magnetotaxis, because new characteristics of the interaction among MTB and the geomagnetic field are arising through the study of the movement as function of the applied magnetic field. There are some evidences that MTB use more than one mechanism to detect the magnetic field direction, and not only through the magnetic torque. So, a new magnetoreception mechanism shall be discovered in MTB in the near future.

Acknowledgements

We acknowledge the microscopy facility Unidade de Microscopia Multiusuário (UniMicro, UFRJ) and financial support from the Brazilian agencies Conselho Nacional de Desenvolvimento Científico e Tecnológico (CNPq), Coordenação de Aperfeiçoamento de Pessoal de Nível Superior (CAPES), and Fundação de Amparo à Pesquisa do Estado do Rio de Janeiro (FAPERJ).

Conflict of interest

The authors declare no conflict of interest.

Author details

Fernanda Abreu[1] and Daniel Acosta-Avalos[2]*

1 Universidade Federal do Rio de Janeiro, Rio de Janeiro, Brazil

2 Centro Brasileiro de Pesquisas Fisicas, Rio de Janeiro, Brazil

*Address all correspondence to: dacosta@cbpf.br

References

[1] Bruce A, Bray D, Lewis J, Raff M, Roberts K, Watson JD. Molecular Biology of the Cell. New York & London: Garland Publishing; 1983. pp. 11-12

[2] Bazylinski DA, Frankel RB. Magnetosome formation in prokaryotes. Nature Reviews. Microbiology. 2004;**2**: 217-230. DOI: 10.1038/nrmicro842

[3] Frankel RB. The discovery of magnetotactic/magnetosensitive bacteria. Chinese Journal of Oceanology and Limnology. 2009;**27**:1-2. DOI: 10.1007/S00343-009-0001-7

[4] Blakemore R. Magnetotactic bacteria. Science. 1975;**190**:377-379. DOI: 10.1126/science.170679

[5] Lin W, Bazylinski DA, Xiao T, Wu LF, Pan Y. Life with compass: Diversity and biogeography of magnetotactic bacteria. Environmental Microbiology. 2014;**16**:2646-2658. DOI: 10.1111/1462-2920.12313

[6] Abreu F, Araujo ACV, Leão P, Silva KT, Marques F, Cunha O, Almeida LGP, Geurinkc C, Farina M, Rondelli D, Jovane L, Pellizari VH, Vasconcelos ATR, Bazylinski DA, Lins U. Culture-independent characterization of novel psychrophilic magnetotactic cocci from Antarctic marine sediments. Environmental Microbiology. 2016;**18**: 4426-4441. DOI: 10.1111/1462-2920.13388

[7] Spring S, Lins U, Amann R, Schleifer KH, Ferreira LC, Esquivel DM, Farina M. Phylogenetic affiliation and ultrastructure of uncultured magnetic bacteria with unusually large magnetosomes. Archives of Microbiology. 1998;**2**:136-147. DOI: 10.1007/s002060050553

[8] Frankel RB, Bazylinski DA, Johnson M, Taylor BL. Magneto-aerotaxis in marine, coccoid bacteria. Biophysical Journal. 1997;**73**:994-1000. DOI: 10.1016/s0006-3495(97)78132-3

[9] Lefèvre CT, Wu LF. Evolution of the bacterial organelle responsible for magnetotaxis. Trends in Microbiology. 2013;**21**(10):534-543. DOI: 10.1016/j.tim.2013.07.005

[10] Amann R, Peplies J, Schüler D. Diversity and taxonomy of magnetotactic bacteria. In: Schüler D, editor. Magnetoreception and Magnetosomes in Bacteria. Berlin Heidelberg: Springer-Verlag; 2007. pp. 25-36. DOI: 10.1007/7171-037

[11] Uebe R, Schüler D. Magnetosome biogenesis in magnetotactic bacteria. Nature Reviews. Microbiology. 2016;**14**: 621-637. DOI: 10.1038/nrmicro.2016.99

[12] Komeili A, Li Z, Newman DK, Jensen GJ. Magnetosomes are cell membrane imaginations organized by the actin-like protein MamK. Science. 2006;**311**:242-245. DOI: 10.1126/science.1123231

[13] Schüler D. Molecular analysis of a subcellular compartment: The magnetosome membrane in *Magnetospirillum gryphiswaldense*. Archives of Microbiology. 2004;**181**:1-7. DOI: 10.1007/s00203-003-0631-7

[14] Staniland S, Ward B, Harrison A, van der Laan G, Telling N. Rapid magnetosome formation shown by real-time X-ray magnetic circular dichroism. Proceedings of the National Academy of Sciences. 2007;**104**:19524-19528. DOI: 10.1073/pnas.0704879104

[15] Lefevre CT, Bazylinski DA. Ecology, diversity, and evolution of magnetotactic bacteria. Microbiology and Molecular Biology Reviews. 2013;**3**: 497-526. DOI: 10.1128/mmbr.00021-13

[16] Bazylinski DA, Frankel RB, Heywood BR, Mann S, King JW, Donaghay PL, Hanson AK. Controlled biomineralization of magnetite (Fe_3O_4) and greigite (Fe_3S_4) in a magnetotactic bacterium. Applied and Environmental Microbiology. 1995;**9**:3232-3239

[17] Lefevre CT, Menguy N, Abreu F, Lins U, Posfai M, Prozorov T, Pignol D, Frankel RB, Bazylinski DA. A cultured greigite-producing magnetotactic bacterium in a novel group of sulfate-reducing bacteria. Science. 2011;**334**: 1720-1723. DOI: 10.1126/science.1212596

[18] Abreu F, Sousa AA, Aronova MA, Kim Y, Cox D, Leapman RD, Andrade LR, Kachar B, Bazylinski DA, Lins U. Cryo-electron tomography of the magnetotactic vibrio *Magnetovibrio blakemorei*: Insights into the biomineralization of prismatic magnetosomes. Journal of Structural Biology. 2012;**181**:162-168. DOI: 10.1016/j.jsb.2012.12.002

[19] Zhang WY, Zhou K, Pan HM, Yue HD, Jiang M, Xiao T, Wu LF. Two genera of magnetococci with bean-like morphology from intertidal sediments of the Yellow Sea, China. Applied and Environmental Microbiology. 2012;**78**: 5606-5611. DOI: 10.1128/aem.00081-12

[20] Ruan J, Kato T, Santini CL, Miyata T, Kawamoto A, Zhang WJ, Bernadac A, Wu LF, NK. Architecture of a flagellar apparatus in the fast-swimming magnetotactic bacterium MO-1. Proceedings of the National Academy of Sciences. 2012;**109**:20643-20648. DOI: 10.1073/pnas.1215274109

[21] Leão P, Chen YR, Abreu F, Wang M, Zhang WJ, Zhou K, Xiao T, Wu LF, Lins U. Ultrastructure of ellipsoidal magnetotactic multicellular prokaryotes depicts their complex assemblage and cellular polarity in the context of magnetotaxis. Environmental Microbiology. 2017;**19**:2151-2163. DOI: 10.1111/1462-2920.13677

[22] Winklhofer M, Abraçado LG, Davila AF, Keim CN, Lins De Barros HGP. Magnetic optimization in a multicellular magnetotactic organism. Biophysical Journal. 2007;**92**:661-670. DOI: 10.1529/biophysj.106.093823

[23] Keim C, Martins JL, Abreu F, Rosado AS, Lins De Barros HGP, Borojevic R, Lins U, Farina M. Multicellular life cycle of magnetotactic prokaryotes. FEMS Microbiology Letters. 2004;**240**:203-208. DOI: 10.1016/j.femsle.2004.09.035

[24] Abreu F, Silva KT, Martins JL, Lins U. Cell viability in magnetotactic multicellular prokaryotes. International Microbiology. 2006;**9**:267-272

[25] Lins U, Keim C, Evans F, Farina M, Buseck P. Magnetite (Fe_3O_4) and greigite (Fe_3S_4) crystals in multicellular magnetotactic prokaryotes. Geomicrobiology Journal. 2007;**24**: 43-50. DOI: 10.1080/0149045060 1134317

[26] ogler C, Schüler D. Genetic and biochemical analysis of magnetosome formation in *Magnetospirillum gryphiswaldense*. In: Bauerelein E, Behrens P, Epple M, editors. Handbook of Biomineralization: Biological Aspects and Structure Formation. Weinheim: Wiley; Vol. 1. 2008. pp. 145-161. DOI: 10.1002/9783527619443.ch9

[27] Ullrich S, Ullrich S, Kube M, Schübbe S, Reinhardt R, Schüler D. A hypervariable 130-kilobase genomic region of *Magnetospirillum gryphiswaldense* comprises a magnetosome island which undergoes frequent rearrangements during stationary growth. Journal of Bacteriology. 2005;**21**:7176-7184. DOI: 10.1128/jb.187.21.7176-7184.2005

[28] Bazylinski DA, Schübbe S. Controlled biomineralization by and applications of magnetotactic bacteria. Advances in Applied Microbiology.

2007;**62**:21-62. DOI: 10.1016/s0065-2164(07)62002-4

[29] Murat D, Quinlan A, Vali H, Komeili A. Comprehensive genetic dissection of the magnetosome gene island reveals the step-wise assembly of a prokaryotic organelle. Proceedings of the National Academy of Sciences. 2010;**12**:5593-5598. DOI: 10.1073/pnas.0914439107

[30] Lohße A, Ullrich S, Katzmann E, Borg S, Wanner G, Richter M, Voigt B, Schweder T, Schüler D. Functional analysis of the magnetosome island in *Magnetospirillum gryphiswaldense*: The mamAB operon is sufficient for magnetite biomineralization. PLoS One. 2011;**6**:e25561. DOI: 10.1371/journal.pone.0025561

[31] Komeili A, Vali H, Beveridge TJ, Newman DK. Magnetosome vesicles are present before magnetite formation, and MamA is required for their activation. Proceedings of the National Academy of Sciences of the United States of America. 2004;**101**:3839-3844. DOI: 10.1073/pnas.0400391101

[32] Uebe R, Junge K, Henn V, Poxleitner G, Katzmann E, Plitzko JM, Zarivach R, Kasama T, Wanner G, Pósfai M, Böttger L, Matzanke B, Schüler D. The cation diffusion facilitator proteins MamB and MamM of *Magnetospirillum gryphiswaldense* have distinct and complex functions, and are involved in magnetite biomineralization and magnetosome membrane assembly. Molecular Microbiology. 2011;**82**: 818-835. DOI: 10.1111/j.1365-2958.2011.07863.x

[33] Quinlan A, Murat D, Vali H, Komeili A. The HtrA/DegP family protease MamE is a bifunctional protein with roles in magnetosome protein localization and magnetite biomineralization. Molecular Microbiology. 2011;**80**:1075-1087. DOI: 10.1111/j.1365-2958.2011.07631.x

[34] Raschdorf O, Müller FD, Pósfai M, Plitzko JM, Schüler D. The magnetosome proteins MamX, MamZ and MamH are involved in redox control of magnetite biomineralization in *Magnetospirillum gryphiswaldense*. Molecular Microbiology;**89**:872-886. DOI: 10.1111/mmi.12317

[35] Scheffel A, Gruska M, Faivre D, Linaroudis A, Plitzko JM, Schüler D. An acidic protein aligns magnetosomes along a filamentous structure in magnetotactic bacteria. Nature. 2006; **440**:110-114. DOI: 10.1038/nature04382

[36] Katzmann E, Scheffel A, Gruska M, Plitzko JM, Schüler D. Loss of the actin-like protein MamK has pleiotropic effects on magnetosome formation and chain assembly in *Magnetospirillum gryphiswaldense*. Molecular Microbiology. 2010;**77**:208-224. DOI: 10.1111/j.1365-2958.2010.07202.x

[37] Yang W, Li R, Peng T, Zhang Y, Jiang W, Li Y, Li J. mamO and mamE genes are essential for magnetosome crystal biomineralization in *Magnetospirillum gryphiswaldense* MSR-1. Research in Microbiology. 2010;**161**: 701-705. DOI: 10.1016/j.resmic.2010.07.002

[38] Siponen MI, Adryanczyk G, Ginet N, Arnoux P, Pignol D. Magnetochrome: A c-type cytochrome domain specific to magnetotactic bacteria. Biochemical Society Transactions. 2012;**40**: 1319-1323. DOI: 10.1042/bst20120104

[39] Kolinko I, Lohße A, Borg S, Raschdorf O, Jogler C, Tu Q, Pósfai M, Tompa E, Plitzko JM, Brachmann A, Wanner G, Müller R, Zhang Y, Schüler D. Biosynthesis of magnetic nanostructures in a foreign organism by transfer of bacterial magnetosome gene clusters. Nature Nanotechnology. 2014; **3**:193-197. DOI: 10.1038/nnano.2014.13

[40] Lowenstam HA. Minerals formed by organisms. Science. 1981;**211**:

1126-1131. DOI: 10.1126/science.7008198

[41] Belov KP. Electronic processes in magnetite (or, 'Enigmas of magnetite'). Physics-Uspekhi. 1993;**36**:380-391. DOI: 10.1070/pu1993v036n05abeh002160

[42] Weaver JC, Wang Q, Miserez A, Tantuccio A, Stromberg R, Bozhilov KN, Maxwell P, Nay R, Heier ST, DiMasi E, Kisailus D. Analysis of an ultra-hard magnetic biomineral in chiton radular teeth. Materials Today. 2010;**13**:42-52. DOI: 10.1016/s1369-7021(10)70016-x

[43] Roberts AP, Chang L, Rowan CJ, Horng CS, Florindo F. Magnetic properties of sedimentary greigite (Fe_3S_4): An update. Reviews of Geophysics. 2011;**49**:RG1002. DOI: 10.1029/2010rg000336

[44] Posfai M, Dunin-Borkowski RE. Sulfides in biosystems. Reviews in Mineralogy and Geochemistry. 2006;**61**: 679-714. DOI: 10.2138/rmg.2006.61.13

[45] Bean CP, Livingston JD. Superparamagnetism. Journal of Applied Physics. 1959;**30**:S120-S129. DOI: 10.1063/1.2185850

[46] Dunin-Borkowski RE, McCartney MR, Frankel RB, Bazylinski DA, Posfai M, Buseck PR. Magnetic microstructure of magnetotactic bacteria by electron holography. Science. 1998;**282**: 1868-1870. DOI: 10.1126/science.282.5395.1868

[47] Kiani B, Faivre D, Klumpp S. Self-organization and stability of magnetosome chains—A simulation study. Plos One. 2018;**13**:e0190265. DOI: 10.1371/journal.pone.0190265

[48] Frankel RB, Blakemore RP, Wolfe RS. Magnetite in freshwater magnetotactic bacteria. Science. 1978; **203**:1355-1356. DOI: 10.1126/science.203.4387.1355

[49] Mann S, Frankel RB, Blakemore RP. Structure, morphology and crystal growth of bacterial magnetite. Nature. 1984;**310**:405-407. DOI: 10.1038/310405a0

[50] Jimenez-Lopez C, Romanek CS, Bazylinski DA. Magnetite as a prokaryotic biomarker: A review. Journal of Geophysical Research. 2010; **115**:G00G03. DOI: 10.1029/2009jg001152

[51] Bazylinski DA, Garratt-Reed AJ, Frankel RB. Electron microscopic studies of magnetosomes in magnetotactic bacteria. Microscopy Research and Technique. 1994;**27**:389-401. DOI: 10.1002/jemt.1070270505

[52] Farina M, Esquivel DMS, Lins de Barros HGP. Magnetic iron-sulphur crystals from a magnetotactic microorganism. Nature. 1990;**343**: 256-258. DOI: 10.1038/343256a0

[53] Frankel RB, Blakemore RP. Navigational compass in magnetic bacteria. Journal of Magnetism and Magnetic Materials. 1980;**15–18**: 1562-1564. DOI: 10.1016/0304-8853(80)90409-6

[54] Kalmijn AJ. Biophysics of geomagnetic field detection. IEEE Transactions on Magnetics. 1981;**17**: 1113-1124. DOI: 10.1109/tmag.1981.1061156

[55] Keim CN, De Melo RD, Almeida FP, Lins de Barros HGP, Farina M, Acosta-Avalos D. Effect of applied magnetic fields on motility and magnetotaxis in the uncultured magnetotactic prokaryote '*Candidatus* Magnetoglobus multicellularis'. Environmental Microbiology Reports. 2018. DOI: 10.1111/1758-2229.12640

[56] Esquivel DMS, Lins de Barros HGP. Motion of magnetotactic microorganisms. The Journal of Experimental Biology. 1986;**121**:153-163

[57] De Melo RD, Acosta-Avalos D. The swimming polarity of multicellular magnetotactic prokaryotes can change during an isolation process employing magnets: Evidence of a relation between swimming polarity and magnetic moment intensity. European Biophysics Journal. 2017;**46**:533-539. DOI: 10.1007/s00249-017-1199-5

[58] Bahaj AS, James PAB, Moeschler FD. An alternative method for the estimation of the magnetic moment of non-spherical magnetotactic bacteria. IEEE Transactions on Magnetics. 1996; **32**(5):5133-5135. DOI: 10.1109/20.539514

[59] Bahaj AS, James PAB, Ellwood DC, Watson JHP. Characterization and growth of magnetotactic bacteria: Implications of clean up of environmental pollution. Journal of Applied Physics. 1993;**73**(10):5394-5396. DOI: 10.1063/1.353743

[60] Petersen N, Weiss DG, Vali H. Magnetic bacteria in lake sediments. In: Lowes F, editor. Geomagnetism and Paleomagnetism. Dordrecht: Kluwer; 1989. pp. 231-241. DOI: 10.1007/978-94-009-0905-2_17

[61] Pan Y, Lin W, Tian L, Zhu R, Petersen N. Combined approaches for characterization of an uncultivated magnetotactic coccus from Lake Miyun near Beijing. Geomicrobiology Journal. 2009;**26**:313-320. DOI: 10.1080/01490450902748633

[62] Wajnberg E, Salvo de Souza LH, Lins de Barros HGP, Esquivel DMS. A study of magnetic properties of magnetotactic bacteria. Biophysical Journal. 1986;**50**:451-455. DOI: 10.1016/s0006-3495(86)83481-6

[63] Rosenblatt C, Torres de Araujo FF, Frankel RB. Birefringence determination of magnetic moments of magnetotactic bacteria. Biophysical Journal. 1982;**40**:83-85. DOI: 10.1016/s0006-3495(82)84461-5

[64] Rosenblatt C, Torres de Araujo FF, Frankel RB. Light scattering determination of magnetic moments of magnetotactic bacteria. Journal of Applied Physics. 1982;**53**(3):2727-2729. DOI: 10.1063/1.330948

[65] Logofatu PC, Ardelean I, Apostol D, Iordache I, Bojan M, Moisescu C, Ionita B. Determination of the magnetic moment and geometrical dimensions of the magnetotactic bacteria using an optical scattering method. Journal of Applied Physics. 2008;**103**:094911. DOI: 10.1063/1.2917404

[66] Zahn C, Keller S, Toro-Nahuelpan M, Dorscht P, Gross W, Laumann M, Gekle S, Zimmermann W, Schuler D, Kress H. Measurement of the magnetic moment of single *Magnetospirillum gryphiswaldense* cells by magnetic tweezers. Scientific Reports. 2017;7: 3558. DOI: 10.1038/s41598-017-03756-z

[67] Lefevre CT, Bennet M, Landau L, Vach P, Pignol D, Bazylinski DA, Frankel RB, Klumpp S, Faivre D. Diversity of magneto-aerotactic behaviors and oxygen sensing mechanisms in cultured magnetotactic bacteria. Biophysical Journal. 2014;**107**: 527-538. DOI: 10.1016/j.bpj.2014.05.043

[68] Nogueira FS, Lins de Barros HGP. Study of the motion of magnetotactic bacteria. European Biophysics Journal. 1995;**24**:13-21. DOI: 10.1007/bf00216826

[69] Crenshaw HC. A new look at locomotion in microorganisms: Rotating and translating. American Zoologist. 1996;**36**:608-618. DOI: 10.1093/icb/36.6.608

[70] Almeida FP, Viana NB, Lins U, Farina M, Keim CN. Swimming behaviour of the multicellular magnetotactic prokaryote '*Candidatus*

Magnetoglobus multicellularis' under applied magnetic fields and ultraviolet light. Antonie Van Leeuwenhoek. 2013; **103**:845-857. DOI: 10.1007/s10482-012-9866-0

[71] Chen YR, Zhang R, Du HJ, Pan HM, Zhang WY, Zhou K, Li JH, Xiao T, Wu LF. A novel species of ellipsoidal multicellular magnetotactic prokaryotes from Lake Yuehu in China. Environmental Microbiology. 2015;**17**: 637-647. DOI: 10.1111/1462-2920.12480

[72] Cui Z, Kong D, Pan Y, Zhang K. On the swimming motion of spheroidal magnetotactic bacteria. Fluid Dynamics Research. 2012;**44**:055508. DOI: 10.1088/0169-5983/44/5/055508

[73] Yang C, Chen C, Ma Q, Wu LF, Song T. Dynamic model and motion mechanism of magnetotactic bacteria with two lateral flagellar bundles. Journal of Bionic Engineering. 2012;**9**: 200-210. DOI: 10.1016/s1672-6529(11)60108-x

[74] Pan Y, Lin W, Li J, Wu W, Tian L, Deng C, Liu Q, Zhu R, Winklhofer M, Petersen N. Reduced efficiency of magnetotaxis in magnetotactic coccoid bacteria in higher than geomagnetic fields. Biophysical Journal. 2009;**97**: 986-991. DOI: 10.1016/j.bpj.2009.06.012

[75] Shapiro OH, Hatzenpichler R, Buckley DH, Zinder SH, Orphan VJ. Multicellular photo-magnetotactic bacteria. Environmental Microbiology Reports. 2011;**3**:233-238. DOI: 10.1111/j.1758-2229.2010.00215.x

[76] Azevedo LV, Lins de Barros H, Keim CN, Acosta-Avalos D. Effect of light wavelength on motility and magnetic sensibility of the magnetotactic multicellular prokaryote 'Candidatus Magnetoglobus multicellularis'. Antonie Van Leeuwenhoek. 2013;**104**:405-412. DOI: 10.1007/s10482-013-9964-7

[77] Azevedo LV, Acosta-Avalos D. Photokinesis is magnetic field dependent in the multicellular magnetotactic prokaryote *Candidatus* Magnetoglobus multicellularis. Antonie Van Leeuwenhoek. 2015;**108**:579-585. DOI: 10.1007/s10482-015-1543-4

[78] De Melo RD, Acosta-Avalos D. Light effects on the multicellular magnetotactic prokaryote '*Candidatus* Magnetoglobus multicellularis' are cancelled by radiofrequency fields: The involvement of radical pair mechanisms. Antonie Van Leeuwenhoek. 2017;**110**:177-186. DOI: 10.1007/s10482-016-0788-0

Chapter 2

Isolation and Purification of Sulfate-Reducing Bacteria

Ivan Kushkevych

This chapter, I dedicate to all scientists who have interests in the research on the physiology and biochemistry of this promising and heterogeneous group of microorganisms, the sulfate-reducing bacteria.

Author.

Abstract

Sulfate-reducing bacteria (SRB) are a widespread group of microorganisms that are often isolated from the anoxygenic environments (lake depths, soil, or swamps), and they are also present in the human and animal intestines. This group is often detected in patients with inflammatory bowel disease, including ulcerative colitis. That is why new rapid methods for their isolation, purification, and identification are important and necessary. In this chapter, the methods of mesophilic SRB isolation from various environments are described. Particular attention is paid to the purification of mesophilic SRB since they can be in close interaction with other microorganisms (*Clostridium*, *Bacteroides*, *Pseudomonas*, etc.), which are their frequent satellites. Moreover, the main methods of mesophilic SRB identification based on their morphological, physiological, biochemical, and genetical characteristics are presented.

Keywords: sulfate-reducing bacteria, sulfate, sulfite agar, hydrogen sulfide, isolation, identification

1. Introduction

Sulfate-reducing bacteria (SRB) are a heterogeneous group of microorganisms which is widespread in anaerobic places where sulfate-containing compounds are present [1–5]. These microorganisms use sulfate ions, which are reduced to hydrogen sulfide in the process called "dissimilatory sulfate reducing" or "sulfate respiration." In this process, sulfate is a terminal electron acceptor [1, 2, 6–8]. For the implementation of dissimilatory sulfate reduction, exogenous electron donors are necessary [3, 4].

Molecular hydrogen is the main electron donor for all SRB, but commonly used electron donors are also lactate, acetate, pyruvate, ethanol, fatty acids, amino acids, dicarboxylic acids, and other organic compounds [9, 10]. Depending on the species of SRB, organic compounds can be oxidized completely to carbon dioxide or incompletely with the formation of acetate [11]. The SRB can also use ammonium salts as nitrogen sources [3, 11]. SRB species can assimilate molecular nitrogen [3]. So, SRB are widespread in the following areas as lake depths, soils, swamps [1, 3], and biogas plant [12–14] and also present in the human and animal intestines [15–19]. The main species of intestinal SRB, *Desulfovibrio* genus, are often isolated from patients with inflammatory bowel disease (IBD) and healthy subjects [15, 20–25]. Other species of SRB, *Desulfomicrobium*, *Desulfobulbus*, *Desulfobacter*, *Desulfomonas*, and *Desulfotomaculum* were also seldom isolated from human and animal feces [1, 23, 26].

An increased number of SRB are often detected in patients with periodontitis [18]; inflammatory bowel diseases, including ulcerative colitis; and many other diseases [27–31]. Some scientists also suggest that SRB may be the cause of some forms of colon cancer, given the fact that these microorganisms produce hydrogen sulfide affecting the intestinal cell metabolism causing various diseases [32, 33]. That is why the isolation of SRB new strains, their purification from other microorganisms, and study of SRB cultural, physiological, biochemical, and genetical properties in detail are necessary.

It should be also noted that many species may be uncultured, so it is important to apply molecular and genetic methods such as Illumina sequencing. This method can give a clear picture of SRB diversity in the detected sample. However, in this chapter, the focus will be on isolation, purification, and cultivation of cultured mesophilic SRB species.

The goal of chapter is to describe:

- Methods of sample selections from water, soil, swamp, and feces of human or animals and from biopsy material

- Media, isolation, purification, and cultivation conditions

- Morphological diversity and physiological and biochemical properties

- Identification based on physiological and biochemical properties and sequence analysis of the 16S rRNA gene

- Generalization of this research

2. Selection of samples

As was noted, the SRB can be present in a sulfate-rich environment. The samples selected from the different ecotopes should be directly placed in anoxic modified Postgate liquid medium [3]. The composition of the medium and conditions of selections is described in Section 3.

2.1 Samples from environment (water, soil, swamp, and environmental surfaces)

One milliliter of water (or 1 g of swamp, soil, and metal rust) should be suspended in 9 ml of anoxic Postgate liquid medium. The tubes should be brim-filled with medium and closed to provide anaerobic conditions. Another option to provide anaerobic conditions is to add in tube 1 ml sterile liquid paraffin. The schema of sampling is presented in **Figure 1**.

2.2 Samples from feces of human or animals

It is thought that the species of SRB, their composition, and the number found in the intestinal lumen differ from that of the composition and number of

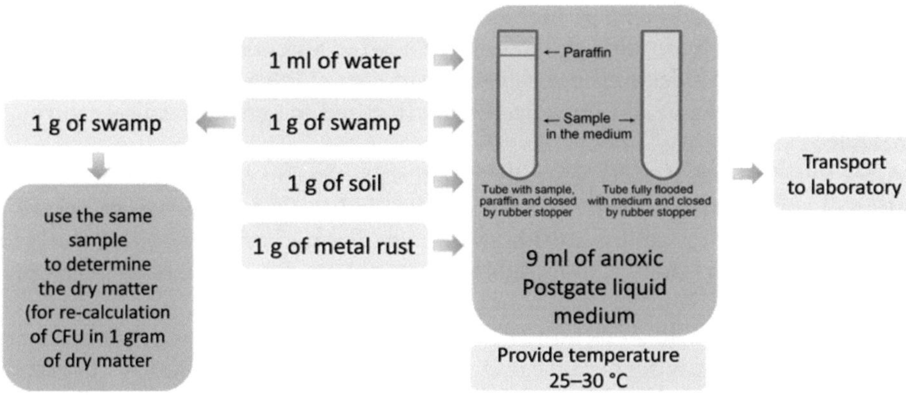

Figure 1.
The scheme of sampling from different environmental biotopes.

microorganisms on the surface of the intestinal mucosa [2, 9, 28, 34]. Similar to environmental samples (see **Figure 1**), fecal samples from human or animals should be fresh and directly suspended in anoxic modified Postgate liquid medium (pH 7.5, *temperature* to +37°C). One gram of feces suspends in 9 ml of the modified Postgate liquid medium [3, 35]. The same quantity of feces should be taken for determining the dry matter and recalculation of colony-forming units (CFU) per 1 g of dry matter. Before this procedure, the medium should be heated in thermostat to +37°C temperature. If the samples from domestic or wild birds (chickens, geese, ducks, etc.), the temperature of medium should be +40°C.

2.3 Samples from intestine (biopsy or sections of the large intestine of animals)

Intestinal SRB with other intestinal bacteria can form biofilms on the surface of the epithelial cells of the large intestine [34]. These biofilms include species of *Desulfovibrio* genus and the species of *Bacteroides*, *Pseudomonas*, *Clostridium*, *Escherichia,* or other intestinal microorganisms. Such biofilms are often resistant to antimicrobial substances [36]; therefore it is an interesting area of the study.

For isolation of SRB from biofilms, 10^{-5} M EDTA (ethylenediaminetetraacetic acid) should be added to the modified Postgate liquid medium for releasing SRB from a biofilm. A fresh piece of biopsy should be weighed, and its approximate square (in cm^2) must be calculated and added to 9 ml of the modified Postgate liquid medium (pH 7.5, *temperature* to +37°C). This calculation of square must be done for recalculation of CFU of SRB released from cm^2 of a biofilm.

The same procedure can be applied for isolation of SRB from sections of the large intestine of animals.

3. Medium and cultivation conditions

The composition of modified liquid Postgate medium [3, 35] is the following (g/l): Na_2SO_4 (0.5); KH_2PO_4 (0.3); K_2HPO_4 (0.5); $(NH_4)_2SO_4$ (0.2); NH_4Cl (1.0); $CaCl_2 \times 6H_2O$ (0.06); $MgSO_4 \times 7H_2O$ (0.1); lactate, $C_3H_5O_3Na$ (2.0); yeast extract (1.0); $FeSO_4 \times 7H_2O$ (0.004); sodium citrate, $C_6H_5O_7Na_3 \times 2H_2O$ (0.3); and distilled water (1 l).

Separated solutions: Mohr's salt solution [$(NH_4)_2Fe(SO_4)_2 \times 6H_2O$] (10%) and $Na_2S \times 9H_2O$ solution (1%) and 10 M solution of NaOH must be sterilized separately.

The modified liquid Postgate medium and solutions of Mohr's salt, sodium sulfide, and sodium hydroxide should be sterilized in autoclave (20 min, at 1 atm.). The sterilization provides sterile conditions and partial release of oxygen from the medium. The solution of sodium sulfide is hydrolyzed to hydrogen sulfide during autoclaving.

After sterilization, 10 ml/l of sterile Mohr's salt solution and 0.05 ml/l of sterile solution of sodium sulfide must be added to the medium. The addition of a small quantity (one drop) of sodium sulfide solution to the medium makes visible a black ring which confirms interactions of hydrogen sulfide and free Fe^{2+} released from Mohr's salt.

A sterile ascorbic acid solution also must be added to the medium, but it cannot be sterilized by autoclaving because it may partially decompose and lose its properties for redox potential. So, 20% ascorbic acid solution should be filtrated through membrane filters (0.2 μm) and added directly to the medium after sterilization. The final concentration of ascorbic acid in the medium should be 0.1 g/l, and the redox potential of the medium must be around −100 mV. Solution of hydrogen sulfide added to medium can also decrease a redox potential [3].

The redox and anaerobic conditions can be controlled by sodium resazurin as an indicator. In addition, FeS reduced and Na_2S contained in the medium provides the necessary redox conditions for SRB cultures. The discoloration of sodium resazurin (redox potential of discoloration Eh = −100 mV) confirms the decrease of redox potential. A pH medium (7.5) provides by the addition of a sterile 10 M solution of NaOH.

The temperature of the media should be +25...+30°C for environmental samples, and + 37°C for intestinal samples (+40°C for samples from birds).

The tubes with samples must be completely filled up to the edges of the test tube with completed medium and closed with rubber stoppers. In another case, tubes can be filled up incompletely, but 1 ml of sterile liquid paraffin must be filled up to the top of the medium and closed with rubber stoppers.

As a control of the quality of the medium, known pure culture of SRB from some collections of microorganisms is recommended to also be used.

Cultivate in the thermostat at +25...+30°C, +37°C, or +40°C, depending on the origin of the sample, during for 1–5 days under anaerobic conditions. SRB from birds, animals, and humans mostly grow faster than environmental species.

Positive growth of SRB is indicated by observing a black FeS precipitate occurred in the bottom of the tube.

4. Isolation and purification of positive SRB samples

As already mentioned above, SRB are in close interactions with other microorganisms and can form biofilms in which they may be in a symbiotic relationship [34, 37]. Such microorganisms cooperating with SRB are often called satellite microorganisms [3]. Among the intestinal microorganisms, the species of the *Bacteroides*, *Pseudomonas*, *Clostridium*, and *Escherichia* genera are most often detected. Phototrophic green sulfur bacteria can make consortium with SRB [38]. On one hand, SRB produce hydrogen sulfide, and on the other hand, green sulfur bacteria oxidize hydrogen sulfide to molecular sulfur in the process of anoxygenic photosynthesis. Molecular sulfur may subsequently be oxidized to sulfate, in which SRB can be used as a final electron acceptor. Such an example of interaction can be consortium *Pelochromatium roseo-viridae* [11]. That is why it is important to purify the mixed cultures of SRB from satellite microorganisms which are very difficult to remove of during this process.

For obtaining pure cultures of SRB colonies, positive SRB samples (mixed SRB cultures) should be diluted (1,9) in a series of tubes (to 10^{-5}) containing the modified Postgate liquid medium. The scheme of the series of tubes is presented in **Figure 2**. Before it, the modified Postgate agar medium of the same composition like liquid should be prepared but in this case adds to the medium additional compounds: Na_2SO_3 (7.5 g/l) and microbiological agar (12 g/l). Sterilize it by autoclaving like Postgate liquid medium. Sodium sulfite in high concentration in medium inhibits most of intestinal species of *Enterobacteriaceae* family, including *Bacteroides*, *Pseudomonas*, and *Clostridium*, *Escherichia*, which can be satellites of SRB. The species of SRB are resisted to sulfite ions and can be used as an alternative electron acceptor [11] in the process of dissimilatory sulfate reduction since they have sulfite reductase activity [39].

The modified Postgate agar medium containing sodium sulfite (Na_2SO_3) after sterilization in autoclave should be cooled to +40°C and 10 ml/l of sterile Mohr's salt solution, 0.05 ml/l of sterile solution of sodium sulfide and ascorbic acid (0.1 g/l) added to the medium. These components must be thoroughly mixed in the flask and a sterile 10 M solution of NaOH added to provide accordingly a pH depending on the samples. To prevent the medium solidation, use a water bath to keep the temperature (+40°C) at a constant level.

In total, 20 ml of warm modified Postgate agar medium spill in Petri palates and add to the medium 100 μl of each diluted suspension of a positive sample, thoroughly mix the suspension with the warm medium. The temperature should be according to the sample from where it was isolated.

Filled with medium and suspension Petri plates introduce into an anaerobic box with oxygen uptake generators for anaerobiosis. Mohr's salt in the agar medium allows to detect black colonies of SRB since as a result, FeS was formed by hydrogen sulfide bacterial production that caused black-colored colonies. Cultivate in the thermostat at the appropriate temperature. The black colonies will be visible in 1–5 days in the deep of agar medium depending on sample and its dilution.

The black colonies obtained from Petri palates cut from agar and suspend in modified liquid Postgate medium. Cultivate in the thermostat at the appropriate temperature. The formation of black sediment (FeS precipitate) will be visible in the tube (about in 1–3 days). This sediment confirms sulfate reduction and

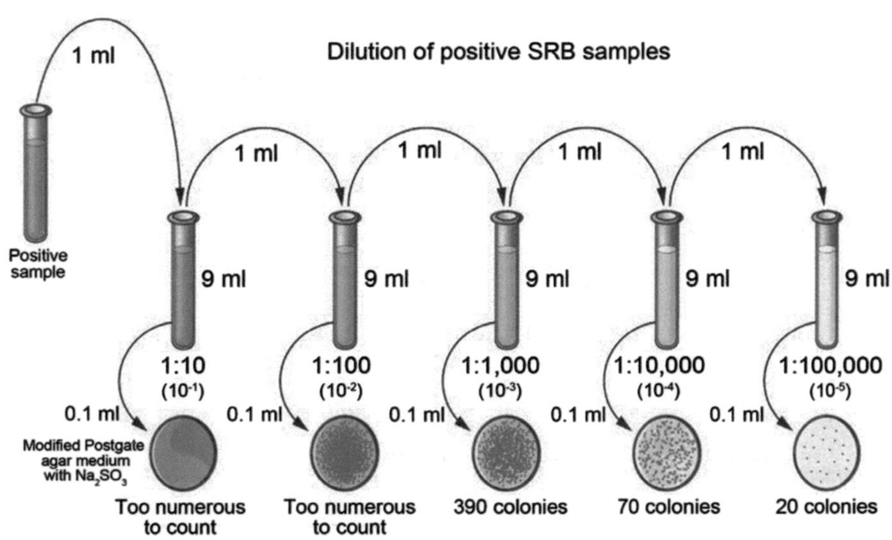

Figure 2.
The scheme of dilution of positive SRB samples (mixed cultures).

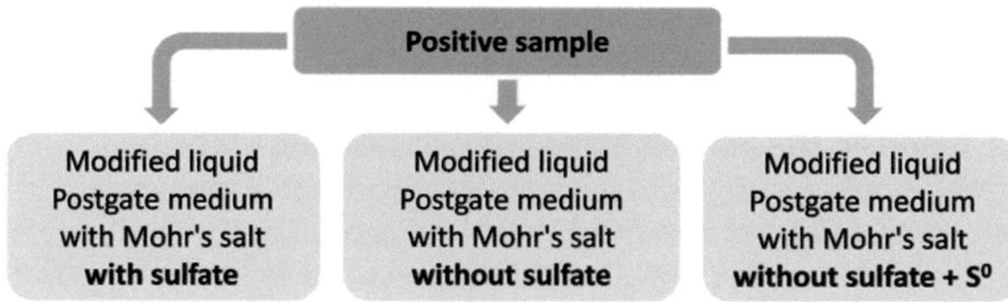

Figure 3.
The scheme of the confirmation that the isolates belong to the SRB.

production of hydrogen sulfide, which interact with Fe^{2+} from Mohr's salt, and FeS precipitate is formed. However, hydrogen sulfide can also be produced by species of *Clostridium*, *Salmonella*, and other intestinal bacteria. Moreover, some sulfur-reducing bacteria in the same case can also use sulfate as an electron acceptor [11]. To be sure that the selected microorganisms are not sulfur reducers or other bacteria capable of hydrogen sulfide production, the liquid media following the composition should be prepared: first modified liquid Postgate medium with sulfate (concentration 3.5 mM), second the same medium but without sulfate, and third the same medium but without sulfate containing molecular sulfur (0.5 g/l). The scheme of the confirmation that the isolates belong to the SRB is presented in **Figure 3**.

The grown black sample should be mixed and 100 μl of bacterial suspension pipetted into Eppendorf tubes (volume 1.5 ml) with 900 μl of liquid media by the scheme (**Figure 3**). Pipette 200 μl of sterile liquid paraffin on the surface of the media with suspension, and close a cap of Eppendorf tubes. Cultivate in thermostat.

If the sample after cultivation forms a black sediment in the modified liquid Postgate medium without sulfate ions that contained molecular sulfur, it means that isolates in a positive sample can belong to the sulfur-reducing bacteria (not SRB).

If the sample after cultivation does not form black sediment (FeS precipitate) in modified liquid Postgate medium without sulfate and the same medium without sulfate ions that contained molecular sulfur, but bacterial growth is observed in the medium with sulfate, it means that isolates in a positive sample belong to the SRB.

The positive sample with SRB culture should be diluted in the modified liquid Postgate medium and again seed each dilution in agar medium containing sodium sulfite (see **Figure 2**). This procedure must be repeated 3–5 times for full purification of SRB from other bacterial satellites.

After that, to check the purity of the SRB cultures from satellites, other additional tests are necessary. These additional tests are bacterial growth on the growth on different nonselective media (meat peptone agar; wort agar; starch-and-ammonia agar; Giltay's, Baalsrud's, and modified Postgate medium). Growth of SRB should be positive only in modified Postgate medium.

5. Morphological diversity: physiological and biochemical properties

The SRB cells are spherical, oval, rod-shaped, spiral, or vibrio-shaped with a diameter of 0.4–3.0 μm. The cells can be either single or in pairs or aggregates also may form a single row of multicellular filaments [1, 3]. Most cells of SRB genera are Gram-negative, although the filamentous and spore-forming microorganisms are Gram-positive. The SRB genera are anaerobes [11]. Morphology of SRB cells can be studied by using the light microscope, phase-contrast microscopy, or electronic microscopy.

Some species of SRB have single flagellum or more flagella depending on the genus. A simple, qualitative, and rapid method for detecting bacterial flagella and their shape, length, curvature, arrangement, and number on the cell is Hardy Diagnostics Flagella Stain (HDFS) [40, 41]. In 1937, Ryu developed this method, and later Kodaka et al. further described it [42, 43]. This test is especially useful in taxonomy and identifying characteristic about SRB motile, and more recently, anaerobic bacteria. Due to their narrow diameter, SRB flagella cannot be seen with a light microscope. The method of flagella stain can provide viewing SRB flagella by employing a crystal violet in an alcoholic solution as the primary stain. The alcoholic solution evaporates and leaves a precipitate around the flagella during the staining procedure and in increasing its apparent size.

In addition to the cell morphology and the presence of flagella, the following physiological characteristic, which is no less important, is also the formation of spores. However, among the heterogeneous quantity of SRB, the species of *Desulfotomaculum* genus can sporulate. To determine the ability of the SRB cells to sporulate, 1 ml of 72 h pure culture of SRB grown in modified liquid Postgate medium should be heated at +80°C for 15 min and then 100 μl of bacterial suspension pipetted into epindorph (volume 1.5 ml) with 900 μl of liquid media, and add 200 μl of sterile liquid paraffin on the surface of the media with suspension, and close the cap of Eppendorf tubes. Cultivate in thermostat. Thermoresistant forms of the *Desulfotomaculum* genus can be observed by FeS precipitate in the Eppendorf tubes. The SRB spores can be also additionally detected by a staining method for endospore. This method was published by Dorner [44] and later modified by Schaeffer and Fulton [45]. The modified process is simpler and faster and commonly used to differentiate bacterial endospores from other vegetative cells. It is also used to differentiate spore-forming bacteria from nonspore-forming [45].

Other physiological and biochemical characteristics which are important for identification are the determination of SRB growth at various pH and temperature, biomass accumulation, sulfate/lactate consumption, hydrogen sulfide and acetate production, catalase test, indole test, nitrate reduction, carbohydrate fermentation, gas production, and desulfoviridin test (**Figure 4**).

The effect of acidity (pH) is one of many important environmental factors which can be used for physiological characteristics of new SRB strains. The decreasing and

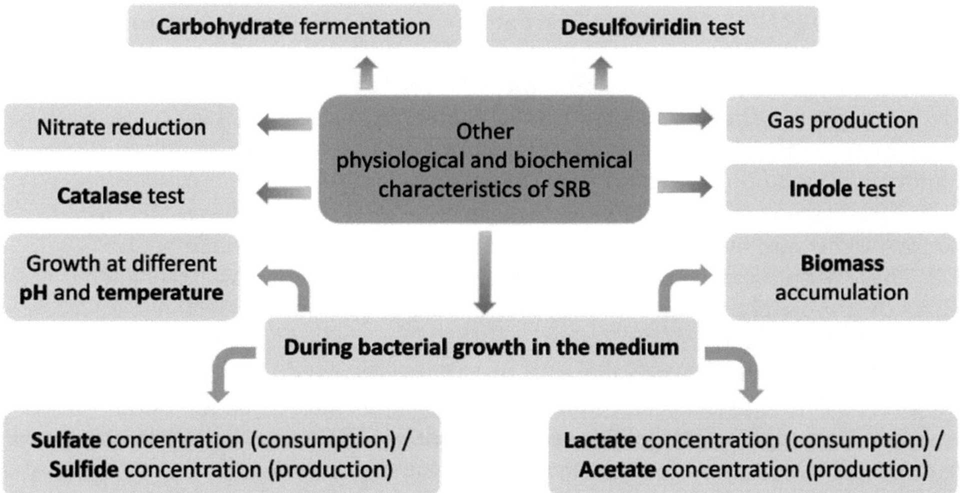

Figure 4.
The scheme of other physiological and biochemical characteristics.

increasing acidity of the medium can lead to the decrease of the SRB growth rate and hydrogen sulfide production [25]. Furutani and Schindler reported that the process of dissimilatory sulfate reduction was significantly slowed at low pH [46]. The increasing of the pH medium until 9.0–10.0 also caused growth inhibition of the studied bacteria [25]. To test the pH effect on the SRB growth, the modified liquid Postgate medium (*without Mohr's salt*) with different pH 3, 4, 5, 6, 7, 8, 9, and 10 is necessary to be prepared. Inoculation (initial concentration) of bacterial cells should be not less than 10%. After cultivation in the thermostat (24–36 h), biomass accumulation can be determined and compared in which the value of pH is optimum for SRB growth.

Most of the species of SRB are mesophilic microorganisms and live at a temperature from +20 to +40°C. Some SRB species can be also thermophilic microorganisms, e.g., *Thermodesulfobacterium* genus (*T. thermophilum*, *T. hveragerdense*, *T. commune,* and others) [11]. However, this chapter is focused on isolation and purification of mesophilic SRB. Similar to the case with test pH effect, the optimum of temperature for SRB growth is necessary to be determined. Inoculate bacterial cells (10%) in the modified liquid Postgate medium (*without Mohr's salt*), and cultivate at different temperatures (+4, +14, +20, +35, +45°C). After cultivation (24–36 h), biomass accumulation can be determined and compared in which the value of the temperature is optimum for SRB growth.

Biomass accumulation of the SRB cells in liquid medium can be measured by the photometric method by using a spectrophotometer, but the medium cannot contain Mohr's salt, since FeS precipitate makes it impossible [26, 47].

The cultivation of SRB in anaerobic, microaerophilic, or aerobic conditions allows testing their viability and resistance to molecular oxygen. However, SRB are anaerobes, but some of them may have high activity of antioxidant enzymes, catalase, and superoxide dismutase [1, 3].

Sulfate consumption as a terminal acceptor and determination of its concentration in the medium during SRB growth is important for observing and understanding more the process of dissimilatory sulfate reduction. The sulfate concentration in the medium (*without Mohr's salt*) can be assayed by the turbidimetric method by precipitation with barium chloride. For stabilizing the suspension, glycerol should be used [48].

The final product of the dissimilatory sulfate reduction process is hydrogen sulfide, which can be measured in the culture medium (*without Mohr's salt*) by a photometric method based on the reaction of sulfide and n-aminodimethylaniline with the methylene blue formation [49]. The concentration of hydrogen sulfide is determined by calibration curve. The data on the concentration of hydrogen sulfide, produced by the isolates, is supposed to help in establishing and assessing a toxicity effect of hydrogen sulfide on the epithelial cells of the human intestine. Such studies might help in predicting the development of diseases in the gastrointestinal tract, by providing further details on the etiology of bowel diseases which are very important for the clinical diagnosis of these disease types.

In the dissimilatory sulfate reduction process, SRB use exogenous electron donors. Molecular hydrogen is a universal electron donor for intestinal SRB [23, 37]. These bacteria are in close interaction with each other. It was established that SRB can completely displace methanogenic microorganisms of the intestine in the process of H_2 competition [9]. This competition for molecular hydrogen between SRB and methanogens largely depends on the presence and quantity of sulfate in the gut [9]. Adding sulfate and sulfated mucopolysaccharides to fecal suspensions which contain metabolically active products of the SRB stimulates the formation of hydrogen sulfide and inhibits the intensity of the methanogenesis [1, 14]. Except H_2, the second important electron donor is lactate, which SRB can oxidize incompletely to acetate or completely to CO_2.

The determination of lactate concentration can be carried out through dehydrogenation of lactate reaction by lactate dehydrogenase in the presence of NAD^+, with formation of pyruvate and NADH. Another method for measurement of lactate concentration is the use of lactate assay kit (Sigma-Aldrich, Catalog Number MAK064). Acetate accumulated during lactate incompletely oxidizing in the process of bacterial growth can be determined by using the acetate assay kit (Colorimetric, Catalog Number KA3764) or by titration.

Simple catalase test on modified Postgate surface agar cultures can be carried out by adding a drop of 10% H_2O_2 solution over the colonies. Another way is adding 5 drops of 10% H_2O_2 solution in 1 ml of a modified liquid Postgate medium. If the culture is catalase positive, the bubbles are formed.

The indole production test can be carried out by using a 24-h liquid culture with nitric acid and isoamylic alcohol reagents (Salkowski's reaction).

Adding sodium nitrate (5%) to modified liquid Postgate medium can be used for testing nitrate reduction. Nitrites can be tested by using a naphthylamine-sulfanilic acid reagent on 24-h cultures.

The ability of SRB strains to metabolize except lactate or H_2 other electron donors and a carbon source is also necessary to test. With this purpose, formate, propionate, pyruvate, fumarate, malate, methanol, citrate, ethanol, acetate, glycerol, glucose, oleate, stearate, and benzoate should be added separately in modified liquid Postgate medium but without electron donor (lactate) and carbon source. A final concentration of each compound should be 1%. Glucose and pyruvate fermentation in the liquid medium can be analyzed by acidity (pH reaction) and pH indicators. This test confirms that SRB isolated strains are capable to *chemolithoheterotrophic* growth. In addition to organic acids with different carbon chain lengths and alcohol, the strains can also assimilate some amino acids.

Gas production can be observed in deep culture Postgate agar in the tubes.

The desulfoviridin production is a very important factor for identification of *Desulfovibrio* and *Desulfomonas* genera. The presence of this protein in bacterial cells can be examined by using ultraviolet (UV) light on Postgate agar surface cultures after treatment with a 1 N NaOH solution. Desulfoviridin will be green in UV light.

6. Identification based on physiological and biochemical properties and sequence analysis of the 16S rRNA gene

Identification of the SRB by morphological, physiological, and biochemical characteristics can be conducted according to Bergey's Manual of Determinative Bacteriology (ninth edition, 1994), where SRB belong to the seventh group and are called "dissimilatory sulfate- or sulfur-reducing bacteria" [11]. This group is divided into four subgroups (**Figure 5**).

However, more modern and complex classification of SRB is published in Bergey's Manual of Systematic Bacteriology (2005), where SRB are divided into different classes, for example, class IV, *Deltaproteobacteria*, including order II, *Desulfovibrionales*; family I, *Desulfovibrionaceae* (genus I. *Desulfovibrio*); or family II, *Desulfomicrobiaceae* (genus I. *Desulfomicrobium*) [10]. For details identification based on physiological and biochemical characteristics is necessary to use both Bergey's manuals.

As was mentioned above, the representatives of *Desulfovibrio* genus are often found in the animal and patients with IBD and healthy subjects, because it is necessary to pay attention to the more detailed steps for identification of the second subgroup where this genus belongs (see **Figure 4**).

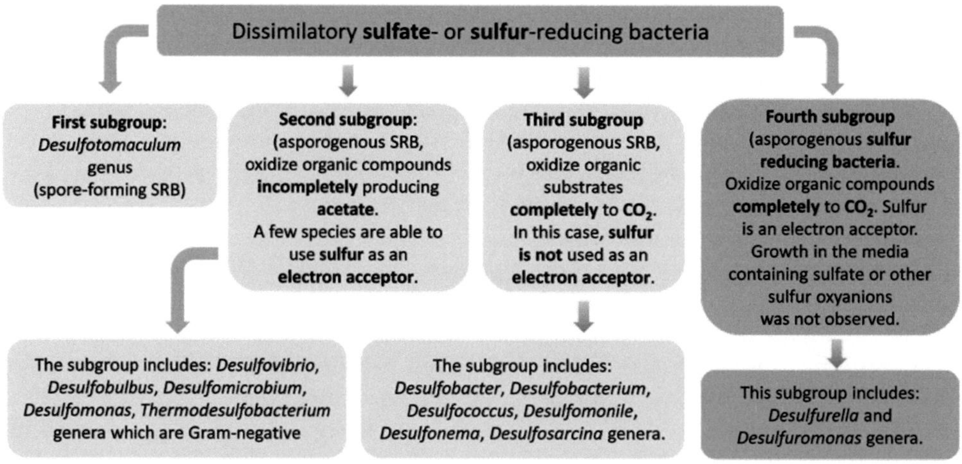

Figure 5.
The scheme of classification of dissimilatory sulfate- or sulfur-reducing bacteria group, according to Bergey's manual of determinative bacteriology (ninth edition, 1994) [11].

Features	Dvi	Dmo	Dbu	Dmi	Tdb
Spiral or vibrio-shaped cells	+	—	—	—	—
Oval or rod-shaped cells	—	+	+	+	+
Movement with the polar flagella	+/–		+/–	+	+/–
Optimal temperature range					
+25...+40°C	+	+	+	+	—
+65...+70°C	—	—	—	—	+
Ability of bacteria to grow in the presence of sulfate					
H_2 + CO_2 + acetate as a carbon source	+	+	+	+	+
Lactate	+	+	+	+	+
Propionate	—	—	+	—	—
Desulfoviridin	+	+	—	—	—

Notes: *the feature is presence "+" or absence "–".*

Table 1.
The differences between features of the second subgroup SRB genera according to Bergey's manual of determinative bacteriology [11].

The second subgroup includes *Desulfovibrio* (Dvi), *Desulfomonas* (Dmo), *Desulfobulbus* (Dbu), *Desulfomicrobium* (Dmi), and *Thermodesulfobacterium* (Tdb) genera (**Table 1**).

Other SRB genera can be identified by Bergey's manuals [10, 11]. However, for complete identification based on morphological, physiological, and biochemical properties, the molecular methods, in particular the sequence analysis of 16S rRNA gene, are also necessary to be applied [50]. Except sequence analysis of 16S rRNA gene, it is important to confirm the SRB species by using primers of functional genes of dissimilatory sulfate-reduction, such as *DsrAB* and A*prBA* (**Table 2**).

Further on the example of one isolate of intestinal SRB, identification based on sequence analysis of 16S rRNA gene by using the universal primers will be described. The schema of this identification is presented in **Figure 6**.

Functional genes	Primer sequence	Amplicon length (pb)
DsrAB gen		
DSR1F	5'-ACSCAYTGGAARCACG-3'	1900
DSR4R	5'-GTGTARCAGTTDCCRCA-3'	
AprBA gen		
aprB-1-FW	5'-TGCGTGTAYATHTGYCC-3'	1200–1350
aprA-5-RV	5'-GCGCCAACYGGRCCRTA-3'	

Table 2.
Primers designed based on functional gens of dissimilatory sulfate reduction, which can be used for amplification.

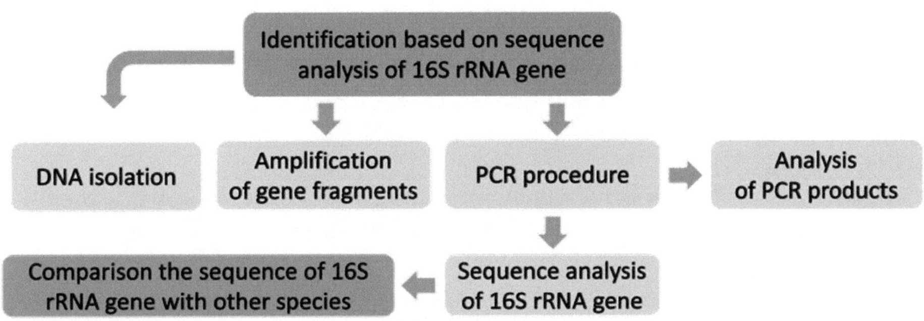

Figure 6.
The scheme of identification based on sequence analysis of 16S rRNA gene.

DNA isolation. Isolation and purification of DNA were carried out with a 72-h culture of SRB by using a "QIAmp DNA Mini Kit (QIAGEN), Cat. No 51304." One single SRB colony was taken from modified Postgate agar medium and suspended in 50 μl of deionized water in a screw cap micro-centrifuge tube. The samples were boiled at 98°C for 5 min prior to being centrifuged for 5 min/14,000 g to settle cell debris. In total 2 μl of supernatant, containing the genomic DNA, were used for PCR amplification.

Amplification of gene fragments. Amplification of 16S rRNA gene fragments was carried out using the universal primers (**Table 3**) according to Weisburg et al. [51] and Persing [52].

PCR procedure. PCR was carried out on DNA isolated from SRB cells in a final volume of 20 μl consisting of 10.0 μl Taq PCR Master Mix Kit (Cat. No 201445), 0.1 μl of each primer, 0.1 μl uracil D-glycosylase (Cat. No. M0280 L), 7.7 μl deionized water, and 2.0 μl of DNA supernatant.

The amplicons were amplified by a preliminary incubation at 94°C for 5 min (initial denaturation) and then 34 cycles of 94°C for 1 min (denaturation), 55°C for 1 min (annealing of primers), and 72°C for 2 min (polymerization), using a

Primers	Sequence	Amplicon length (pb)
8FPL	5'-AGTTTGATCCTGGCTCAG-3' position 8–27	Approximately 1500
1492RPL	5'-GGTTACCTTGTTACGACTT-3' position 1510–1492	
806R	5'-GGACTACCAGGGTATCTAAT-3' position 806–787	Approximately 800

Table 3.
Universal primers for amplification of 16S rRNA gene fragments.

thermocycler (model MJ Research PTC-200, USA). After the last amplification cycle, the samples were incubated further at 72°C for 2 min for complete elongation of the final PCR products and cooled at 10°C.

Analysis of PCR products. Analysis of PCR products was carried out by electrophoresis in 1.5% agarose gel, with field strengths of 5 V/cm. Electrophoresis time was 40 min. The 100 bp ladder (Malamité, Czech Republic) was used as a size standard and molecular weight markers. Isolation and purification of fragments from agarose were performed by centrifugation of gel strips containing DNA through aerosol filters. For purification of SRB amplicons, the commercial kit from QIAGEN "MinElute Gel Extraction Kit" was used. The sequence was carried out using a "genetic analyzer" and reagents "BigDye Terminator v3.1 Cycle Sequencing Kit." Search homologous deposited in the GenBank nucleotide sequence encoding the 16S rRNA gene, was performed using BLASTN and Blast2 programs.

The 16S rRNA gene amplicons which were used for sequence analysis were obtained by using the PCR method. The PCR products were separated by electrophoresis (**Figure 7**). Before sequence analysis the absorbance of amplicons (8FPL/806R, amplicon I about 800 bp; 8FPL/1492RPL, amplicon I about 1500 bp; 8FPL/806R, amplicon II about 800 bp; 8FPL/1492RPL, amplicon II about 1500 bp) was determined [50].

By comparison of individual sequencing data from the amplicons 1–5, the following gene for 16S rRNA sequence of the total length 1370 bp was completed:

TTCGGTCCCGAGTAAAGTGGCGCACGGGTGAGTAACACGTGGATGATCTGCCTCTATGATGGGATA
ACAGTTGGAAACGACTGCTAATACCGAATACGCTCATGATGAACGTTGTGAGGAAAGGTGGCCTCTG
CTTGCAAGCTATCGCATAGAGATGAGTCCGCGTCCCATTAGCTCGTTGGTGGGGTAACGGCCTACCA
AGGCAACGATGGGTAGCCGATCTGAGAGGATGATCGGCCACACTGGAACTGAAACACGGTCCAGAC
TCCTACGGGAGGCAGCAGTGGGGAATATTGCGCAATGGGCGAAAGCCTGACGCAGCGACGCCGCGT
GAGGGATGAAGGTCTTCGGATCGTAAACCTCTGTCAGAAGGGAAGAAACTGGGGTGTTCTAATCAG
CATCCCACTGACGGTACCTTCAAAGGAAGCACCGGCTAACTCCGTGCCAGCAGCCGCGGTAATACGG
AGGGTGCAAGCGTTAATCGGAATCACTGGGCGTAAAGCGCACGTAGGCTGTTATGTAAGTCAGGGG
TGAAAGCCCACGGCTCAACCGTGGAACTGCCCTTGATACTGCACGACTCGAATCCGGGAGAGGGTGG
CGGAATTCCAGGTGTAGGAGTGAAATCCGTAGATATCTGGAGGAACATCAGTGGCGAAGGCGGCCA
CCTGGACCGGTATTGACGCTGAGGTGCGAAAGCGTGGGGAGCAAACAGGATTAGATACCCTGGTAG
TCCACGCCGTAAACGATGGATGCTAGATGTCGGGATGTATGTCTCGGTGTCGTAGTCAACGCGTTAA
GCATCCCGCCTGGGGAGTACGGTCGCAAGGCTGAAACTCAAAGAAATTGACGGGGGCCCGCACAAG
CGGTGGAGTATGTGGTTTAATTCGATGCAACGCGAAGAACCTTACCTGGGTTTGACATCTGGGGAAT
CCTCCCGAAAAGGAGGAGTGCCCTTCGGGGAGCCCCAAGACAGGTGCTGCATGGCTGTCGTCAGCTC
GTGTCGTGAGATGTTGGGTTAAGTCCCGCAACGAGCGCAACCCCTATGCATAGTTGCCAGCAGGTAG
AAGCTGGGCACTCTATGCAGACTGCCCGGGTTAACCGGGAGGAAGGTGGGGACGACGTCAAGTCAT
CATGGCCCTTACACCCAGGGCTACACACGTACTACAATGGCACGCACAAAGGGCAGCGATACCGTGA
GGTGGAGCCAATCCCAAAAAACGTGTCCCAGTCCGGATTGCAGTCTGCAACTCGACTGCATGAAGTC
GGAATCGCTAGTAATTCGAGGTCAGCATACTCGGGTGAATGCGTTCCCGGGCCTTGTACACACCGCC
CGTCACACCACGAAAGTCGGTTTACCCGAAGCCGGGGG

Black nucleotides are a summary from 1, 2, 3, and 5: the best sequence data (totally four sequencing). **Green nucleotides** are a summary from 3 and 4: good sequence data (two sequencing). **Red nucleotides** are the rest of 4: the worse sequence data because they were received from one sequencing only, but its quality was excellent. The obtained sequence BLASTN was analyzed. The highest homology of SRB colony was identified with *Desulfovibrio piger* ATCC 29098 from GenBank.

The obtained sequence results of SRB isolated colony were also compared by BLASTN analysis with the nucleotide sequences of 16S rRNA gene of other strains (**Table 4**).

Thus, the nucleotide sequence of the 16S rRNA gene of SRB has the highest homology (99%) compared to deposited nucleotide sequence *D. piger* ATCC 29098 (AF192152) in the GenBank database.

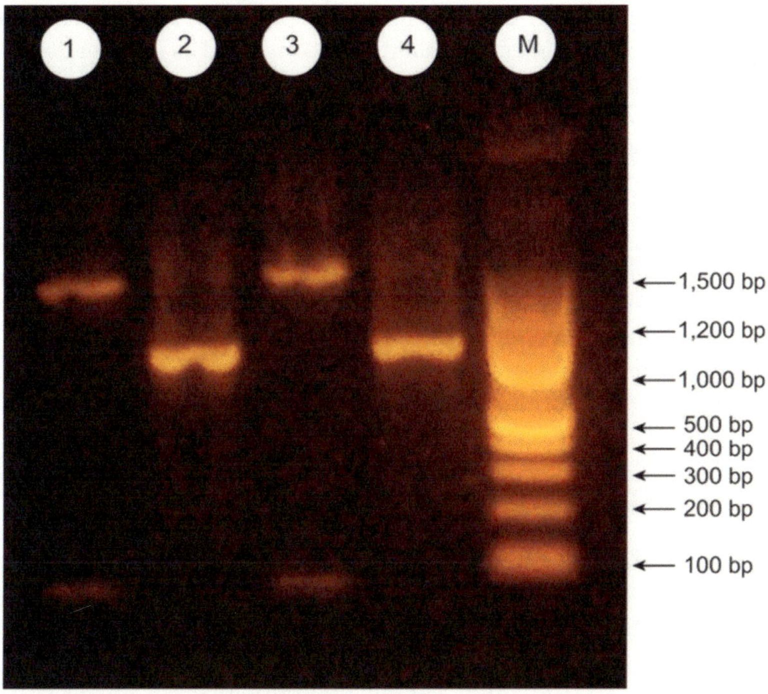

Figure 7.
The results of electrophoresis 16S rRNA PCR products (amplicons): 8FPL/1492RPL, 1500 bp (1); 8FPL/806R, 800 bp (2); 8FPL/1492RPL, 1500 bp (3); 8FPL/806R, 800 bp (4); ***M*** *is marker (100 bp ladder).*

SRB strains	Acc. No	Identities	Identity (%)
Desulfovibrio piger ATCC 29098	AF192152	1352/1368	99
Desulfovibrio fairfieldensis ATCC700045	U42221	1313/1374	96
Desulfovibrio desulfuricans Essex 6	AF192153	1299/1369	95
Desulfovibrio intestinalis DSM 11275	Y12254	1289/1373	94
Desulfovibrio desulfuricans MB	AF192154	1293/1373	94
Bilophila wadsworthia ATCC49260	L35148	1242/1369	91
Desulfovibrio vulgaris subsp. Oxamicus DSM 1925	AJ295677	1089/1195	91
Desulfovibrio longreachensis ACM 3958	Z24450	1253/1374	91
Desulfovibrio termitidis DSM 5308	X87409	1237/1372	90
Lawsonia intracellularis NCTC 12656	U30147	1215/1374	88
Desulfovibrio vulgaris subsp. vulgaris DSM 644	M34399	726/834	87

Table 4.
Comparing the resulting sequence of 16S rRNA gene with other Desulfovibrio *species.*

The *D. piger* belongs to sulfate-reducing bacteria which are usually considered as a commensal bacterium in humans [5, 10, 50]. More recently, *D. piger* has attracted more interest as it was found to be the most prevalent species of SRB in feces of patients with inflammatory bowel disease [20, 21, 25, 31]. The obtained bacterial strains have such phenotypic features as the presence of desulfoviridin, cytochrome c_3, and menaquinone MK-6. They oxidize organic compounds incompletely to acetate [10, 11].

Moore W.E. found SRB for the first time in people's feces and identified it as *Desulfomonas pigra* [53], which subsequently is reclassified as *Desulfovibrio piger* [16]. The described bacterial strains are similar to that of Moore et al. [53] except for the G–C content of the DNA, which is 64 mol%. Obligate anaerobic, sulfate-reducing, non-saccharolytic, non-proteolytic, nonspore-forming, and Gram-negative bacteria that are straight and vibrio-like and have rounded ends (0.8–1.0 × 2.5–10.0 μm) [10]. These microorganisms use lactate, pyruvate, ethanol, and hydrogen as electron donors for sulfate reduction. They oxidize lactate and pyruvate incompletely to acetate. The optimum temperature for growth is +37°C. Growth is not affected by 20% bile. Colonies on anaerobic blood agar are translucent, 2 mm in diameter, circular, and non-hemolytic. Cells contain desulfoviridin and cytochrome c_3. These bacteria isolated from human specimens (feces, peritoneal fluids, and intra-abdominal collections). The type strain, isolated from human feces, is ATCC 29098 [10].

7. Generalization of the research

Taking into consideration all research described in the chapter, it is necessary to generalize that isolation of mesophilic SRB from environmental samples (water, soil, swamp, etc.) and intestinal samples can be similar, although swamps and feces are required to determine dry matter of the samples. It is important to purify a positive sample of SRB from other satellite microorganisms such as *Clostridium*, *Bacteroides*, *Pseudomonas*, etc. With this aim, obtained SRB mixed cultures should be 3–5 times repassed to the modified Postgate agar medium with sulfite which inhibit the growth other microorganisms. Ability to growth of SRB mixed culture with high sulfite concentration allows to eliminate (purify SRB) from other microorganisms which can be in close interactions with SRB. The cultivation conditions depend on sample from where it is isolated. The key criteria for identification based on physiological and biochemical characteristics are the morphology of bacterial

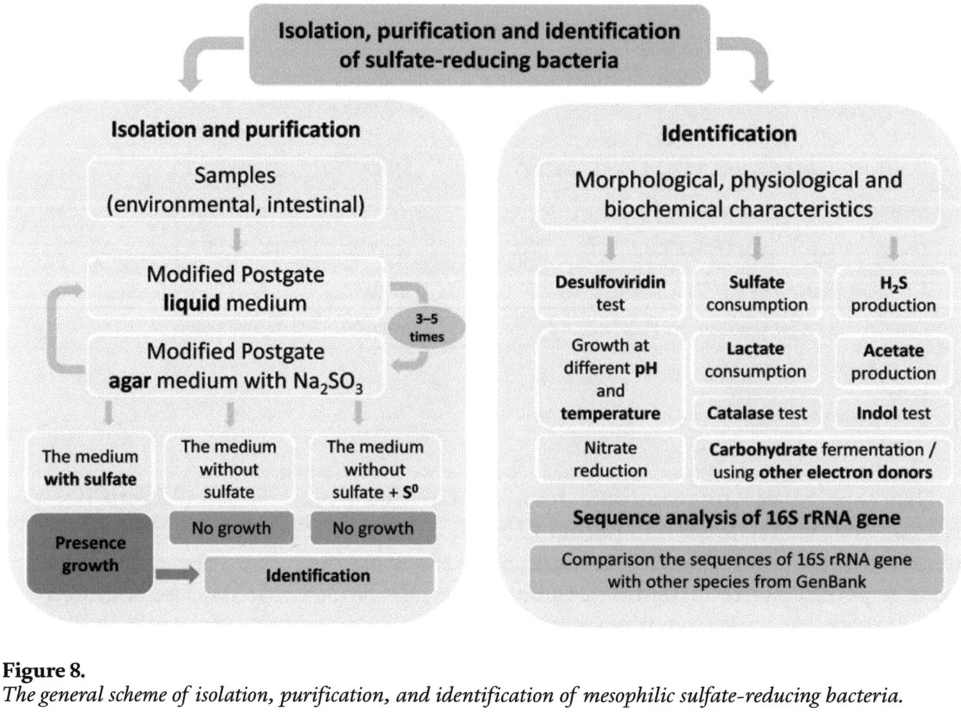

Figure 8.
The general scheme of isolation, purification, and identification of mesophilic sulfate-reducing bacteria.

cells, ability to form of spores, sulfate reduction to hydrogen sulfide, lactate oxidation to acetate or CO_2, use of other organic compounds as an electron donor and carbon sources, etc. The general scheme of isolation, purification, and identification of mesophilic sulfate-reducing bacteria is presented in **Figure 8**.

For identifications of SRB based on morphological, physiological, and biochemical characteristics, two Bergey's Manuals [10, 11] are recommended. Moreover, all isolated SRB species should be confirmed by the sequence analysis of the 16S rRNA gene by using universal primers or primers of functional genes of dissimilatory sulfate-reduction, such as *DsrAB* and A*prBA*.

8. Conclusions

The methods of sample selections from water, soil, swamp, and feces of humans or animals and from biopsy material and the process of SRB isolation and purifications are similar, although cultivation conditions may differ. Identification based on physiological and biochemical properties is a complex process, and many other factors must be considered. For this identification, Bergey's Manuals are recommended to be used. The sequence analysis of the 16S rRNA gene should confirm the identification process based on physiological and biochemical properties.

It is of vital importance to obtain new strains of the SRB from various ecotopes and identify them and study their growth and physiological and biochemical properties. Aside from that, the process of dissimilatory sulfate reduction by SRB and the production of hydrogen sulfide should be investigated in order to clarify the etiological role of these bacteria in the nature and in the development of various diseases.

Acknowledgements

This study was supported by Grant Agency of the Masaryk University (MUNI/A/0902/2018).

Conflict of interest

The authors declare no conflict of interest.

Author details

Ivan Kushkevych
Department of Experimental Biology, Faculty of Science,
Masaryk University in Brno, Czech Republic

*Address all correspondence to: kushkevych@mail.muni.cz

References

[1] Barton LL, Hamilton WA. Sulphate-reducing bacteria. In: Environmental and Engineered. 1st ed. Cambridge University Press; 2007. 552 p

[2] Kushkevych IV. Dissimilatory sulfate reduction in the intestinal sulfate-reducing bacteria. Studia Biologica. 2016;**10**(1):197-228. DOI: 10.30970/sbi.1001.560

[3] Postgate JR. The Sulfate-Reducing Bacteria. 2nd ed. Cambridge University Press; 1984. 199 p

[4] Kushkevych I, Kollar P, Suchy P, Parak K, Pauk K, Imramovsky A. Activity of selected salicylamides against intestinal sulfate-reducing bacteria. Neuroendocrinology Letters. 2015;**36**:106-113. Available at: https://www.ncbi.nlm.nih.gov/pubmed/26757109

[5] Kushkevych I, Fafula R, Parak T, Bartos M. Activity of Na^+/K^+-activated Mg^{2+}-dependent ATP hydrolase in the cell-free extracts of the sulfate-reducing bacteria *Desulfovibrio piger* Vib-7 and *Desulfomicrobium* sp. Rod-9. Acta Veterinaria Brno. 2015;**84**:3-12. DOI: 10.2754/avb201585010003

[6] Kushkevych IV. Activity and kinetic properties of phosphotransacetylase from intestinal sulfate-reducing bacteria. Acta Biochemica Polonica. 2015;**62**:1037-1108. DOI: 10.18388/abp.2014_845

[7] Kushkevych IV. Kinetic properties of pyruvate ferredoxin oxidoreductase of intestinal sulfate-reducing bacteria *Desulfovibrio piger* Vib-7 and *Desulfomicrobium* sp. Rod-9. Polish Journal of Microbiology. 2015;**64**:107-114. Available at: https://www.ncbi.nlm.nih.gov/pubmed/26373169

[8] Kushkevych I, Kollar P, Ferreira AL, Palma D, Duarte A, Lopes MM, et al. Antimicrobial effect of salicylamide derivatives against intestinal sulfate-reducing bacteria. Journal of Applied Biomedicine. 2016;**14**(2):125-130. DOI: 10.1016/j.jab.2016.01.005

[9] Gibson GR, Macfarlane GT, Cummings JH. Sulphate-reducing bacteria and hydrogen metabolism in the human large intestine. Gut. 1993;**34**:437-439. Available at: https://www.ncbi.nlm.nih.gov/pmc/articles/PMC1374298/

[10] Brenner DJ, Krieg NR, Staley JT, Garrity GM. Bergey's Manual of Systematic Bacteriology. Volume 2: The Proteobacteria, Part C: The Alpha-, Beta-, Delta-, and Epsilonproteobacteria. 2nd ed. United States of America; 2005. 1388 p

[11] Holt JG, Krieg NR, Sneath PH, Staley JT, Williams ST. Bergey's Manual of Determinative Bacteriology. 9th ed. Vol. 1. Baltimore: William & Wilkins; 1994. 426 p

[12] Kushkevych I, Vítězová M, Vítěz T, Kováč J, Kaucká P, Jesionek W, et al. A new combination of substrates: Biogas production and diversity of the methanogenic microorganisms. Open Life Sciences. 2018;**13**:119-128. DOI: 10.1515/biol-2018-0017

[13] Kushkevych I, Kováč J, Vítězová M, Vítěz T, Bartoš M. The diversity of sulfate-reducing bacteria in the seven bioreactors. Archives of Microbiology. 2018;**200**:945-950. DOI: 10.1007/s00203-018-1510-6

[14] Kushkevych I, Vítězová M, Vítěz T, Bartoš M. Production of biogas: Relationship between methanogenic and sulfate-reducing microorganisms. Open Life Sciences. 2017;**12**:82-91. DOI: 10.1515/biol-2017-0009

[15] Kováč J, Vítězová M, Kushkevych I. Metabolic activity of sulfate-reducing bacteria from rodents with colitis.

Open Medicine. 2018;**13**:344-349. DOI: 10.1515/med-2018-0052

[16] Rowan FE, Docherty NG, Coffey JC, O'Connell PR. Sulphate-reducing bacteria and hydrogen sulphide in the aetiology of ulcerative colitis. British Journal of Surgery. 2009;**96**:151-158. DOI: 10.1002/bjs.6454

[17] Beerens H, Romond C. Sulfate-reducing anaerobic bacteria in human feces. The American Journal of Clin Nutr. 1977;**30**:1770-1776. DOI: 10.1093/ajcn/30.11.1770

[18] Langendijk PS, Kulik EM, Sandmeier H, Meyer J, van der Hoeven JS. Isolation of *Desulfomicrobium orale* sp. nov. and *Desulfovibrio* strain NY682, oral sulfate-reducing bacteria involved in human periodontal disease. International Journal of Systematic and Evolutionary Microbiology. 2001;**51**(3):1035-1044. DOI: 10.1099/00207713-51-3-1035

[19] Kushkevych I, Vítězová M, Fedrová M, Vochyanová Z, Paráková L, Hošek J. Kinetic properties of growth of intestinal sulphate-reducing bacteria isolated from healthy mice and mice with ulcerative colitis. Acta Vetrinaria Brno. 2017;**86**:405-411. DOI: https://doi.org/10.2754/avb201786040405

[20] Loubinoux J, Bronowicji JP, Pereira IA, Mougenel JL, Faou AE. Sulphate-reducing bacteria in human feces and their association with inflammatory diseases. FEMS Microbiology Ecology. 2002;**40**:107-112. DOI: 10.1111/j.1574-6941.2002.tb00942.x

[21] Loubinoux J, Mory F, Pereira IA, Le Faou AE. Bacteremia caused by a strain of *Desulfovibrio* related to the provisionally named *Desulfovibrio fairfieldensis*. Journal of Clinical Microbiology. 2000;**38**:931-934

[22] Kushkevych I, Vítězová M, Kos J, Kollár P, Jampílek J. Effect of selected 8-hydroxyquinoline-2-carboxanilides on viability and sulfate metabolism of *Desulfovibrio piger*. Journal of Applied Biomedicine. 2018;**16**:241-246. DOI: 10.1016/j.jab.2018.01.004

[23] Gibson GR, Cummings JH, Macfarlane GT. Growth and activities of sulphate-reducing bacteria in gut contents of health subjects and patients with ulcerative colitis. FEMS Microbiology Ecology. 1991;**86**:103-112. DOI: 10.1111/j.1574-6941.1991.tb01742.x

[24] Kushkevych I, Dordević D, Kollár P. Analysis of physiological parameters of Desulfovibrio strains from individuals with colitis. Open Life Sciences. 2018;**13**(1):481-488. DOI: 10.1515/biol-2018-0057

[25] Kushkevych I, Dordević D, Vítězová M. Analysis of pH dose-dependent growth of sulfate-reducing bacteria. Open Medicine. 2019;**14**(1): 66-74. DOI: 10.1515/med-2019-0010

[26] Kushkevych IV. Identification of sulfate-reducing bacteria strains of human large intestine. Studia Biologica. 2013;**7**(3):115-124. DOI: 10.30970/sbi.0703.312

[27] Cummings JH, Macfarlane GT, Macfarlane S. Intestinal bacteria and ulcerative colitis. Current Issues in Intestinal Microbiology. 2003;**4**:9-20 Available at: https://www.ncbi.nlm.nih.gov/pubmed/12691258

[28] Zinkevich VV, Beech IB. Screening of sulfate-reducing bacteria in colonoscopy samples from healthy and colitic human gut mucosa. FEMS Microbiology Ecology. 2000;**34**: 147-155. DOI: 10.1111/j.1574-6941.2000.tb00764.x

[29] Kushkevych IV. Etiological role of sulfate-reducing bacteria in the development of inflammatory bowel diseases and ulcerative colitis. American Journal of Infectious Diseases and

Microbiology. 2014;**2**:63-73. DOI: 10.12691/ajidm-2-3-5

[30] Kushkevych I, Dordević D, Vítězová M, Kollár P. Cross-correlation analysis of the *Desulfovibrio* growth parameters of intestinal species isolated from people with colitis. Biologia. 2018;**73**:1137-1143. DOI: 10.1158/1541-7786.MCR-05-0126

[31] Kushkevych I, Kos J, Kollar P, Kralova K, Jampilek J. Activity of ring-substituted 8-hydroxyquinoline-2-carboxanilides against intestinal sulfate-reducing bacteria *Desulfovibrio piger*. Medicinal Chemistry Research. 2018;**27**:278-284. DOI: 10.1007/s00044-017-2067-7

[32] Kushkevych I, Dordević D, Vítězová M. Toxicity of hydrogen sulfide toward sulfate-reducing bacteria *Desulfovibrio piger* Vib-7. Archives of Microbiology. 2019;**201**(3):389-397. DOI: 10.1007/s00203-019-01625-z

[33] Pitcher MC, Cummings JH. Hydrogen sulphide: A bacterial toxin in ulcerative colitis? Gut. 1996;**39**:1-4. DOI: 10.1136/gut.39.1.1

[34] Macfarlane S, Dillon JF. Microbial biofilms in the human gastrointestinal tract. Journal of Applied Microbiology. 2007;**102**:1187-1196. DOI: 10.1111/j.1365-2672.2007.03287.x

[35] Kováč J, Kushkevych I. New modification of cultivation medium for isolation and growth of intestinal sulfate-reducing bacteria. Proceeding of International PhD Students Conference MendelNet. 2017:702-707

[36] Casals-Pascual C, Vergara A, Vila J. Intestinal microbiota and antibiotic resistance: Perspectives and solutions. Human Microbiome Journal. 2018;**9**:11-15. DOI: 10.1016/j.humic.2018.05.002

[37] Gibson GR, Macfarlane S, Macfarlane GT. Metabolic interactions involving sulfate-reducing and methanogenic bacteria in the human large intestine. FEMS Microbiology Ecology. 1993;**12**:117-125. DOI: 10.1111/j.1574-6941.1993.tb00023.x

[38] Overmann J, van Gemerden H. Microbial interactions involving sulfur bacteria: Implications for the ecology and evolution of bacterial communities. FEMS Microbiology Reviews. 2000;**24**(5):591-599. DOI: 10.1111/j.1574-6976.2000.tb00560.x

[39] Lie TJ, Godchaux W, Leadbetter ER. Sulfonates as terminal electron acceptors for growth of sulfite-reducing bacteria (*Desulfitobacterium* spp.) and sulfate-reducing bacteria: Effects of inhibitors of sulfidogenesis. Applied and Environmental Microbiology. 1999;**65**(10):4611-4617. Available at: https://www.ncbi.nlm.nih.gov/pubmed/10508097

[40] Jorgensen JH, Pfaller MA, Carroll KC. Manual of Clinical Microbiology. 11th ed. Washington, DC: American Society for Microbiology; 2015

[41] Koneman EW et al. Color Atlas and Textbook of Diagnostic Microbiology. Philadelphia, PA: J.B. Lippincott Company; 2016

[42] Ryu E. A simple method of staining bacterial flagella. The Kitasato Archives of Experimental Medicine. 1937;**14**:218-219

[43] Kodaka H, Armfield AY, Lombard GL, Dowell VR. Practical procedure for demonstrating bacterial flagella. Journal of Clinical Microbiology. 1982;**16**(5):948-952 Available at: https://www.ncbi.nlm.nih.gov/pubmed/6185531

[44] Dorner W. Ein neues verfahren für isolierte sporenfärbung. Landwirtschaftliches Jahrbuch der Schweiz. 1922;**36**:595-597

[45] Schaeffer AB, Fulton MD. A simplified method of staining endospores. Science. 1933;**77**(1990):194-194

[46] Furutani A, Schindler DW. Effects of lake acidification on rates of organic matter decomposition. Limnology and Oceanography. 1984;**29**:687-694. DOI: 10.4319/lo.1984.29.4.0687

[47] Sutton S. Measurement of microbial cells by optical density. Journal of Validation Technology. 2011;**17**(1):46-49

[48] Kolmert A, Wikstrom P, Hallberg KB. A fast and simple turbidimetric method for the determination of sulfate in sulfate-reducing bacterial cultures. Journal of Microbiol Methods. 2000;**41**:179-184. DOI: 10.1016/S0167-7012(00)00154-8

[49] Sugiyama M. Reagent composition for measuring hydrogen sulfide and method for measuring hydrogen. U.S. Pat. 6340596 B1 USA. Int. Cl. G 01 N 33/00; 2002

[50] Kushkevych I, Bartos M, Bartosova L. Sequence analysis of the 16S rRNA gene of sulfate-reducing bacteria isolated from human intestine. International Journal of Current Microbiology and Applied Sciences. 2014;**3**:239-248

[51] Weisburg WG, Barns SM, Pelletier DA, Lane DJ. 16S ribosomal DNA amplification for phylogenetic study. Journal of Bacteriology. 1991;**173**:697-703. DOI: 10.1128/jb.173.2.697-703.1991

[52] Persing DH. Molecular Microbiology: Diagnostic Principles and Practice. 2nd ed. ASM Press; 2011. 960 p

[53] Moore WE, Johnson JL, Holdeman LV. Emendation of *Bacteroidaceae* and *Butyrivibrio* and descriptions of *Desulfomonas* gen. Nov. and ten new species of the genera *Desulfomonas*, *Butyrivibrio*, *Eubacterium*, *Clostridium* and *Ruminococcus*. International Journal of Systematic Bacteriology. 1976;**26**:238-252. DOI: 10.1099/00207713-26-2-238

Section 2

Bacteria in Nature

Chapter 3

First Insights into the Resilience of the Soil Microbiome of a Tropical Dry Forest in Puerto Rico

Michelle Jovanne Rivera-Rivera and Elvira Cuevas

Abstract

This study evaluated the effect that tree species traits and wet/dry periods display on soil microbial communities in a tropical dry forest in Puerto Rico. Understanding the ecological role of soil microorganisms in tropical dry forests and how they relate to different tree species is necessary to protect these fragile forest ecosystems. Thus, by using 454 pyrosequencing, we explored how microbial diversity was affected by dominant tree species during the wettest and driest periods at the Guánica Dry Forest. We found that 9 out of 17 phyla were more abundant during the dry period demonstrating that soil communities have adapted to historically low rainfall patterns. The most abundant phyla during both periods were Proteobacteria, Actinobacteria, and Bacteroidetes. During the dry period, Actinobacteria increased significantly ($p < 0.0001$), whereas Proteobacteria and Bacteroidetes decreased significantly ($p < 0.0001$; $p < 0.001$). Canonical correspondence analysis (CCA) also demonstrated that soil microbes are shaped by wet and dry periods, thus axis 1 of CCA explained 80% of the variation. This study offers baseline information in order to help elucidate how microbial diversity is affected by climate change in tropical areas and extrapolate this information to agricultural areas in order to develop better management practices.

Keywords: historical rainfall patterns, bacterial resilience, soil microbiome, soil microbial ecology, soil enzyme activity, Guánica Dry Forest, Puerto Rico, bacterial diversity, DNA sequencing

1. Introduction

Arid and semiarid ecosystems comprise almost 1/3 of the Earth's surface, and it is expected that these ecosystems will increase their total coverage area due to anthropogenic activities and climate change [1]. In tropical dry forests, seasonality and rainfall distribution fluctuate more often than in other ecosystems. Dry periods can extend for many months, and in some cases, they are accompanied by pulsed rainfall that can last from hours to days. These fluctuations control temporal growth patterns, productivity, turnover of organic matter, and other forest soil functional traits [2]. After a dry period, the first pulse of rainfall causes abrupt changes in soil moisture and water potential leading to microbial physiological stress and the reawakening of soil microbial communities. Seasonally tropical dry forests are already towards the extreme of water availability. Climate model predictions for

IntechOpen

the Caribbean point towards progressively drier periods, with precipitation loss between −10 and −50% [3]. There is limited information regarding the diversity of soil microbial communities in these ecosystems, and it needs to be assessed in order to establish baseline information that is crucial to help elucidate the degree of the ecosystems resilience to the proposed precipitation changes that are affecting these ecosystems.

The intrinsic effects of vegetation are strong influencers of soil properties. Due to the confounding factor of plant species and plant roots sharing the same area, there is very little information on the effect of specific plant species on microbial diversity and soil enzymatic activities [4, 5]. Seasonally dry ecosystem, such as the Guánica Dry Forest, can serve as a model system to better understand the impact of seasonal variations and tree species effect on microbial community composition and activity. In this forest the trees are growing in the cracks of the calcareous platform, forming individual islands of leaf litter and organic matter under similar climatic conditions [5, 6]. The canopies of the trees are close to ground level limiting the transfer of leaf litter between neighboring trees, thus forming individual islands of fertility. These individual islands of fertility prevent belowground competition for resources [6]. Given that plant species vary in their effects on soil properties [7–9], one of our objectives was to determine how responsive the soil microbial diversity is to plant species effects. Our second objective was to understand to what degree the soil microbial diversity shows resilience to rainfall variability and dry periods. The study was conducted during the rainiest period of 2011 (August) and the driest period of 2012 (January). We selected three tree species (a pantropical species and two native species) that are highly distributed in the forest [10]. We hypothesize that both rainfall and plant species will regulate of modify the soil microbiome.

2. Materials and methods

2.1 Study site and sampling

The study was carried out at the Guánica Dry Forest which is situated in the semiarid region of Southwestern Puerto Rico (**Figure 1A–C**). Trees in this area are dwarfed, and the vegetation is located between 0 and 150 m from the coastline. The mean annual precipitation of this zone is 750 mm and exhibits a bimodal distribution. Even though monthly rainfall is highly erratic, most studies have documented that almost 50% of the annual precipitation occurs between September and November [11, 12]. Historical data also demonstrates that the forest also exhibits two periods of predominant drought that start in January and June. The data presented here correspond to samples that were taken during July 2011 (wet) and January 2012 (dry) representing the months of higher (wet) and lower (dry) rainfall [13], therefore allowing us to measure maximum and minimum response of the substrates microbiome.

The soils of the area are described as isohyperthermic Calcic Lithic Petrocalcids of the Pitahaya-Limestone outcrop-La Covana Association which consist of shallow, well-drained, very slowly permeable soils that formed or were deposited in material that weathered from limestone bedrock (USDA NRCS 2008). The depth of the substrate varies according to ground relief and among seasons [14]. The low stature woody vegetation grows on a rocky calcareous substrate where plants establish their roots in the holes, cracks, and crevices, of the rocky material accumulating water and very shallow soil substrates. Due to the high variability of the surface area soil

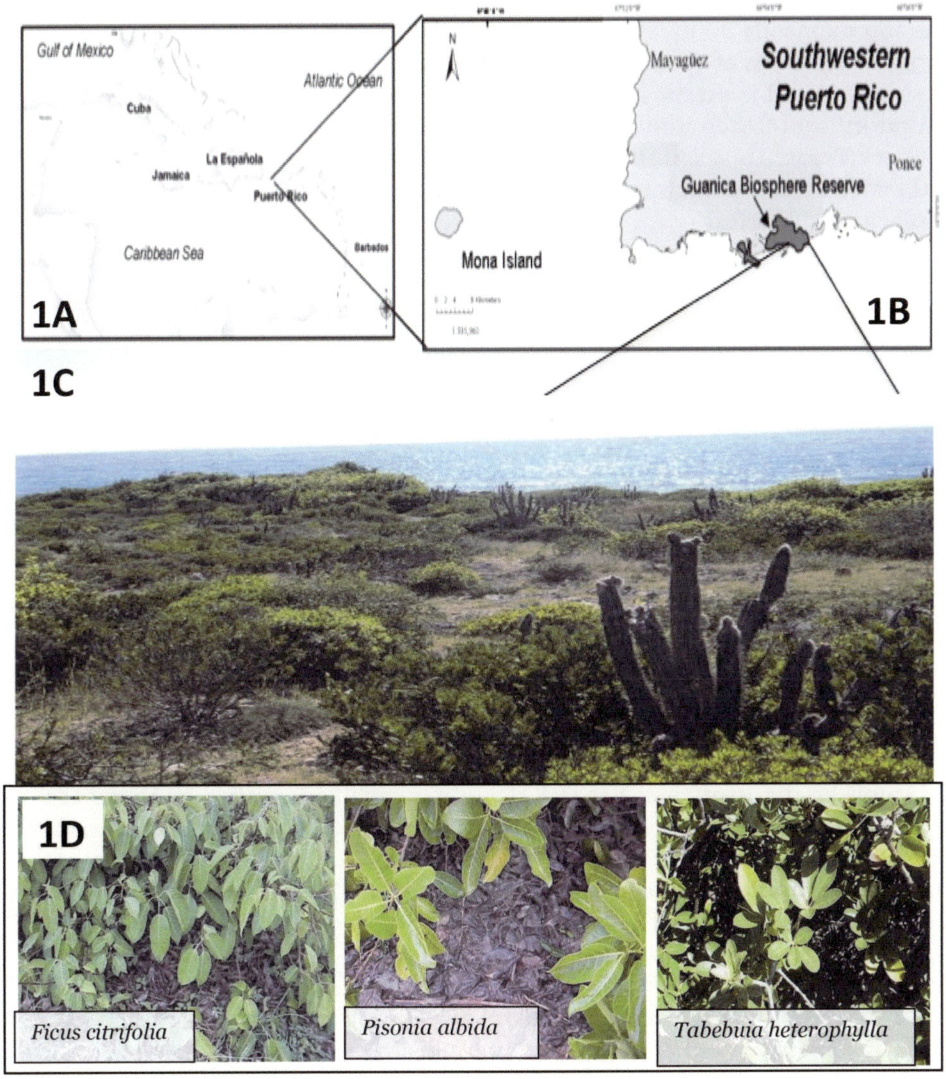

Figure 1.
Location and description of the study site. (A) The Island of San Juan Puerto Rico forms part of the Greater Antilles and is bordered by the Caribbean Sea. (B) The Guánica Dry Forest Biosphere Reserve found in the southwestern area of Puerto Rico. (C) Picture of the landscape of the study site. Here we observe the coastal area of the forest where trees are dwarfed and have established their growth in the cracks and crevices of the rocky substrate. (D) Representation of tree species used in this study.

sampling depth was not fixed; it ranged from 0 to 8 cm. We selected three (3) trees from three species, previously tagged and studied, that grow from 100 m to approximately 300 m from the coast. The tree species selected complied with the following characteristics: (a) they were interspersed within the study area, (b) each tree formed an island that was isolated from other trees by exposed rock, and (c) that their litter and belowground substrate originated from their own residue decomposition [14]. The three species selected were *Tabebuia heterophylla (DC.)* Briton (facultative deciduous), *Pisonia albida* (Heimerl) Briton ex Standal, (obligate deciduous) and *Ficus citrifolia* Mill., (facultative deciduous) (**Figure 1D**).

Each tree was used as a sampling unit supported by the very high heterogeneity in the vegetation structure of the site and the actual physical separation of the trees.

The minimum distance between any two trees was about 1 m, and the maximum was approximately 30 m. We collected one soil sample of each sampling unit (tree) during the months of the study. Soil samples were sieved in the field with a 2 mm mesh and placed in plastic bags. Samples were then placed on ice, taken to the laboratory, and frozen until they were sent to the Molecular Research Facility at Lubbock Texas. Total soil DNA extraction and 454 pyrosequencing were completed at the Molecular Research Facility. The molecular research facility reported all results as OUT tables.

2.2 Soil enzyme activities

Enzyme activities were performed as described in [5, 15]. The activities of enzymes relevant in C cycling (β-glucosidase), C and N cycling (β-glucosaminidase), P cycling (alkaline phosphatase, acid phosphatase, phosphodiesterase), and the S cycle (arylsulfatase) were assayed using 0.5 g of air-dried soil (<2 mm). Duplicate replicates and one control were used for all the soils that were tested; furthermore, the appropriate substrate was used for each assay, and reactions were incubated for 1 h at 37°C at their optimal pH as described in [5]. For the controls, the substrate was added after the 1 h incubation period and subtracted from a sample control value. Enzyme activity is expressed in mg p-nitrophenol (PN) released in kg^{-1} soil h^{-1}.

2.3 Pyrosequencing data processing and analysis

Amplicon pyrosequencing (bTEFAP) was originally described by Dowd et al. (2008) and has been utilized in describing a wide range of environmental and health-related microbiomes including the intestinal populations of a variety of sample types and environments, including cattle [16–18]. The 16S universal eubacterial primers (F = AGRGTTTGATCMTGGCTCAG, R = GTNTTACNGCGGCKGC TGG) were used for PCR amplification. A single-step 30 cycle PCR using HotStarTaq Plus Master Mix Kit (Qiagen, Valencia, CA) was used under the following conditions: 94°C for 3 minutes, followed by 28 cycles of 94°C for 30 seconds; 53°C for 40 seconds and 72°C for 1 minute; and after which a final elongation step at 72°C for 5 minutes was performed. Following PCR, all amplicon products from different samples were mixed in equal concentrations and purified using Agencourt AMPure beads (Agencourt Bioscience Corporation, MA, USA). Samples were sequenced utilizing Roche 454 FLX titanium instruments and reagents and following manufacturer's guidelines.

The sequence data derived from the sequencing analysis was processed using a proprietary analysis pipeline (www.mrdnalab.com, MR DNA, Shallowater, TX). Sequences were depleted of barcodes and primers, and all sequences shorter than <200 bp were removed. Sequences with ambiguous base calls were removed, and sequences with homopolymer runs exceeding 6 bp were also removed. All sequences were then denoised and chimeras were removed. Operational taxonomic units were defined after the removal of singleton sequences, clustering at 3% divergence (97% similarity) [16–20]. OTUs were then taxonomically classified using BLASTn against a curated Greengenes database [20] and compiled into each taxonomic level into both "counts" and "percentage" files. Operational taxonomic unit tables (OTU) tables reported by the Molecular Research Facility were used to complete all statistical analysis. Bacterial diversity was estimated by using the Shannon-Wiener (H′) and Equitability (J′) indexes; both were calculated using the

PAST3 statistical program [22]. Nonparametric Kruskal-Wallis was calculated using the JMP10 statistical software to evaluate differences between diversity indexes as affected by tree species and rainfall. In the book "Microbial Source Tracking: Methods applications and case studies," Cao et al. (2011), page 278, discusses that "common multivariate techniques used for the examination of microbial community structure include cluster analysis, principle components analysis (PCA), correspondence analysis (CA), and nonmetric multidimensional scaling (NMDS). All of these techniques belong to a group called indirect gradient analysis, which aims to reveal community similarities among sites or samples through grouping or ordering the sites or samples into either dendrograms or on a two (2D) or three-dimensional (3D) plot." On the other hand, they also mention that "direct gradient analysis such as canonical correspondence analysis (CCA), aims to correlate the overall multivariate community profile with environmental variables." To identify the influence of soil physicochemical characteristics on the bacterial community, canonical correspondence analysis (CCA) was performed using the OTU tables of each community and the soil physicochemical characteristics. Canonical correspondence analysis is a site/species matrix where each site has given values for one or more environmental variables. The ordination axes are linear combinations of the 169 environmental variables. It is a gradient analysis that shows species abundances as a response to an environmental gradient. Environmental variables are plotted as correlations with site scores. I reported two types of scaling. Type 1 emphasizes the relationship between sampling sites and environmental variables, and Type 2 emphasizes relationships between species and environmental variables [21, 22]. Indicator species analysis (ISA) was performed to identify the bacterial species responsible for changes in soil microbial communities between tree species and sampling periods. ISA analysis was completed in R using the IndVal script, where R calculates the indicator value d of species as the product of the relative frequency and relative average abundance [23].

3. Results

3.1 OTU abundance at the Guánica Dry Forest as affected by sampling period and tree species

Sequencing data revealed 17 predominant bacterial phyla for this forest (**Table 1** and **Figure 2**). The phyla with the highest relative abundance were *Proteobacteria*

Phyla	Dry	S.D.	Wet	S.D.	p value
			Bacterial phyla		
Bacteroidetes	**2.80**	**0.94**	**7.60**	**23.26**	**0.01**
Chloroflexi	**2.60**	**0.99**	**0.75**	**0.58**	**0.01**
Cyanobacteria	0.08	0.15	0.00	0.00	0.47
Actinobacteria	**37.65**	**6.57**	**16.93**	**6.77**	**<0.0001**
Verrucomicrobia	**0.32**	**0.30**	**0.08**	**0.15**	**0.02**
Spam (candidate division)	0.07	0.15	0.00	0.00	0.47
Planctomycetes	**2.26**	**0.68**	**0.68**	**0.40**	**0.01**
Nitrospirae	**0.56**	**0.39**	**0.18**	**0.21**	**0.03**

Phyla	Dry	S.D.	Wet	S.D.	p value
			Bacterial phyla		
Acidobacteria	**2.28**	**1.18**	**0.73**	**0.60**	**0.01**
Gemmatimonadetes	**1.48**	**0.75**	**0.45**	**0.44**	**0.01**
Firmicutes	**5.32**	**2.86**	**2.49**	**1.37**	**0.01**
Tenericutes	0.00	0.00	0.02	0.07	>0.9999
ws3 (candidate division)	0.11	0.24	0.00	0.00	0.47
Proteobacteria	**44.35**	**5.87**	**69.99**	**7.78**	**<0.0001**
op3 (candidate division)	0.04	0.12	0.00	0.00	>0.9999
tm6 (candidate division)	0.02	0.07	0.01	0.02	>0.9999
tm7 (candidate division)	0.05	0.10	0.08	0.13	0.86

Bold numbers represent significant differences ($p < 0.05$).

Table 1.
Kruskal-Wallis analysis of the relative abundance (%) of bacteria at the Guánica Dry Forest as affected by wet and dry periods ($n = 9$).

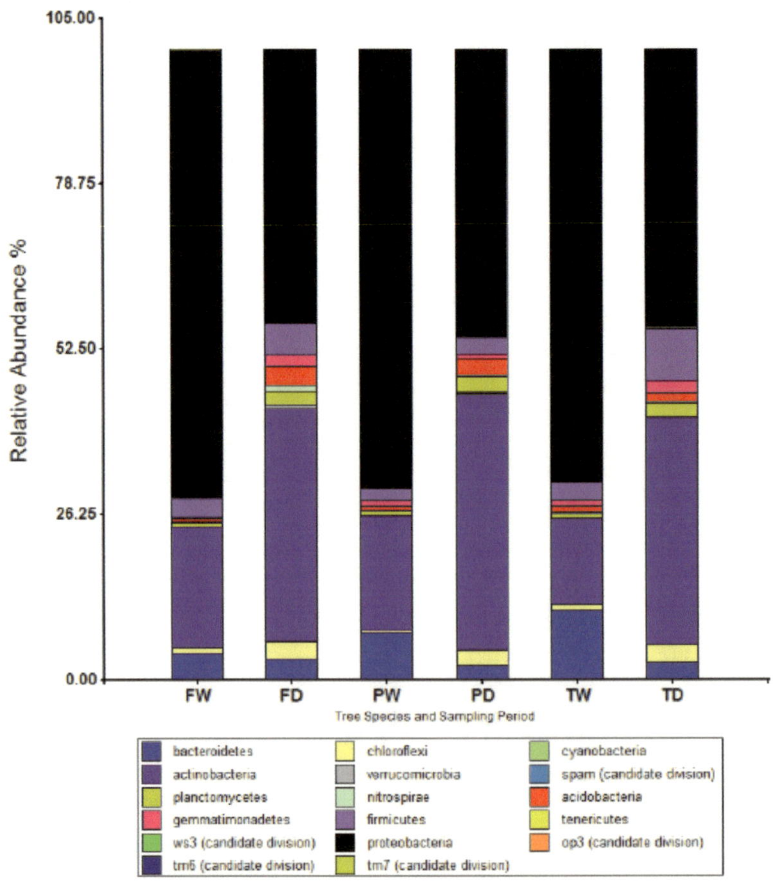

Figure 2.
Relative abundance (%) of bacterial phyla at the Guánica Dry Forest during the wet (W) and dry (D) period under three different tree species (F = Ficus citrifolia, *P =* Pisonia albida, *T =* Tabebuia heterophylla*).*

Bacterial phyla	*F. citri.*	S.D.	*P. albida*	S.D.	*T. heter.*	S.D.	p
			Tree Species				
Bacteroidetes	3.71	1.27	4.97	3.45	6.92	4.89	0.69
Chloroflexi	1.91	1.39	1.29	1.24	1.84	1.19	0.70
Cyanobacteria	0.05	0.13	0.06	0.15	0.00	0.00	0.59
Actinobacteria	27.95	11.7	29.16	14.5	24.77	13.07	0.85
Verrucomicrobia	0.30	0.39	0.25	0.17	0.07	0.14	0.18
Spam (candidate division)	0.00	0.00	0.07	0.17	0.04	0.10	0.59
Planctomycetes	1.46	1.00	1.65	1.12	1.30	0.96	0.85
Nitrospirae	0.53	0.49	0.22	0.17	0.36	0.35	0.65
Acidobacteria	1.65	1.39	1.67	1.40	1.20	0.96	0.85
Gemmatimonadetes	1.03	0.91	0.59	0.63	1.28	0.80	0.20
Firmicutes	3.97	1.75	2.24	1.87	5.51	3.22	0.17
Tenericutes	0.04	0.09	0.00	0.00	0.00	0.00	0.37
ws3 (candidate division)	0.00	0.00	0.00	0.00	0.16	0.29	0.12
Proteobacteria	57.23	17	57.77	15.4	56.51	14.77	0.98
op3 (candidate division)	0.06	0.14	0.00	0.00	0.00	0.00	0.37
tm6 (candidate division)	0.04	0.09	0.00	0.00	0.01	0.03	0.59
tm7 (candidate division)	0.09	0.15	0.07	0.11	0.03	0.07	0.77

Bold numbers represent significant differences ($p < 0.05$).

Table 2.
Kruskal-Wallis analysis of the relative abundance (%) of bacteria phyla at the Guánica Dry Forest as affected by tree species (Ficus citrifolia, Pisonia albida, and Tabebuia heterophylla).

	Tree Species			Period		
	ChiSquare	DF	Prob>ChiSq	ChiSquare	DF	Prob>ChiSq
			Bacteria			
S_Taxa	0.9297	2	0.6282	3.7974	1	**0.0513**
Shannon index (H)	1.1345	2	0.5671	2.3879	1	0.1223
Equitability (J)	0.3158	2	0.8539	5.0702	1	**0.0243**

Significant differences are found in bold.

Table 3.
Nonparametric ANOVA for bacterial alpha diversity in soils as affected by tree species and sampling period at the Guánica Dry Forest.

(44%) and *Actinobacteria* (37%). These dominant bacterial phyla presented the same pattern during the wet and dry period. During the wet period, relative abundance of *Actinobacteria* decreased almost four times when compared to the dry period, and the relative abundance of *Proteobacteria* almost duplicated (**Table 1** and **Figure 2**). Relative abundance of *Planctomycetes*, *Acidobacteria*, *Gemmatimonadetes*, and *Firmicutes* was 1–2 orders of magnitude higher during the dry period when compared to wet period, whereas only *Proteobacteria* and *Bacteroidetes* were 1–2 orders of magnitude higher during the wet period when compared to the dry period. The most predominant bacterial phyla for all tree species during the wet and dry periods were *Proteobacteria*, *Actinobacteria*, and *Bacteroidetes* (**Table 2**). Kruskal-Wallis test did not demonstrate significant differences in bacterial relative

Bacteria Indicator Species	group	indval	p-value	freq
Acetobacteraceae spp	Ficus Wet	0.9036854	0.018	4
Brevundimonas wangchunensis	Ficus Wet	0.7891088	0.019	6
Brevundimonas asdaezzhy	Ficus Wet	0.7146215	0.011	8
Brevundimonas spp.	Ficus Wet	0.6583452	0.029	12
Bosea thiooxidans	Ficus Wet	0.6415078	0.042	8
Devosia spp.	Ficus Wet	0.6211101	0.009	14
Pseudomonas fluorescens	Ficus Wet	0.7762022	0.046	6
Pseudomonas stutzeri	Ficus Wet	0.6999245	0.016	8
Pedobacter ginsengisoli	Ficus Wet	0.5858814	0.049	8
Delftia lacustris	Ficus Wet	0.9448138	0.013	4
Streptomyces platensis malvinus	Pisonia Wet	1	0.014	3
Streptomyces thermolineatus	Pisonia Wet	0.7624343	0.03	4
Hyphomicrobium zavarzinii	Pisonia Wet	0.6518947	0.015	7
Aeromicrobium kwangyangensis	Pisonia Wet	0.651858	0.022	4
Mycobacterium bavariae	Pisonia Wet	0.6077151	0.004	10
Mycobacterium moriokaense	Pisonia Wet	0.5624626	0.035	10
Defluviicoccus spp.	Pisonia Wet	0.5216104	0.031	8
Nitrospira spp.	Pisonia Wet	0.4342276	0.021	14
Nocardioides spp.	Pisonia Wet	0.3847424	0.048	16
Rhodospirillaceae spp.	Pisonia Wet	0.3789393	0.05	18
Streptomyces sakaiensis	Pisonia Wet	1	0.008	3
Sciscionella spp.	Pisonia Wet	1	0.01	3
Phenylobacterium spp.	Pisonia Wet	0.8661456	0.011	5
Bradyrhizobium sp.	Pisonia Wet	0.7317982	0.011	8
Mycobacterium avium complex	Pisonia Wet	0.6628777	0.006	8
Frankia spp.	Pisonia Wet	0.6492538	0.008	12
Bradyrhizobium spp.	Pisonia Wet	0.6320272	0.045	18
Crossiella spp.	Pisonia Wet	0.6190261	0.038	16
Candidatus entotheonella	Pisonia Wet	0.569092	0.025	13
Opitutus spp	Pisonia Wet	0.5654218	0.035	6
Hyphomonadaceae spp.	Pisonia Wet	0.5261401	0.004	13
Mycobacterium celatum	Pisonia Wet	0.4244481	0.006	15
Streptomyces aureus	Pisonia Wet	0.966957	0.014	4
Nocardioides albus	Pisonia Wet	0.8408999	0.026	5
Bacillus spp.	Pisonia Wet	0.8086407	0.011	9
Nitrospinaceae spp.	Pisonia Wet	0.601469	0.02	12
Sphingomonas spp.	Pisonia Wet	0.5841869	0.013	18
Microlunatus spp.	Pisonia Wet	0.5766045	0.006	17
Acidimicrobiales spp.	Pisonia Wet	0.5665658	0.011	18
Mycobacterium asiaticum	Pisonia Wet	0.4652796	0.008	9
Paenibacillus spp.	Pisonia Wet	0.3844302	0.017	18
Bradyrhizobiaceae spp.	Pisonia Wet	0.3625066	0.035	18

Table 4.
Bacterial indicator species analysis (ISA) at the Guánica Dry Forest.

abundance due to tree species (**Table 2**). There were no significant differences in any of the diversity indexes with regard to tree species (**Table 3**). Sampling period exhibited an effect on bacterial species richness (P = 0.05) and on equitability (P = 0.0243). but not on the Shannon index. We found higher bacterial species richness during the wet period and higher equitability during the dry period (**Table 3**).

3.2 Evaluation of indicator species analysis (ISA) for this forest ecosystem

The identification of species associated or indicative of groups of samples is a common aspect of ecological research [24]. Indicator species analysis (ISA) identified several bacterial (**Table 4**) species responsible for changes in soil microbial communities. Out of 185 bacterial OTUs, 10 served as indicator species for *Ficus citrifolia* during the wet period (**Table 4**). A total of 31 bacterial OTUs were identified as indicator species for *Pisonia albida* during the wet period. No bacterial indicator species were found for the dry period (**Table 4**).

3.3 Relationship between the relative abundance of taxa and enzyme activities

Axis 1 of the canonical correspondence analysis for bacterial community explained 80% of the variation (**Figure 3**). Two groups were segregated with regard

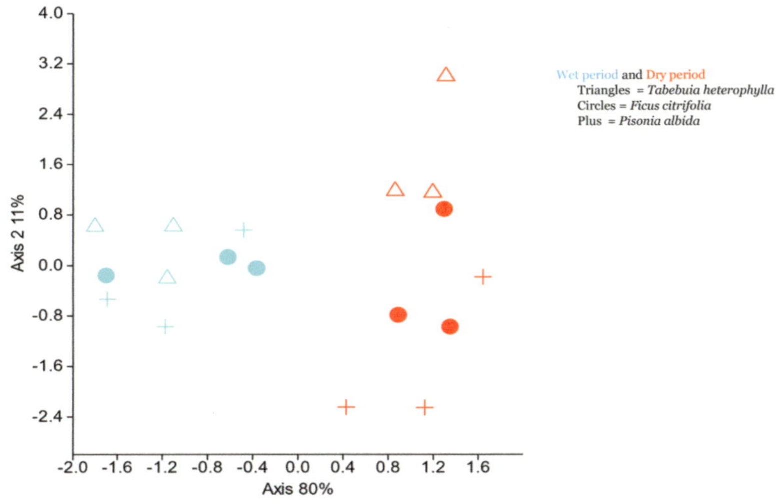

Figure 3.
Canonical correspondence analysis (CCA) of bacterial phyla demonstrating the effect of wet/dry periods at the Guánica Dry Forest. Blue symbols represent wet periods and red symbols represent dry periods. Triangles, circles and plus sign represent tree species (Tabebuia heterophylla, Ficus citrifolia and Pisonia albida), respectively.

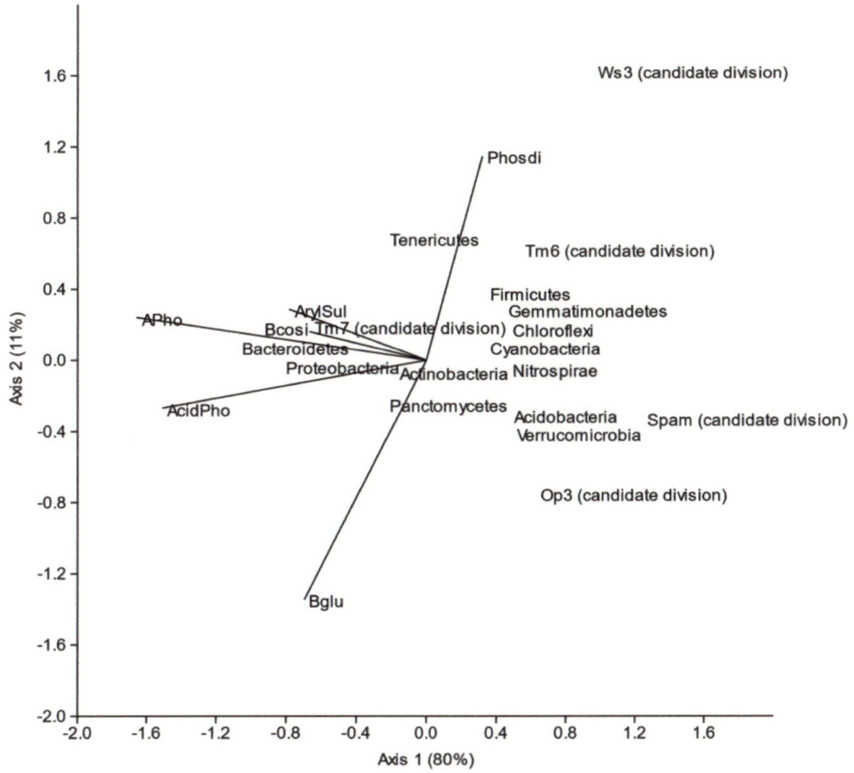

Figure 4.
Canonical correspondence analysis (CCA) of bacterial phyla and soil enzyme activities (Phosdi = phosphodiesterasae, ArylSul = aryl sulphatase, APho = alkaline phosphatase, AcidPho = acid phosphatase, Bcosi = β-glucosaminidase, and Bglu = β-glucosidase). Vectors represent enzyme activities.

to dry and wet periods (**Figure 3**). Samples that correspond to the wet period were associated with acid phosphatase, alkaline phosphatase, β-glucosaminidase, β-glucosidase, and arylsulfatase (**Figure 4**), whereas samples corresponding to the dry period are associated with phosphodiesterase (**Figure 4**). Microbial taxa associated with wet samples were Bacteroidetes, Proteobacteria, and Tenericutes. Microbial taxa associated with dry sampling points were *Actinobacteria*, *Planctomycetes*, *Acidobacteria*, *Verrucomicrobia*, *Cyanobacteria*, and *Chloroflexi*.

4. Discussion

4.1 Response of microbial diversity to wet and dry periods at the Guánica Dry Forest

Historical rainfall patterns contribute to the acclimatization and resilience of soil bacterial communities to low and high rainfall events. Bacterial Shannon index (3.9) was similar to values reported by Žifčáková et al. (2016) for a Norway spruce forest (S = 3.5) and lower than the one reported for a hardwood forest (S = 6–6.5) or dry heath in a tundra (S = 7.5) [25]. Our study demonstrates that soil bacterial richness, diversity, and equitability were impacted by rainfall patterns and not by tree species. Both bacterial richness and equitability were higher during the dry period, but bacterial diversity was not impacted by rainfall regime. Our trends imply that a total number of bacterial species do not change during low rainfall events in this forest, but changes occur in the quantity of each species and in their distribution, indicating that soil bacterial communities have adapted to low rainfall at the Guánica Dry Forest. There is indirect evidence that microbial communities do become resistant and function optimally under their historical rainfall regime [26–28]. Cruz-Martínez et al. (2009) [29] found that soil microbial communities were more resilient to long-term changes in rainfall after a 7-year rainfall amendment study. They stated that after 7 years soil microbial communities developed a degree of robustness or acclimatization to the rainfall amendments. Additionally, other studies have reported acclimatization of soil heterotrophic communities to experimental warming and seasonal variation [30]. Compositional changes exhibit historical legacy with regard to moisture regimes [27] suggesting that microbial communities will be shaped in part by the historical climate to which they are exposed [30].

In this study the bacterial communities under all tree species were dominated by *Proteobacteria* (57%), *Actinobacteria* (24–29%), and *Bacteroidetes* (4–7%). All of these bacterial phyla are ubiquitous and have been identified in desert soils [31], agroecosystems, and other types of forest environments [32–35]. *Proteobacteria* are one of the largest bacterial divisions within the prokaryotes and account for most of the known Gram-negative bacteria [36]. Although *Proteobacteria* are one of the top bacterial phyla that are found colonizing many soil types their high predominance at the GDF could be associated with high accumulation of litter and SOM that is reported in these soils. An interesting detail is that Proteobacteria dominated during both periods, with the exception that during the dry period they reduced their relative abundance, suggestion there association with soil moisture as found in other studies [28].

Actinobacteria, one of the largest bacterial phyla known, was the second most abundant phylum found in this study. This group is mainly composed of Gram-positive bacteria [32–35] and are known to withstand in harsh environments, due to their metabolic, physiologic, and morphological diversity [36]. Soils of the area under study have an accumulation of organic matter mainly composed of new and old leaf litter due to the harsh environmental conditions; this bacterial phylum is associated with the degradation of more complex substrates of SOM including lignin, which could explain their predominance in this forest soils [37–39]. Our data suggest that *Actinobacteria* were highly associated with phosphodiesterase at the Guánica forest. Other studies have also found that *Actinobacteria* populations (characterized by TRLFP) correlated with P content in semiarid environments [40]. Other studies have also demonstrated the production of phosphatases by cultivable soil *Actinobacteria* [41–43] serving as a further evidence of the trend observed in this study. I also found that Actinobacteria were the most abundant phyla during the dry period which is consistent with other reports. A similar trend was also found for a semiarid, high desert grassland north of Flagstaff, Arizona [44] and in another study, which evaluated African tropical forest soils and Chinese forest soils [45].

Bacteroidetes, the third most abundant phylum in this forest, are widely distributed in different habitats ranging from Antarctic ice, lakes, the gut of animals, and terrestrial environments [46]. Additionally, this phylum has the ability to withstand extreme desiccation conditions including droughts and UV light [47]. They have been found in fine dust traveling thousands of kilometers [48] and inside microaggregates; authors explain that in semiarid agroecosystems this could be a protection strategy employed by these microbes to endure extreme environmental conditions [48]. I found that *Bacteroidetes* were highly associated with the activities of two soil enzymes evaluated (acid phosphatase and alkaline phosphatase). These enzymes mineralize organic P forms into inorganic P forms. Another study provided evidence suggesting that environmental *Bacteroidetes* specialize in the mineralization of high molecular weight organic matter making them a key compartment for carbon fluxes and budgets in ecosystems [49]. *Bacteroidetes* are oligotrophic and are commonly associated with substrates rich in organic matter [50] as is the forest area where I based my study. The high abundance of *Bacteroidetes* could be associated with the rich organic matter present in the sampling area.

4.2 Influence of tree species on the bacterial populations in this forest

Even though the most abundant bacterial phyla identified under all tree species were the same (*Proteobacteria*, *Actinobacteria*, and *Bacteroidetes*), some were specific for each tree species. Indicator species analysis (ISA) revealed that specific bacterial species were present under *Ficus citrifolia* and *Pisonia albida* only during the wet period. This trend suggests that these species assemblages may play an important role in the soil ecosystem processes under these specific tree species. One species was *Devosia* spp. whose OUT frequency was 14 (ind val = 0.62; p = 0.009) in soils collected under *Ficus citrifolia*. *Devosia* spp. forms part of the α-*Proteobacteria*; this genus is a non-rhizobia nodulating, nitrogen fixer [51]. Three different species of *Brevundimonas* spp. also occurred in high frequency under *Ficus citrifolia* during the wet period as indicated by ISA. This genus is actually known to produce phosphodiesterase, a group of enzymes involved in the degradation of organophosphorus [52]. During the wet period, 32 different OTUs were identified exclusively for *Pisonia albida*, and at least 7 of them had a frequency of 18. This high frequency suggests that indicator species may be playing an important role in the soil dynamics of *Pisonia albida*. For instance, three of the most frequent species are nitrogen fixers (*Microlunatus* sp., *Rhodospirillacease* spp., and *Paenibacillus*), and *Pisonia albida* was the tree species with the highest total available nitrogen in this study (data not shown here). Even though we did not cultivate the indicator species nor have information regarding the physiology of the indicator species identified for

Ficus citrifolia and *Pisonia albida*, we can infer that they contribute with relevant functions in this soil system. The dwarfed tree species selected for this study influenced the structure and diversity of specific bacterial populations.

4.3 Potential disadvantages and bias with 454-pyrosequencing

Amplicon-based pyrosequencing methods have major advantages over the tools that have been used in the past to study microbial community structure. Although the results presented in this chapter have a similar pattern as the results presented Rivera et al. (2018), it important to acknowledge certain biases that have been described for amplicon-based pyrosequencing. Even though 454 pyrosequencing has a higher resolving power than Sanger sequencing or EL-FAME analysis in 454 pyrosequencing, there are some sequencing errors and chimeras that can be retained in the datasets that can inflate the estimated richness of the sample. Bias can also occur with primer selection as the primers used can select for the most predominant DNA present in the sample underestimating the rare DNA in the sample [53]. Using inappropriate primers consequently can lead to questionable biological conclusions. Another concern is that the techniques used for processing amplicon pyrosequencing data can result in the detection of several hundred "false" OTUs, mostly at low abundance, rising the concern that species abundance can be overestimated [54]. More stringent techniques such as shotgun sequencing, Ion Torrent sequencing, and Illumina platforms have been developed that help mitigate some of the concerns with pyrosequencing, but these stringent technologies have biases of their own.

5. Conclusions

Soil bacterial communities have adapted to low rainfall at the Guánica Dry Forest; this could be a response to historical rainfall patterns encountered at the Guánica Dry Forest. The fact that 9 out of the 17 bacterial phyla identified were higher during the dry period supports this conclusion. For this forest, bacterial diversity did not change as a response to rainfall; however, equitability and richness changed demonstrating bacterial resilience. We are seeing how the same three bacterial phyla (*Proteobacteria*, *Actinobacteria*, and *Bacteroidetes*) are predominant during both dry and rainfall periods. Even though predominant bacterial phyla were the same during both periods under all tree species, differences were found at a finer scale. For instance, *Pisonia albida* had the soil with the most bacterial indicator species present. It is evident that this tree is shaping the soil microbiome in different ways. The general trend for predominant phyla found for Guánica is similar to the predominant phyla found in other terrestrial ecosystems even though the conditions of Guánica are unique. In the future, it would be nice to compare the sequences obtained in this study with other terrestrial environments but to the species level. This will help elucidate which are the species that could be playing important roles in ecosystem function and resilience.

Acknowledgements

The project was funded by NSF Grant HRD-0734826 and is a contribution of the Centre of Applied Tropical Ecology and Conservation of the University of Puerto Rico. We appreciate the support of Mr. Larry Diaz, laboratory coordinator, and students from the Ecosystems Processes and Function laboratory of the University of Puerto Rico, Río Piedras Campus, that participated in the field collection of soil samples. Special thanks to Dr. Verónica Acosta-Martínez for offering her comments and suggestions in the preparation of this manuscript.

Conflict of interest

No conflict of interest.

Author details

Michelle Jovanne Rivera-Rivera* and Elvira Cuevas
Department of Biology and Center for Applied Tropical Ecology and Conservation, University of Puerto Rico, Rio Piedras, Puerto Rico

*Address all correspondence to: yersiniamjr@hotmail.com

References

[1] Vargas-Gastélum L, Romero-Olivares AL, Escalante AE, Rocha-Olivares A, Brizuela C, Riquelme M. Impact of seasonal changes on fungal diversity of a semi-arid ecosystem revealed by 454 pyrosequencing. FEMS Microbiology Ecology. 2015;**91**(5): fiv044

[2] Medina E, Cuevas E. Propiedades fotosinteticas y eficiencia de uso de agua de plantas leñosas del bosque deciduo de Guánica: Consideraciones generales y resultados preliminares. Acta Cientifica. 1990;**4**:25-36

[3] Jennings LN, Douglas J, Treasure E, González G. Climate change effects in El Yunque National Forest, Puerto Rico, and the Caribbean region. General Technical Report-Southern Research Station, USDA Forest Service (SRS-193); 2014

[4] Tremont O, Cuevas E. Carbono orgánico, nutrientes y cambios estacionales de la biomasa microbiana en las principales especies de dos tipos de bosques tropicales. Multiciencias. 2006;**4**:1-4

[5] Rivera-Rivera MJ, Acosta-Martínez V, Cuevas E. Tree species and precipitation effect on the soil microbial community structure and enzyme activities in a tropical dry forest reserve. In: Extremophilic Microbes and Metabolites-Diversity, Bioprespecting and Biotechnological Applications. IntechOpen; 2018

[6] Medina E, Cuevas E, Molina S, Lugo A, Ramos O. Structural variability and species diversity of a dwarf Caribbean dry forest. Caribbean Journal of Science. 2012;**46**:1-13

[7] Waldrop M, Firestone M. Microbial community seasonal dynamics. Microbial Ecology. 2006;**52**:470-479

[8] Ayres E, Steltzer H, Berg S, Wallenstein MD, Simmons BL, Wall DH. Tree species traits influence soil physical, chemical, and biological properties in high elevation forests. PLoS ONE. 2009;**4**(6):e5964

[9] Augusto L, De Schrijver A, Vesterdal L, Smolander A, Prescott C, Ranger J. Influences of evergreen gymnosperm and deciduous angiosperm tree species on the functioning of temperate and boreal forests. Biological Reviews. 2015;**90**(2):444-466

[10] Medina E, Cuevas E, Molina S, Lugo AE, Ramos O. Structural variability and species diversity of a dwarf Caribbean dry forest. Caribbean Journal of Science. 2010;**46**(2–3): 203-215

[11] Medina E, Garcia V, Cuevas E. Sclerophylly and oligotrophic environments: Relationships between leaf structure, mineral nutrient content, and drought resistance in tropical rain forests of the upper Rio Negro region. Biotropica. 1990;**22**:51-64

[12] Govender Y, Cuevas E, Sternberg LDS, Jury MR. Temporal variation in stable isotopic composition of rainfall and groundwater in a tropical dry forest in the northeastern Caribbean. Earth Interactions. 2013; **17**(27):1-20

[13] Lugo AE, Gonzalez-Liboy JA, Cintron B, Dugger K. Structure, productivity, and transpiration of a subtropical dry forest in Puerto Rico. Biotropica. 1978;**10**:278-291

[14] Barberena M. Single tree species effects on temperature nutrients and arthropod diversity in liter and humus in the Guánica Dry Forest [Thesis Dissertation]. Biology Department, University of Puerto Rico, Rio Piedras Campus; 2008

[15] Acosta-Martínez V, Bell CW, Morris B, Zak J, Allen VG. Long-term soil microbial community and enzyme activity responses to an integrated cropping-livestock system in a semi-arid region. Agriculture, Ecosystems and Environment. 2010;**137**:231-240

[16] Dowd SE, Sun Y, Wolcott RD, Domingo A, and Carroll JA. Bacterial tag-encoded FLX amplicon pyrosequencing (bTEFAP) for microbiome studies: bacterial diversity in the ileum of newly weaned *Salmonella*-infected pigs. Foodborne Pathogens and Disease. 2008;**5**:459-472

[17] Callaway TR, Dowd SE, Edrington TS, Anderson RC, Krueger N, Bauer N, et al. Evaluation of bacterial diversity in the rumen and feces of cattle fed different levels of dried distillers grains plus solubles using bacterial tag-encoded FLX amplicon pyrosequencing. Journal of Animal Science. 2010;**88**(12):3977-3983

[18] Capone KA, Dowd SE, Stamatas GN, Nikolovski J. Diversity of the human skin microbiome early in life. Journal of Investigative Dermatology. 2011;**131**(10):2026-2032

[19] Swanson KS, Dowd SE, Suchodolski JS, Middelbos IS, Vester BM, Barry KA, et al. Phylogenetic and gene-centric metagenomics of the canine intestinal microbiome reveals similarities with humans and mice. The ISME Journal. 2011;**5**(4):639-649

[20] DeSantis TZ, Hugenholtz P, Larsen N, Rojas M, Brodie EL, Keller K, et al. Greengenes, a chimera-checked 16S rRNA gene database and workbench compatible with ARB. Applied and Environmental Microbiology. 2006; **72**(7):5069-5072

[21] Legendre P, Legendre L. Numerical Ecology. 2nd English ed. Amsterdam: Elsevier; 1998. p. 853

[22] Hammer Ø, Harper DA, Ryan PD. PAST: paleontological statistics software package for education and data analysis. Palaeontologia Electronica. 2001;**4**:1-275

[23] Dufrene M, Legendre P. Species assemblages and indicator species: The need for a flexible asymmetrical approach. Ecological Monographs. 1997; **67**(3):345-366

[24] Bakker JD. Increasing the utility of indicator species analysis. Journal of Applied Ecology. 2008;**45**(6): 1829-1835

[25] Yu Z, Yang J, Yu X, Liu L, Tian Y. Aboveground vegetation influences belowground microeukaryotic community in a mangrove nature reserve. Wetlands. 2014;**34**(2):393-401

[26] Schwinning S, Sala OE, Loik ME, Ehleringer JR. Thresholds, memory, and seasonality: understanding pulse dynamics in arid/semi-arid ecosystems. Oecología. 2004;**141**:191-193

[27] Evans SE, Wallenstein MD. Soil microbial community response to drying and rewetting stress: Does historical precipitation regime matter? Biogeochemistry. 2012;**109**(1–3): 101-116

[28] Evans SE, Wallenstein MD. Climate change alters ecological strategies of soil bacteria. Ecology Letters. 2014;**17**(2): 155-164

[29] Cruz-Martínez K, Suttle KB, Brodie EL, Power ME, Andersen GL, Banfield JF. Despite strong seasonal responses, soil microbial consortia are more resilient to long-term changes in rainfall than overlying grassland. The ISME Journal. 2009;**3**(6):738-744

[30] Strickland MS, Keiser AD, Bradford MA. Climate history shapes contemporary leaf litter decomposition. Biogeochemistry. 2015;**122**(2–3):165-174

[31] Makhalanyane TP, Valverde A, Gunnigle E, Frossard A, Ramond JB, Cowan DA. Microbial ecology of hot desert edaphic systems. FEMS Microbiology Reviews. 2015;**39**(2): 203-221

[32] Jenkins JR, Viger M, Arnold EC, Harris ZM, Ventura M, Miglietta F, et al. Biochar alters the soil microbiome and soil function: Results of next-generation amplicon sequencing across Europe. GCB Bioenergy. 2017;**9**(3): 591-612

[33] Acosta-Martínez V, Cotton J, Gardner T, Moore-Kucera J, Zak J, Wester D, et al. Predominant bacterial and fungal assemblages in agricultural soils during a record drought/heat wave and linkages to enzyme activities of biogeochemical cycling. Applied Soil Ecology. 2014;**84**:69-82

[34] Lladó S, López-Mondéjar R, Baldrian P. Forest soil bacteria: Diversity, involvement in ecosystem processes, and response to global change. Microbiology and Molecular Biology Reviews. 2017;**81**(2):e00063-e00016

[35] Pathan SI, Žifčáková L, Ceccherini MT, Pantani OL, Větrovský T, Baldrian P. Seasonal variation and distribution of total and active microbial community of β-glucosidase encoding genes in coniferous forest soil. Soil Biology and Biochemistry. 2017;**105**:71-80

[36] Gao B, Gupta RS. Phylogenetic framework and molecular signatures for the main clades of the phylum Actinobacteria. Microbiology and Molecular Biology Reviews. 2012;**76**(1): 66-112

[37] Větrovský T, Steffen KT, Baldrian P. Potential of cometabolic transformation of polysaccharides and lignin in lignocellulose by soil Actinobacteria. PLoS ONE. 2014;**9**(2):e89108

[38] Woo HL, Hazen TC, Simmons BA, DeAngelis KM. Enzyme activities of aerobic lignocellulolytic bacteria isolated from wet tropical forest soils. Systematic and Applied Microbiology. 2014;**37**(1):60-67

[39] Pisani O, Lin LH, Lun OO, Lajtha K, Nadelhoffer KJ, Simpson AJ, et al. Long-term doubling of litter inputs accelerates soil organic matter degradation and reduces soil carbon stocks. Biogeochemistry. 2016;**127**(1):1-14

[40] Pasternak Z, Al-Ashhab A, Gatica J, Gafny R, Avraham S, Minz D, et al. Spatial and temporal biogeography of soil microbial communities in arid and semiarid regions. PLoS ONE. 2013;**8**(7): e69705

[41] Yamamura H, Ohkubo SY, Nakagawa Y, Ishida Y, Hamada M, Otoguro M, et al. *Nocardioides iriomotensis* sp. nov., an actinobacterium isolated from forest soil. International Journal of Systematic and Evolutionary Microbiology. 2011;**61**(9): 2205-2209

[42] Balakrishna G, Shiva Shanker A, Pindi PK. Isolation of phosphate solubilizing actinomycetes from forest soils of Mahabubnagar district. IOSR Journal of Pharmacy. 2012;**2**(2):271-275

[43] Ghorbani-Nasrabadi R, Greiner R, Alikhani HA, Hamedi J, Yakhchali B. Distribution of actinomycetes in different soil ecosystems and effect of media composition on extracellular phosphatase activity. Journal of Soil Science and Plant Nutrition. 2013;**13**(1): 223-236

[44] McHugh TA, Schwartz E. A watering manipulation in a semiarid grassland induced changes in fungal but not bacterial community composition. Pedobiologia. 2016;**59**:121-127

[45] Zhou X, Fornara D, Ikenaga M, Akagi I, Zhang R, Jia Z. The resilience of

microbial community under drying and rewetting cycles of three forest soils. Frontiers in Microbiology. 2016;7(1101): 1-12

[46] Gupta RS. The phylogeny and signature sequences characteristics of Fibrobacteres, Chlorobi, and Bacteroidetes. Critical Reviews in Microbiology. 2004;**30**(2):123-143

[47] Ordoñez OF, Flores MR, Dib JR, Paz A, Farías ME. Extremophile culture collection from Andean lakes: Extreme pristine environments that host a wide diversity of microorganisms with tolerance to UV radiation. Microbial Ecology. 2009;**58**(3):461-473

[48] Gardner T, Acosta-Martínez V, Calderón FJ, Zobeck TM, Baddock M, Van Pelt RS, et al. Pyrosequencing reveals bacteria carried in different wind-eroded sediments. Journal of Environmental Quality. 2012;**41**(3): 744-753

[49] Nagata T. Organic matter-bacteria interactions in seawater. In: Kirchman DL, editor. Microbial Ecology of the Oceans. Second ed. 2008. pp. 207-241. DOI: 10.1002/9780470281840.ch7

[50] Fierer N, Bradford MA, Jackson RB. Toward an ecological classification of soil bacteria. Ecology. 2007;**88**(6): 1354-1364

[51] Balachandar D, Raja P, Kumar K, Sundaram SP. Non-rhizobial nodulation in legumes. Biotechnology and Molecular Biology Reviews. 2007;**2**(2): 49-57

[52] Singh BK. Organophosphorus-degrading bacteria: Ecology and industrial applications. Nature Reviews Microbiology. 2009;**7**(2):156-164

[53] Hert DG, Fredlake CP, Barron AE. Advantages and limitations of next-generation sequencing technologies: A comparison of electrophoresis and non-electrophoresis methods. Electrophoresis. 2008;**29**(23):4618-4626

[54] Kunin V, Engelbrektson A, Ochman H, Hugenholtz P. Wrinkles in the rare biosphere: Pyrosequencing errors can lead to artificial inflation of diversity estimates. Environmental Microbiology. 2010;**12**(1):118-123

Chapter 4

Tropical Crops and Microbes

Barachel Odaro-Junior Umukoro

Abstract

Sustaining crop production in order to meet the growing demand of the teeming populace in the tropics has been one of the utmost goals of Scientists nowadays since the conversion of the tropical ecosystems to other uses has posed serious threat to it. Crops that were either introduced or adopted to the tropical nations by the European conquerors are referred to as tropical crops. The ubiquitous nature of microorganisms has made the soil to be one of their habitats or reservoirs. Microorganisms belonging to bacteria, fungi, protozoa, micro-algae, and viruses inhabit the soil. In crop production, beneficial soil microbes have been used as biofertilizers, biopesticides, and phytostimulators and also increase resilience in plants. Biofertilizers obtained from effective and indigenous microorganism have been used to improve and maintain the biological, chemical, and physical properties of cropland soils, which in turn improve crop growth and yields. Plants also contribute to the population of microbes in the soil by supplying them carbon from their photosynthates. The mutual relationship between beneficial microbes and plants cannot be underestimated in improving crop growth and yields in threatened tropical ecosystem.

Keywords: soil, tropical crops, beneficial microorganisms, mutual relationship, crop production

1. Introduction

In the era of sustainable crop production, the interaction between plants and soil microbes play an important role in the transformation, solubilization, mobilization, etc. of nutrients from a limited nutrient pool and make it available for the uptake of plants in order to realize their full genetic potential. Microorganisms perform numerous metabolic functions which are essential for their own maintenance and can directly or indirectly be beneficial to the biosphere through environmental detoxification, soil health improvement, nutrient recycling, waste water treatment, etc. [1].

For more than three centuries, endophytic microorganisms which colonize and reside in plant roots have been known to be in existence. Though their value in increasing crop yields and buffering environmental conditions have become appreciated in recent decades. When describing the formation of galls on roots in 1967, Malpighi reported the symbiotic association between microorganisms and plants. His report was not seen as scientific rather borne out of curiosity. After two centuries, Hellriegel and Wilfath, showed that these galls are nodules composed of both bacterial (Rhizobiaceae) and plant cells which fix N_2 from the atmosphere, providing leguminous plants with an essential nutrient known as ammonia (NH_3) [2]. In 1882, fungi which are presently known as arbuscular mycorrhizal fungi

(AMF) were found to symbiotically improve plants' productivity by colonizing their roots [3]. In the 1920s and 30s, *Trichoderma,* the commonly known soil-inhabiting fungi were found to biologically control pathogenic fungi, thus having the potential in protecting agricultural crops [4]. In addition to protecting plants against pathogenic fungi, some strains of *Trichoderma* have been found to induce multiple benefits to plants when they colonize their roots [5]. Another group of fungi known as *Piriformospora indica* that beneficially colonizes and inhabits plant roots was discovered in the 1990s [6]. Once these microorganisms colonize and inhabit plant roots, they induce physiological changes and modify the expression of genes in the plant they reside in, thus improving plants' productivity and resilience.

2. Tropical agriculture

Geographically, the tropical region is the region of the earth that centers in the equator and limited by the tropics of cancer (23.5°N) and capricorn (23.5°S). All the parts of the earth where the sun reaches at an altitude of 90° and move between the two tropics during the average length of the year are contained in the tropical region. The sun's position makes this region not to experience notable changes in temperature (seasons), and during the wet seasons, water evaporation produces abundant rainfall in this "torrid" region due to constant daily radiation. A dry season which ranges from a month to over 6 months also occurs at different times and regions depending on the sun's position during the year and the region. A bimodal or unimodal distribution may be presented by these dry and wet (raining) seasons during the year. However, tropical ecosystems vary considerably from deserts to rainforests, and the concept of vertical geography which ranges from hot lowlands to snow-capped mountains within a few hundred kilometers can change the temperature drastically [7].

In terms of crop, tropical agriculture is usually described as those crops that were either introduced or adopted by European conquerors in the tropical nations that are under their dominion. Tropical agriculture is often dominated by crops [8]. A large number of plants use in agriculture today were originated and domesticated in the tropics, mountainous area where ecoclines often overlap [9–11]. This is due to the wide range of microclimate, temperatures and rainfalls in the mountainous tropics thereby increasing genetic diversification through selection, mutation and adaptation [12, 13]. A major reservoir of plant and animal biodiversity is the tropical ecosystems which play essential roles in global climate regulation and biogeochemical cycling [14, 15]. The exact yield potential for almost all tropical fruit crops still remains unclear. Though some industrialized crops such as banana, oil palm and citrus have very high production efficiencies, this is exceptional to the norm. For most tropical crops, the maximum recorded yields are much higher than the average yields over large area. Poor soil and water management, pests and diseases, low commodity prices, shortage of skilled and productive labor and failure of the market to provide incentive to growers are some of the reasons for the yield gap [16]. Crop production in the tropical regions is highly diversified compared to the large acre crop farming system in the temperate regions. In the tropics, food and industrial crops may be cultivated either in small holdings, plantations or in mixed gardens. Plantation crops may be large and continuous but they may be owned either by major corporations with uniform cultural practices or by a number of contiguous small farmers with varying practices [17]. Large number of pathogenic bacteria, fungi, viruses and other pests, especially insects often destroy tropical crops. The harsh winter conditions of the tropical regions do not reduce the pathogens population as in the temperate regions. The availability of plants, which serve as hosts to the pathogens during the year, maintains the populations of a myriad of pests at a damaging level in tropical regions. The tropical ecosystems have been seriously threatened due to its conversion to other uses [18].

3. Soil

Soil is gradually formed from various parent materials, which is modified by climate, time, micro- and macro-organisms, vegetation and topography. It is a complex mixture of minerals, organic compounds and living organisms that interact continuously in response to natural, biologically, chemically and physically imposed changes [19]. In addition to root anchorage, the soil serves as a natural medium for plant growth and habitat for a wide range of microbes [20]. The growth of soil microbes and plants may be directly and indirectly influenced through a variety of interaction. These interactions may result in positive growth through mutual benefits, negative growth through antagonism, or no growth stimulation through neutral effects [21].

Cropland soil can be classified into three types and these include spermosphere, rhizosphere and bulk soil. Spermosphere is the portion of soil that surrounds germinating seeds. It has been described as the short-lived, rapidly changing and microbiologically dynamic zone of soil that surround germinating seed [22]. Rhizosphere has been described as a narrow soil zone that surrounds the roots of leguminous plant which stimulates intense bacterial activities [23]. It is a huge reservoir of microbial diversity. The release of exudates from the roots of plant into the rhizoenvironment initiates the establishment of rhizosphere. The exudates include plant mucilage, mucigel, root secretions and lysates [24]. Thus, exudates are the most vital factors that contribute towards the dynamics of rhizosphere. The rhizosphere is the most important niche that affects diverse aspects of plant life. Bulk soil is composed of soil outside the spermosphere and rhizosphere. With regard to microbial activities, it is considered to be the least dynamic. Out of the three groups of cropland soils, it occupies the largest portion of cropland soil. A large population of micro- and meso-organisms that include surviving propagules of soil inhabiting plant pathogens are harbored in the bulk soil [25].

3.1 Soil microorganisms

Microorganisms are small microscopic organisms that cannot be seen with naked eyes. They inhabit animal intestine, food, soil, water and other different environment. They belong to any of the following group of organisms: bacteria, fungi, protozoa, micro-algae and viruses [26]. The largest proportion of the earth's biodiversity has been reported to be microorganisms and they play an integral role in the processes of ecosystem thereby providing functions that eventually sustain all forms of life [27, 28].

Soil microbial biomass is the living component of soil organic matter. Soils with high organic substances tend to have a higher microbial biomass contents as well as their activities since organic matters are the preferred energy source for microbes. The surface horizon of the soil has the highest microbial activities when compared to the deeper horizon [29]. The soil microbial biomass helps in the enzymatic transformation of soil organic matter into humus, carbon and other nutrients which are utilized by microorganisms for their own growth [30]. Soil microbial biomass and its enzymatic activities are strongly influenced by seasonal changes in soil temperature, moisture and available residue [31]. Soil enzymes may originate from animals, plants or microbes and can either exist in bound or free form within the soil. Soil enzymes play a vital role in the biochemical functioning of soils [32] including nutrients cycling [33], soil structure maintenance [34] and decomposition of organic residue [35]. The activities of soil enzymes are controlled by many factors such as soil microbial community [36], soil physio-chemical properties [37], vegetation type [38] and ecological disturbances [39]. Prior to the utilization of complex organic matter by microorganisms as their source of energy, they produce a quite number of extra cellular enzymes in order to decompose them [40]. Soil enzymes are specific in the types of reactions they participate. For example, a starch hydrolyzing enzyme known as amylase hydrolyses α1-4D glucosidic linkage of amylase and amylopectin and consist of α-amylase and β-amylase. α-amylase is synthesized by animals, plants and microorganisms while β-amylase is primarily synthesize by plants [41]. To a large extent, soil microbial activities is dependent on the quantity of available carbon and this is shown by dehydrogenase activity [29]. Dehydrogenase is involved in the biological oxidation of soil organic matter, and also responsible in oxidizing organic matter by transferring hydrogens and electrons from substrates to acceptors [42]. Phosphatase originate from root exudates and microorganisms, it cleaves the phosphate from organic substrates and also involved in P cycle in soil [43]. It has been evidently suggested by Ushio *et al.* [44, 45] that plant species significantly have more direct impacts on the composition of soil microbial community and their activities in addition to soil physicochemical properties. Plants' rhizosphere has been reported by Vyas and Gupta [46] to have profound effect on microbial population and activities. From the study of Islam and Borthakur [47], increase in microbial biomass and enzyme activities indicates high rate of release of nutrients by rice crops which aid microbial activities.

4. Indigenous microorganisms

Indigenous microorganisms are naturally occurring microorganisms that have adapted to the environmental conditions where they are found thus being capable of accelerating decomposition of organic materials found in that environment [48]. They contain mainly Lactobacillus and sometimes Rhizobium with a few other species [49]. Effective microorganisms are composed of mixed cultures of beneficial and naturally occurring microorganisms which are applied to the soil in order to increase the soil microbial diversity and the growth of plants [50]. This concept was first discovered by Higa [51]. It is used as a means of improving crops' efficiency in utilizing organic matter. There are three main families of over 80 different species contained in effective microorganisms [52]. In agriculture, microorganisms are of great importance because they promote decomposition, cycling and circulation of plant nutrients and reduce the need for chemical fertilizers [53]. From the study of Desire et al. [53] the use of biofertilizers obtained from indigenous and effective microorganisms significantly improved and maintained the chemical, physical and biological properties of the soil, and thus increased the yield of potato in terms of number and weight of tubers when compared to untreated (control) soil.

5. Endophytic microorganism

In 1886, Anton de Bary, a German Botanist and father of plant pathology coined the term endophyte and described it as microorganisms that colonize internal tissues of stem and leaves of plants. Endophytic microorganisms are microorganisms that inhabit at least a period of their life cycle in the interior parts of plants especially leaves, branches and stems, showing no apparent harm to the host [54].

They are also capable in colonizing the roots and shoots of plants and may not remain as endophyte throughout their life cycle [55]. They include both bacteria and fungi that colonize almost every plant species [56]. Endophytic fungal appear to be symbiotically associated with almost all plants in natural ecosystem and constitute important components of plant micro-ecosystems. They have impacts on the composition of plant communities by increasing their tolerance to biotic and abiotic stress, biomass and decreasing water consumption or altering allocation of resources [57]. Endophytic microbes produce a plethora of secondary metabolites, including toxins, enzymes, anti-inflammatory, antibiotics, anticancer and antifungal compounds in order to colonize plants and also compete with other microorganisms [58]. Zhao *et al.* [59] reported that endophytic fungi produce good bioactive compound paclitaxel (taxol) and many other bioactive molecules such as terpenoids, alkaloids, steroids, lignans, phenols, quinones and lactones.

Endophytic bacteria have been detected inside the stems, leaves and inside the reproductive organs of different host plants [60]. Several endophytic bacteria produce low molecular weight compounds, phytohormones, enzymes, antimicrobial substances and siderophores which support the growth of plants and also increase their nutrient uptake. Endophytic bacteria in combination with the plants they are in association with, produce some metabolites which plants cannot produce alone [61]. *Enterobacter, Pseudomonas, Burkholderia, Bacillus, Erwinia* and *Xanthomonas* are the most commonly isolated genera of endophytic bacteria. Eleven culturable bacterial strains belonging to the genera, *Rahnella, Pseudomonas, Rhodanobacter, Enterobacter, Stenotrophomonas, Phyllobacterium* and *Xanthomonas* have been isolated from the stems of sweet potato. Among these isolates, *Pseudomonas, Enterobacter* and *Rahnella* produced higher amount of indole acetic acid (IAA) which promote plant growth, and *Rahnella sp.* which is resilient to stress like cold shock, antibiotics and UV radiation [62].

Webber [63] was probably the first researcher to report plant protection given by an endophytic fungus, *Phomopsis oblonga* which protected elm trees against *Physocnemum brevilineum*. He suggested that the endophytic fungus protected elm tree against the Dutch disease caused by *Ceratocystis ulmi* by reducing its spread and controlling the vector, *P. brevilineum*. It was reported by Claydon *et al.* [64] that endophytic fungi belonging to the family *Xylariaceae* synthesize secondary metabolites in the hosts of the genus *Fagus* which affect beetle larvae. Stress-related genes in *Oryza sativa* such as aquaporin, dehydrin and malondialdehyde have been found to be upregulated by *Trichoderma harzianum* responses. *Trichoderma harzianum* used in treating *Brassica juncea* improved oil content affected with sodium chloride was found to increase its vital nutrients uptake, improve aggregation of osmolytes and antioxidants, and also reduces its NaCl uptake [65]. Brotman *et al.* [66] reported that *T. harzianum* synthesize 1-aminocyclopropane-1-carboxylate (ACC) deaminase to ameliorate salinity stress. *Acinetobacter sp.* and *Pseudomonas sp.* have also been reported to increase indole acetic acid and ACC deaminase production in oats and barley under salinity stress [67].

Beneficial endophytic bacteria and fungi can be used as inoculant in roots and other plant tissues for many tuberous crops to enjoy the mutualist benefits confer to their original host plants. Many growth promoting endophytes can also be applied as a potential bio-fertilizers in tuber crops with minimal environmental risks [56]. Endophytic microorganisms have frequently been reported to be associated with crop plants such as *Triticum aestivum, Glycine max, Zea mays, Hordeum brevisubulatum and Hordeum bogdanii* [68]. The growth of tomato plants in a salinity stress soil have been improved by *Streptomyces sp.* strain PGPA39 by alleviating the salinity stress [69]. PsJN strain of *Burkholderia phytofirmans* have been reported to combat drought stress in maize and wheat, and also salinity stress in *Arabidopsis thaliana* [70].

5.1 Fungi - plants association

Fungi symbiotic relationship with plants are present in a broad range of terrestrial ecosystems which include a large proportion of plant taxa [71]. It has been established that at least 85% of plant species have been able to establish a symbiotic relationship with fungi, of which those belonging to the phylum Glomeromycota account for 70% of the association [72]. Because of the wide geographical distribution of mycorrhiza and the large proportion of plant taxa involved, mycorrhizal associations are extremely important for terrestrial ecosystem. Due to the development of specialized structures such as proteoid roots, carnivorism or parasitism on other plants, some families of plants have lost their ability to associate with mycorrhizal fungi throughout evolution [71]. For a long time, plant species belonging to the Cyperaceae family was believed not to be able to associate with the mycorrhizal fungi [71] though Bohlen [73] study has evidently shown otherwise. Plant species belonging to the Cyperaceae family are able to associate with arbuscular mycorrhizal fungi and dark septate endophytes (DSE), but the intensity of root colonization intensity may vary depending on the environment in which the samples were collected and phenological stage of the plant [72]. Mycorrhizal associations play an important role in determining the composition of plant communities, since plants that establish this type of association can obtain competitive advantages [74] or facilitate the establishment of other species [75]. van der Heijden *et al.* [76] experimental study evidently suggests the coexistence of different plants. They showed that plants inoculated with AMF grew on the average of 11.8 times more than those not inoculated, and that the distribution of phosphorus and nitrogen between plant species varied depending on the presence of AMF. They further said that AMF can redistribute resources among different species of plant thus allowing their coexistence. The final composition of AMF species varies greatly depending on the plant species cultivated in a soil [77]. The diversity of AMF was much smaller in areas dominated with the invasive species than in areas dominated by native species. Thus, the composition of plant communities and AMF are influenced by feedback interactions in each communities [78, 79].

5.1.1 Arbuscular mycorrhizal fungi (AMF)

Arbuscular mycorrhizae are formed by non-septate phycomycetous fungi belonging to the genera Glomus, Acaulospora and Sclerocystics in the family Endogonaceae of the order Mucorales which are not specialized in host range [80]. The arbuscular mycorrhizal (AM) symbiosis is the association between fungi of the order Glomales (Zygomycetes) and the roots of terrestrial plants [81]. Arbuscular Mycorrhizal Fungi (AMF) also known as Vesicular Arbuscular Mycorrhizal (VAM) are widespread in terrestrial ecosystems and form mutually beneficial association with nearly 80% of higher plants [82]. According to Voko *et al.* [83], the population of AMF, frequency of occurrence and distribution varied with site.

During the formation of AM symbiosis, the fungus forms a haustoria-like structure (arbuscules) that interface with the host cytoplasm by penetrating the cortical cell wall of the root [84]. They penetrate the living cells of plants without harming them and their hyphae can range far into the bulk soil establishing equally intimate contact with the microbiota of soil aggregates and micro-sites [85]. From the fixed photosynthates of the plant, it supplies carbon to the fungus while the fungus in turn assist the plant in the uptake of phosphorus and other mineral nutrients from the soil [86]. It has been demonstrated that plants can receive up to 100% of the phosphorus through mycorrhizal pathway, and 4 to 20% of plant carbon can be transferred to fungi [87].

The transference of these resources between plants and fungi have profound effect on plant nutrition, growth and ecology [88]. The activities of AMF improve crop growth and yield by increasing nutrients availability and increasing root proliferation [82] as well as altering some physiological processes in the plant that result in increased yield [89]. This might also be as a result of modification of host hormonal relations [90] and soil structure [91]. AMF can alter the pattern of gene expression, cellular programming and organ development of the host crop [92]. AMF can improve both plant growth under low fertility conditions, improve plant water balance and help in the establishment of plants in new environment [93]. AMF are useful in the cultivation of cassava in the tropics where rainfall is erratic and may seize for 2–3 months giving rise to drought-prone water deficit-stress condition [94]. AMF enhances plant resistance to drought by building up macroporous structure in soil that allows water and air to penetrate and also prevent erosion thus improving photosynthesis and reducing micropropagation stresses [95]. The mutualistic association between AMF and cassava in AMF-inoculated cassava stimulated the production of more leaf chlorophyll which increased their photosynthetic potential and enhanced growth [94].

5.1.2 Dark septate endophytes and crop plants

Dark Septate Endophytes (DSEs) are another important group of soil microorganism that have the capacity to associate with the roots of several plant species [96]. They sometimes colonize roots containing AMF [96]. The increasing severity of environmental conditions increase the importance of DSE. The associations of plants with DSE in high-stress environment is more frequent than their associations with AMF [97]. AMF and DSE have appeared to have similar and complimentary roles in various terrestrial ecosystem [98]. Grunig *et al.* [99] said that since DSEs can alter the performance of colonized plants, they can also play a vital role in determining the composition of plant communities. In the study of Barrow and Osuna [100], some plants colonized by DSE were more advantageous in the absorption of phosphorus from the soil and production of biomass when compared to those not inoculated. Though DSE is advantageous to plants, its colonization of roots can be of disadvantage to plants, such as decreases in the production of biomass [99]. Thus, the interaction of DSEs with plants seems to vary from mutualism to parasitism and may alter the competitive relations between plants [99].

5.2 Beneficial soil bacteria and crop production

Apart from fungi, there are several groups of soil bacteria that are important to plant growth. Some bacteria have the ability to fix atmospheric nitrogen and form symbiotic relationship with plants [101]. In tropical soils, phosphate-solubilizing microorganisms indirectly provide phosphorus for plants by solubilizing phosphorus precipitated with iron, aluminum and calcium thus making it important for plant growth and development [102].

5.2.1 Root colonization by bacteria

Root colonization is the microbial attachment to and proliferation on roots. It is an essential factor in the beneficial interactions used for biofertilization, microbiological control, phytoremediation and phytostimulation as well as in plant pathogenesis of soil borne microbes [103]. PGPR may colonize the rhizosphere, root surface, or even superficial intercellular spaces [104].

Howie *et al*. [105] hypothesized two phase processes in which bacterium can attach itself to the plant and soil. In the phase I, bacteria on the seed are attached to the emerging root tip where they are passively transported into the soil. During root growth, some bacteria cells remain associated with the tip while others are left behind on the older parts of the root and the rhizosphere. In phase II, bacteria deposited along the root zone multiply and form microcolonies in nutrient-rich microsites, where they compete with indigenous microflora in order to avoid displacement. Both phases occur simultaneous on different root parts [106]. Root colonization can be influenced by both biotic (such as genetic traits of the host plant and the colonizing organism) and abiotic (such as soil humidity, growth substrate, soil and rhizosphere pH and temperature) factors. Changes in plant physical and chemical composition in the rhizosphere can strongly influence root colonization and competence [107]. Root exudates and mucilage-derived nutrients attract beneficial and neutral bacteria as well as harmful bacteria allowing them to colonize and reproduce in the rhizosphere [108].

5.2.2 Plant growth promoting rhizobacteria (PGPR)

In developing a sustainable crop production system, the use of plant growth promoting rhizosphere has played a potential role [109] though its mechanisms of enhancing plant growth and yields have not been fully understood [110]. PGPR play an important role in plant growth through different mechanisms [111]. The relationship of PGPR differs with different host plants. Rhizospheric and endophytic relationships are the two major classes of PGPR relationships. PGPRs that colonize root surfaces or superficial intercellular spaces of the host plant forming root nodules are known to have Rhizospheric relationships. A microbe belonging to the genus *Azospirillum* is the dominant species in the rhizosphere [112]. PGPRs that inhabit and grow within the apoplastic spaces of the host plants are known to have endophytic relationships [107]. Some researchers have shown that inoculation of plants with PGPR help in increasing their nutrient contents [113, 114] and resistivity to pathogens [115, 116]. PGPR colonize plant by interacting with the host plant thus enhancing its nutrient uptake by fixing nitrogen biologically, increasing the availability of nutrients in the rhizosphere, inducing increases in the root surface area, enhancing other beneficial symbioses of the host, and combining the modes of action [107]. PGPR help to solubilize mineral phosphates and other nutrients, stabilize soil aggregates, improve soil structure and organic matter content, and increase plant resistivity to stress. It retains more soil organic nitrogen and other nutrients in the plant–soil system, thereby reducing their need for nitrogen and phosphorus fertilizer and enhancing the release of nutrients.

In addition to increasing plant nutrient contents, PGPR capable of producing phytohormones produce hormones such as cytokinins, ethylene, gibberellins, auxins and abscisic acid. Some of the bacterial genera belonging to the PGPR produce indole-3-acetic acid (IAA), a compound belonging to auxins which promote plant growth. Some PGPR function as a sink for 1-aminocyclopropane-1-carboxylate (ACC), the immediate precursor of ethylene in higher plants, by hydrolyzing it into α–ketobutyrate and ammonia, thereby promoting root growth by lowering indigenous ethylene levels in the micro-rhizo environment [117]. In different ecosystems, bacteria can also play a core role in the composition of plant communities by specifically acting on certain plant species and also participating in key environmental processes. In addition to increasing plant nutrient content, it is capable of increasing the population of other beneficial microorganisms and controlling the population of harmful ones in the rhizosphere [111].

5.2.3 Biological nitrogen fixation (BNF)

The process of fixing nitrogen biologically by soil microbes is an economically attractive and ecologically sound method to reduce external nitrogen input and enhance the quality and quantity of internal resources [118]. Soil microbe can be considered as a living component of soil organic matter because the biomass comprises all soil organisms with a volume approximately less than 5×10^3 μm^3 apart from plant tissue [119]. This process accounts for 65% of nitrogen that are currently used in agriculture, and will continuously be of importance in the sustenance of crop production systems in the future [120]. In most terrestrial ecosystems, BNF is their largest source of new nitrogen [121]. The rates of BNF in tropical forests (15 to 36 kg N/ha/yr) are higher than/similar to their temperate counterparts (7–27 kg N/ha/yr), which are subjected to strong nitrogen limitation [122]. In the tropics, diazotrophs could have been favored because they receive enough quantity of nitrogen to maintain higher extracellular phosphatase activity, which is prerequisite for overcoming phosphorus limitation and also they have optimum temperature for their activities [123]. Important biochemical reactions of BNF occur mainly through symbiotic relationship of N_2-fixing microbes (especially bacteria) with legumes that convert atmospheric nitrogen (N_2) into ammonia (NH_3) [124].

Symbiotic and non-symbiotic microorganisms in the soil rhizosphere can assist in fixing atmospheric nitrogen in crops and non-crop plants. Over the years, it has been accepted generally that legumes (and the non-legumes genus *Parasponia*) are exclusively nodulated by member of the *Rhizobiaceae* Family in the α-proteobacteria, which includes the genera *Bradyrhizobium, Sinorhizobium, Azorhizobium, Mesorhizobium* and *Rhizobium* [125]. Recently, other species of α-proteobacteria such as *Methylobacterium*, *Blastobacter denitrificans, Devosia* have been reported to nodulate *Crotalaria, Aeschynomene indica* and *Neptunia natans*, respectively [126–128]. *Ralstonia taiwanensis* and *Burkholderia spp.* belonging to the β-proteobacteria have been found in the nodules of some tropical legumes [129, 130].

Generally, PGPR are classified as biofertilizers, biopesticides and phytostimulators [131]. The biofertilizers help to promote plant growth by supplying nutrients to the host, and these include *Allorhizobium spp., Pseudomonas fluorescens, Rhizobium spp.* and *Trichoderma spp.* (e.g. *T. asperellum* and *T. hamatum*) [132]. The symbiotic association of Rhizobacteria with soil introduces $50–70 \times 10^6$ tons of nitrogen into agricultural soils thus reducing the use of inorganic fertilizers [133]. The phytostimulators produce phytohormones such as indole acetic acid, gibberellin and cytokinins which alter root architecture and promote plant development [134] and these include *Bacillus, Azospirillum, Azotobacter, Enterobacter, Pantoea, Pseudomonas, Streptomyces* and *Rhizobium spp.* The biopesticides inhibit the proliferation of pathogen and help in plant growth, and these include *Pseudomonas spp.*, *Streptomyces spp.* and *Bacillus spp.* (e.g. *B. subtilis*) [135]. In addition to these three groups, there are other PGPRs that induce tolerance in plants to abiotic stress. Those in this group include *Paenibacillus polymyxa, Achromobacter piechaudii* and *Rhizobium tropici* [136].

The nitrogen fixed by symbiotic *Rhizobia* in legumes can be beneficial to associated non-leguminous crops through direct transfer of biologically fixed nitrogen to cereals growing in intercrops [137] or to subsequent crops rotated with symbiotic leguminous crops [138]. In many low input grassland systems, the grasses depend on the nitrogen fixed by their legume counterparts for their nitrogen nutrition and protein synthesis, which is much needed for forage quality in livestock production [117]. *Rhizobium* and *Bradyrhizobium* species of *Rhizobia* produce molecules such as auxins, abscisic acids, riboflavin, cytokinins, vitamins and lipochitooligosaccharides that promote plant growth in addition to fixing atmospheric nitrogen [139]. Other PGPR traits of *Rhizobia* and *Bradyrhizobia* assist in the production of phytohormones [140], release of siderophore [141], solubilization of inorganic phosphorus [142] and also act as antagonist against plant pathogenic microbes [143]. In the study of Kennedy *et al.* [144], a several number of non-symbiotic PGPR significantly increase the vegetative growth and grain yield of C_3 and C_4 plants such as rice, maize, wheat, cotton and sugarcane due to their interactions. The application of *Azotobacter* increased the yield of rice, cotton and wheat [145, 146]. In a field trial experimental study, Tran Van *et al.* [147] used *Burkholderia vietnamiensis* to inoculate rice and found out that it significantly increased the grain yields up to 8 t/ha. It has been reported that the species belonging to genus *Burkholderia* can produce substances that are antagonistic to nematodes [148].

5.2.4 Phosphorus-solubilizing bacteria

Strains of bacteria belonging to the genera of *Pseudomonas, Rhizobium, Agrobacterium, Flavobacterium, Bacillus, Burkholderia, Aerobacter, Achromobacter, Erwinia* and *Micrococcus* have been found to have the ability to solubilize insoluble inorganic phosphate compounds such as rock phosphate, dicalcium phosphate, tricalcium phosphate and hydroxyl apatite [149, 150]. Tricalcium phosphate and hydroxyl apatite have been reported to be more degradable substrates than rock phosphate while the most powerful phosphate solibilizers are strains belonging to the genera *Pseudomonas, Bacillus* and *Rhizobium* [151].

6. Conclusion

The soil is a good reservoir or habitat for microorganisms which might be beneficial to both plants and animals. Soil microbes help in aerating the soil by increasing the pore sizes thus increasing the rate of percolation. The relevance of soil microorganisms to crop growth and productivity cannot be overemphasize. They enhance crop growth and productivity both in stress, low fertile and fertile soil by facilitating transformation, solubilization and mobilization of nutrients, and altering the physiological processes of plants.

In order to increase agricultural production in terms of cropping in the tropics without polluting or degrading the environment, most especially cropland soils, the use of biofertilizers and biopesticides composed of beneficial microbes should be encouraged among peasant and large scale farmers instead of synthetic fertilizers. Since they do not only improve the soil fertility but also assist the roots of plants in the absorption and uptake of nutrients from the soil.

Author details

Barachel Odaro-Junior Umukoro
Department of Botany, University of Ibadan, Ibadan, Nigeria

*Address all correspondence to: umukorobacheloj@gmail.com

References

[1] Sengupta A, Gunri SK. Microbial intervention in agriculture: An overview. African Journal of Microbiology Research. 2015; **9**(18):1215-1226

[2] Hirsch AM. Brief History of the Discovery of Nitrogen-Fixing Organisms; 2009

[3] Berch SM, Massicotte HB, Tackaberry LE. Republication of a translation of "the vegetative organs of *Monotropa hypopitys* L." published by Kamienski, F. in 1882, with an update on Monotropa mycorrhizas. Mycorrhiza. 2005;**15**(5):323-332

[4] Weindling R. Studies on a lethal principle effective in the parasitic action of *Trichoderma lignorum* on *Rhizoctonia solani* and other soil fungi. Phytopathology. 1934;**24**:1153-1179

[5] Harman GE, Howell CR, Viterbo A, Chet I, Lorito M. Trichoderma species—Opportunistic, avirulent plant symbionts. Nature Reviews. Microbiology. 2004;**2**(1):43-56

[6] Varma A, Verma S, Sudha SN, Butehorn B, Franken P. *Piriformospora indica*, a cultivable plant growth-promoting root endophyte. Applied and Environmental Microbiology. 1999;**65**:2741-2744

[7] Morale FJ. Introduction to tropical agriculture and outlook for tropical crops in a globalized economy. Tropical Biology and Conservation Management. n.d; 3. 27 p

[8] Simmonds NW. Diseases of tropical crops: Problems and controls. Botanical Journal of Scotland. 1994;**47**(1):129-137

[9] Vavilor NI. Theoretical basis for plant breeding. Origin and geography of cultivated plants. In: Love D, editor. The Phytogeographical Basis for Plant Breeding. Vol. 1. Cambridge, UK: Cambridge University Press; 1935. pp. 316-366

[10] Harlan JR. Agricultural origins: Centers and non centers. Science. 1971;**174**:468-474

[11] Harlan JR. Diseases as a factor in plant evolution. Annual Review of Phytopathology. 1975;**14**:31-51

[12] Hawkes JG. The taxonomy of cultivated plants. In: Frankel OH, Bennet E, editors. Genetic Resources in Plants: Their Exploration and Conservation. Oxford, UK: Blackwell; 1970. pp. 69-85

[13] Zeven AC, de Wet JMJ. Dictionary of Cultivated Plants and their Regions of Diversity. Wageningen: Neth.: Cent. Agric. Publ. Doc; 1982

[14] Gibson L, Lee TM, Koh LP, Brook BW, Gardner TA, Barlow J. Primary forests are irreplaceable for sustaining tropical biodiversity. Nature. 2011;**478**:378-381

[15] Townsend AR, Cleveland CC, Houlton BZ, Alden CB, White JWC. Multi-element regulation of the tropical forest carbon cycle. Frontiers in Ecology and the Environment. 2011;**9**:9-17

[16] Herforth A. Access to adequate nutritious food: New indicators to track progress and information. In: Sahn DE, editor. Fight against Hunger and Malnutrition. Oxford, UK: Oxford University Press; 2015. pp. 139-162

[17] Geering ADW, Randles JW. Virus diseases of tropical crops. In: eLS. Chichester: John Wiley and Sons, Ltd.; 2012. pp. 1-14

[18] Stork NE, Coddington JA, Colwell RK, Chazdon RL, Dick CW, Peres CA, et al. Vulnerability and

resilience of tropical forest species to land-use change. Conservation Biology. 2009;**23**:1438-1447

[19] Allan DL, Adriano DC, Bezdicek DF, Cline RG, Coleman DC, Doran JW, et al. Soil Science Society of America: Statement on soil quality. Agronomy New, Madison, Wisconsin: ASA; 1995;7

[20] SSSA. Glossary of Soil Science Terms. Madison: Soil Science Society of America; 1984

[21] Lartey RT, Conway KE. Novel considerations in biological control of plant pathogens: Microbial interactions. In: Lartey RT, Ceasar AJ, editors. Emerging Concepts in Plant Health Management. Trivandrum, India: Research Signpost; 2004. pp. 141-157

[22] Nelson EB. Microbial dynamics and interactions in the spermatosphere. Annual Review of Phytopathology. 2004;**42**:271-309

[23] Hiltner L. On new findings and problems in the area of soil bacteriology with special consideration of green manure and fallow. Abhandlungen der Naturforschenclen Gesell. 1904;**98**:59-78

[24] Rovira AD, Foster RC, Martin JK. Note on terminology: Origin, nature and nomenclature of the organic materials in the rhizosphere. In: Harley JC, Scott-Russellm R, editors. The Soil-Root Interface. London: Academic Press; 1979

[25] Nielsen TH, Sorensen J. Production of cyclic lipopeptides by *Pseudomonas fluorescens* strains in bulk soil and in the sugar beet rhizosphere. Applied and Environmental Microbiology. 2003;**69**:861-868

[26] Mosttafiz S, Rahman M, Rahman M. Biotechnology: Role of microbes in sustainable agriculture and environmental health. The Internet Journal of Microbiology. 2012;**10**(1):1-6

[27] Falkowski PG, Fenchel T, Delong EF. The microbial engines that drive earth's biogeochemical cycles. Science. 2008;**320**:1034-1039

[28] Prosser JI. Ecosystem processes and interactions in a morass of diversity. FEMS Microbiology Ecology. 2012;**81**:507-519

[29] Januszek K. The enzyme activity of the forest soils of Southern Poland as a measure of soil quality. Electronic Journal of Polish Agricultural Universities. 2011;**14**(2):1

[30] Anderson TH, Domsch KH. Ratios of microbial biomass carbon to total organic-C in arable soils. Soil Biology and Biochemistry. 1980;**21**:471-479

[31] Diaz-Ravina M, Acea MJ, Carballas T. Seasonal changes in microbial biomass and nutrient flush in forest soils. Biology and Fertility of Soils. 1995;**19**:220-226

[32] Wang C, Wang G, Liu W, Wu P. The effects of plant-soil-enzyme interactions on plant composition, biomass and diversity of Alpine meadow in the Qinghai-Tibetan plateau. International Journal of Ecology. 2011;**2011**:1-10

[33] Makoi JHJR, Ndakidemi PA. Soil enzymes: Examples of their potential roles in the ecosystem. African Journal of Biotechnology. 2008;**7**:181-191

[34] Balota EL, Kanashiro M, Filho AC, Andrade DS, Dick RP. Soil enzyme activities under long term tillage and crop rotation systems in subtropical agro-ecosystems. Brazilian Journal of Microbiology. 2004;**35**:300-306

[35] Lei T. Understanding soil organic matter mineralization in agro ecosystems: Soil enzyme perspectives. Ph.D Thesis. United State: North Carolina State University. 2011

[36] Kourtev PS, Ehrenfeld JG, Haggblom M. Exotic plant species alter

the microbial community structure and function in the soil. Ecology. 2002;**83**:3152-3166

[37] Amador JA, Glucksman AM, Lyons JB, Gorres JH. Spatial distribution of soil phosphatase activity within a riparian forest. Soil Science. 1997;**162**:808-825

[38] Sinsabaugh RL, Carreiro MM, Repert DA. Allocation of extracellular enzymatic activity in relation to litter composition, N deposition and mass loss. Biogeochemistry. 2002;**60**:1-24

[39] Boerner REJ, Decker KLM, Sutherland EK. Prescribed burning effects on soil enzyme activity in a southern Ohio hardwood forest: A landscape-scale analysis. Soil Biology and Biochemistry. 2000;**32**:899-908

[40] Lalitha S, Santhaguru K. Improving soil physical properties and effect on tree legume seedlings growth under barren soil. Agricultural Science Research Journal. 2012;**2**:126-130

[41] Pazur JH. Enzymes in the synthesis and hydrolysis of starch. In: Whistler R, Paschall EF, editors. Starch Chemistry and Technology, Fundamental Aspects. New York, U.S.A: Academic Press; 1965. pp. 133-175

[42] Maurya BR, Singh V, Dhyani PP. Enzymatic activities and microbial population in agric-soils of Almora District of central Himalaya as influenced by altitudes. International Journal of Soil Science. 2011;**6**:238-248

[43] Huang W, Liu J, Zhou G, Zhang D, Deng Q. Effects of precipitation on soil acid phosphatase activity in three succession forests in southern China. Biogeosciences. 2011;**8**:1901-1910

[44] Ushio M, Wagai R, Balser TC, Kitayama K. Variations in the soil microbial community composition of a tropical montane forest ecosystem: Does tree species matter? Soil Biology and Biochemistry. 2008;**40**:2699-2702

[45] Ushio M, Kitayama K, Balser TC. Tree species effects on soil enzyme activities through effects on soil physicochemical and microbial properties in a tropical montane forest on Mt. Kinabalu, Borneo. Pedobiologica. 2010;**53**:227-233

[46] Vyas D, Gupta RK. Effect of edaphic factors on the diversity of VAM fungi. Tropical Plant Research. 2014;**1**(1):14-25

[47] Islam NF, Borthakur. Effect of different growth stages on rice crop on soil microbial and enzyme activities. Tropical Plant Research. 2016;**3**(1):40-47

[48] Singh A, Sharma S. Effect of microbial inoculants on mixed solid waste composting, vermicomposting and plant response. Compost Science & Utilization. 2003;**11**:190-199

[49] Hiddink GA, van Bruggen AHC, Termorshuizen AJ, Raaijmakers JM, Semenov AV. Effect of organic management of soils on suppressiveness to *Gaeumannomyces graminis* Var. *tritici* and its antagonist, *Pseudomonas fluorescens*. European Journal of Plant Pathology. 2005;**113**(4):417-435

[50] Muthaura C, Musyimi DN, Ogur JA, Okello SV. Effective microorganisms and their influence on growth and yield of pigweed (*Amarantus dubains*). Journal of Agricultural and Biological Sciences. 2010;**5**:17-22

[51] Suthamathy N, Seran TH. Residual effect of organic manure EM bokashi applied to proceeding crop of vegetable cowpea (*Vigna unguiculata*) on succeeding crop of radish (*Raphus sativus*). Research Journal of Agriculture and Forestry Sciences. 2013;**1**:2-5

[52] Daly MJ, Stewart DPC. Influence of effective microorganisms (EM) on vegetable production and carbon mineralization- a preliminary investigation. Journal of Sustainable Agriculture. 1999;**14**(2-3):15-25

[53] Desire TV, Fosah MR, Desire MH, Fotso. Effect of indigenous and effective microorganism fertilizers on soil microorganisms and yield of Irish potato in Bambill, Cameroon. African Journal of Microbiology Research. 2018;**12**(15):345-353

[54] Azevedo JL. Microorganismos endofíticos. In: Melo IS, Azevedo JL, editors. Ecologia Microbiana. Jaguariuna, São Paulo, Brazil: Editora EMBRAPA; 1998. pp. 117-137

[55] Porras-Alfaro A, Bayman P. Hidden Fungi, emergent properties: Endophytes and microbiomes. Annual Review of Phytopathology. 2011;**49**:291-315

[56] Shubhransu N, Archana M, Soma S. Endophytic microorganisms of tropical tuber crops: Potential and perspectives. Journal of Applied and Natural Science. 2017;**9**(2):860-865

[57] Rodriguez RJ, White JF Jr, Arnold AE, Redman RS. Fungal endophytes: diversity and functional roles. New Phytologist. 2009:1-17

[58] De Melo FMP, Fiore MF, DeMoraes LAB, Stenico MES, Scramin S, Teixeira MA, et al. Antifungal compound produced by the cassava endophyte *Bacillus pumilus* MAIIIM4A. Scientia Agricola (Piracicaba, Brazil). 2009;**66**(5):583-592

[59] Zhao J, Zhou L, Wang J, Shan T, Zhong L, Liu X, et al. Endophytic fungi for producing bioactive compounds originally from their host plants. In: Current Research, Technology and Education Topics in Applied Microbiology and Microbial Biotechnology. Formatex, Spain; 2010. pp. 567-576

[60] Rosenblueth M, Martinez-Romero E. Bacterial endophytes and their interactions with hosts. Molecular Plant-Microbe Interactions. 2006;**19**(8):827-837

[61] Brader G, Compant S, Mitter B, Trognitz F, Sessitsch A. Metabolic potential of endophytic bacteria. Current Opinion in Biotechnology. 2014;**27**:30-37

[62] Khan Z, Doty SL. Characterization of bacterial endophytes of sweet potato plants. Plant and Soil. 2009;**322**(1):197-207

[63] Webber J. A natural control of Dutch elm disease. Nature (London). 1981;**292**:449-451

[64] Claydon N, Grove JF, Pople M. Elm bark beetle boring and feeding deterrents from *Phomopsis oblonga*. Phytochemistry. 1985;**24**:937-943

[65] Ahmad P, Hashem A, Abd-Allah EF, Alqarawi AA, John R, Egamberdieva D. Role of *Trichoderma harzianum* in mitigating NaCl stress in Indian mustard (*Brassica juncea* L.) through antioxidative defense system. Frontiers in Plant Science. 2015;**6**:868

[66] Brotman Y, Landau U, Cuadros-Inostroza A, Tohge T, Fernie AR, Chet I. *Trichoderma*-plant root colonization: Escaping early plant defense responses and activation of the antioxidant machinery for saline stress tolerance. PLoS Pathogens. 2013;**9**(3):1-15

[67] Chang P, Gerhardt KE, Huang XD, Yu XM, Glick BR, Gerwing PD. Plant growth promoting bacteria facilitate the growth of barley and oats in salt impacted soil: Implications for phytoremediation of saline soils. International Journal of Phytoremediation. 2014;**16**:1133-1147

[68] Yuan Z, Zhang C, Lin F, Kubicek CP. Identity, diversity, and molecular phylogeny of the endophytic mycobiota in the roots of rare wild rice (Oryza granulate) from a nature Reserve in Yunnan, China. Applied and Environmental Microbiology. 2010;**76**(5):1642-1652

[69] Palaniyandi SA, Damodharan K, Yang SH, Suh JW. *Streptomyces sp.* strain PGPA39 alleviates salt stress and promotes growth of 'micro tom' tomato plants. Journal of Applied Microbiology. 2014;**117**:766-773

[70] Pinedo I, Ledger T, Greve M, Poupin MJ. *Burkholderia phytofirmans* PsJN induces long-term metabolic and transcriptional changes involved in *Arabidopsis thaliana* salt tolerance. Frontiers in Plant Science. 2015;**6**:466

[71] Brundrett M. Mycorrhizal associations and other means of nutrition of vascular plants: Understanding the global diversity of host plants by resolving conflicting information and developing reliable means of diagnosis. Plant and Soil. 2009;**320**(1):37-77

[72] Wang B, Qiu YL. Phylogenetic distribution and evolution of mycorrhizas in land plants. Mycorrhiza. 2006;**16**(5):299-363

[73] Bohlen PJ. Biological invasions: Linking the aboveground and belowground consequences. Applied Soil Ecology. 2006;**32**(1):1-5

[74] Shah MA, Reshi Z, Rashid I. Mycorrhizal source and neighbour identity differently influence *Anthemis cotula* L. invasion in the Kashmir Himalaya, India. Applied Soil Ecology. 2008;**40**(2):330-337

[75] Chen X et al. Effects of weed communities with various species numbers on soil features in a subtropical orchard ecosystem. Agriculture, Ecosystems and Environment. 2004;**102**(3):377-388

[76] van der Heijden MGA, Wiemken A, Sanders IR. Different arbuscular mycorrhizal fungi alter coexistence and resource distribution between co-occurring plant. The New Phytologist. 2003;**157**(3):569-578

[77] Eom AH, Hartnett DC, Wilson GWT. Host plant species effects on arbuscular mycorrhizal fungal communities in tallgrass prairie. Oecologia. 2000;**122**(3):435-444

[78] Bever JD. Soil community feedback and the coexistence of competitors: Conceptual frameworks and empirical tests. The New Phytologist. 2003;**157**(3):465-473

[79] Hart MM, Reader RJ, Klironomos JN. Plant coexistence mediated by arbuscular mycorrhizal fungi. Trends in Ecology & Evolution. 2003;**18**(8):418-423

[80] Sridevi S, Ramakrishnan K. Effect of inoculation with NPK fertilizer and arbuscular mycorrhizal fungi on growth and yield of cassava. International Journal of Developmental Research. 2013;**3**(9):46-50

[81] Harley JL, Smith SE. Mycorrhizal Symbiosis. London and New York: Academic Press; 1983. 483 p

[82] Smith SE, Facelli E, Pope S, Smith FA. Plant performance in stressful environments. Interpreting new and established knowledge of the roles of arbuscular mycorrhizas. Plant and Soil. 2010;**326**(1-2):3-20

[83] Voko DR, Nandjui J, Sery JD, Fotso B, Amoa JA, Kouadio MA, et al. Abundance and diversity of arbuscular mycorrhizal fungal (AMF) communities associated with cassava (*Manihot esculenta* Crantz.) rhizosphere in Abengourou, east

cote d'Ivoire. Journal of Ecology and the Natural Environment. 2013;**5**(11):360-370

[84] Smith SE, Read DJ. Mycorrhizal symbiosis. San Diego: Academic Press; 1997. p. 605

[85] Ibiremo OS, Ogunlade MO, Oyetunji OJ, Adewale BD. Dry matter yield and nutrient uptake of cashew seedlings as influenced by arbuscular mycorrhizal inoculation, organic and inorganic fertilizers in two soils in Nigeria. ARPN Journal of Agricultural and Biological Science. 2012;7(3):196-205

[86] Gadkar V, David-Schwartz R, Kunik T, Kapulnik Y. Arbuscular mycorrhizal fungal colonization. Factors involved in host recognition. Plant Physiology. 2001;**127**:1493-1499

[87] Cavagnaro TR, Langley AJ, Jackson LE, Smukler SM, Koch GW. Growth, nutrition and soil respiration of a mycorrhiza-defective tomato mutant and its mycorrhizal wild-type progenitor. Functional Plant Biology. 2008;**35**:228-235

[88] Smith SE, Read DJ. Mycorrhizal Symbiosis. 3rd edition. London: Academic Press; 2008. 800p

[89] Oyetunji OJ, Ekanayeke IJ, Osonubi O. The influence of yield of yam in an agroforestry system in South Western Nigeria. Maurik Bulletin. 2003;**6**:75-82

[90] Yao Q, Zhu HH, Chen JZ. Growth responses and endogenous IAA and iPAS changes of litchi (*Litchi chinensis* Sonn.) seedlings induced by arbuscular mycorrhizal fungal inoculation. Science Horticulture. 2005;**105**:145-151

[91] Rilling MC. Arbuscular mycorrhizae, glomalin and soil aggregation. Canadian Journal of Soil Science. 2004;**84**:355-363

[92] Poulton JL, Bryla DR, Koide RT, Stephenson AG. Effect of mycorrhizal infection and soil phosphorus availability on leaf area measurements in tomato. In: Lynch JP, Deikman J, editors. Phosphorus in Plant Biology: Regulatory Roles in Molecular, Cellular, Organism and Ecosystem Processes. Current Tropics in Plant Physiology. Maryland, USA: ASPB Series; 1998. p. 19

[93] Jha A, Kumar A, Saxena RK, Kamalvanshi M, Chakravarty N. Effect of arbuscular mycorrhizal inoculations on seedling growth and biomass productivity of two bamboo species. Indian Journal of Microbiology. 2011;**52**:281-285

[94] Ekanayake IJ, Oyetunji OJ, Osonubi O, Lyasse O. The effects of arbuscular mycorrhizal fungi and water stress on leaf chlorophyll production of cassava (*Manihot esculenta* Crantz). Food, Agriculture and Environment. 2004;**2**(2):190-196

[95] Ranveer K, Yogendra SG, Vivek K. Interaction and symbiosis of AM fungi, actinomycetes and plant growth promoting rhizobacteria with plants: Strategies for the improvement of plants health and defense system. International Journal of Current Microbiology and Applied Sciences. 2014;**3**(7):564-585

[96] Weishampel P, Bedford B. Wetland dicots and monocots differ in colonization by arbuscular mycorrhizal fungi and dark septate endophytes. Mycorrhiza. 2006;**16**(7):495-502

[97] Postma JWM, Olsson PA, Falkengrengrerup U. Root colonisation by arbuscular mycorrhizal, fine endophytic and dark septate fungi across a pH gradient in acid beech forests. Soil Biology and Biochemistry. 2007;**39**(2):400-408

[98] Jumpponen A. Dark septate endophytes - are they mycorrhizal? Mycorrhiza. 2001;**11**(4):207-211

[99] Grünig CR et al. Dark septate endophytes (DSE) of the Phialocephala fortinii s.l. – Acephala applanata species complex in tree roots: Classification, population biology, and ecology. Botany. 2008;**86**(12):1355-1369

[100] Barrow JR, Osuna P. Phosphorus solubilisation and uptake by dark septate fungi in fourwing saltbush, *Atriplex canescens* (Pursh) Nutt. Journal of Arid Environments. 2002;**51**:449-459

[101] Franche C, Lindström K, Elmerich C. Nitrogen-fixing bacteria associated with leguminous and non-leguminous plants. Plant and Soil. 2009;**321**(1):35-59

[102] Gyaneshwar P et al. Role of soil microorganisms in improving P nutrition of plants. Plant and Soil. 2002;**245**(1):83-93

[103] Chin-A-Woeng TFC, Lugtenberg BJJ, Root TFC, Woeng CA. Root colonisation following seed inoculation. In: Varma A, Abbott L, Werner D, Hampp R, editors. Plant Surface Microbiology. Berlin, Heidelberg: Springer; 2008

[104] McCully ME. Niches for bacterial endophytes in crop plants: A plant biologist's view. Functional Plant Biology. 2001;**28**:983-990

[105] Howie WJ, Cook RJ, Weller DM. Effects of soil matric potential and cell motility on wheat root colonization by fluorescent pseudomonads suppressive to take-all. Phytopathology. 1987;**77**:286-292

[106] Weller DM, Thomashow LS, O'Gara E, Dowling DN, Boesten B. Current challenges in introducing beneficial microorganisms into the rhizosphere. In: O'Gara F, Dowling DN, Boesten B, editors. Molecular Ecology of Rhizosphere Microorganism. VCH Weinheim; 1994. p. 1-18

[107] Vessey JK. Plant growth promoting rhizobacteria as biofertilizers. Plant and Soil. 2003;**255**:571-586

[108] Walker TS, Bais HP, Grotewold E, Vivanco JM. Root exudation and rhizosphere biology. Plant Physiology. 2003;**132**:44-51

[109] Shoebitz M, Ribaudo CM, Pardo MA, Cantore ML, Ciampi L, Curá JA. Plant growth promoting properties of a strain of *Enterobacterludwigii* isolated from *Loliumperenne rhizosphere*. Soil Biology and Biochemistry. 2009;**41**(9):1768-1774

[110] Dey R, Pal KK, Bhatt DM, Chauhan SM. Growth promotion and yield enhancement of peanut (*Arachis hypogaea* L.) by application of plant growth promoting rhizobacteria. Microbiological Research. 2004;**159**:371-394

[111] Saharan B, Nehra V. Plant growth promoting rhizobacteria: A critical review. Life Sciences and Medicine Research. 2011;**21**:1-30

[112] Bioemberg GV, Lugtenberg BJJ. Molecular basis of plant growth promotion and biocontrol by rhizobacteria. Current Opinion in Plant Biology. 2001;**4**:343-350

[113] Orhan E et al. Effects of plant growth promoting rhizobacteria (PGPR) on yield, growth and nutrient contents in organically growing raspberry. Scientia Horticulturae. 2006;**111**(1):38-43

[114] Karthikeyan B et al. Effect of root inoculation with plant growth promoting rhizobacteria (PGPR) on plant growth, alkaloid content and nutrient control of *Catharanthus roseus* (L.) G. Don. Nature Croatica. 2010;**19**(1):205-212

[115] Saravanakumar D et al. PGPR-induced defense responses in the tea plant against blister blight disease. Crop Protection. 2007;**26**(4):556-565

[116] Maksimov I, Abizgil'dina R, Pusenkova L. Plant growth promoting rhizobacteria as alternative to chemical crop protectors from pathogens (review). Applied Biochemistry and Microbiology. 2011;**47**(4):333-345

[117] Hayat R, Ali S. Nitrogen fixation of legumes and yield of wheat under legumes-wheat rotation in Pothwar. Pakistan Journal of Botany. 2010;**42**(4):2317-2326

[118] Ademir SFA, Luiz FCL, Bruna DFI, Mario DAL Jr, Gustavo RX, Marcia DVBF. Microbiological process in agroforestry systems. A review. Agronomy for Sustainable Development. 2012;**32**:215-226

[119] Brookes PC. The soil microbial biomass: Concept, measurement and applications in soil ecosystem research. Microbes and Environments. 2001;**16**:131-140

[120] Matiru VN, Dakora FD. Potential use of rhizobial bacteria as promoters of plant growth for increased yield in landraces of African cereal crops. African Journal of Biotechnology. 2004;**3**(1):1-7

[121] Galloway JN, Dentener F, Capone D. Nitrogen cycles: past, present, and future. Biogeochemistry. 2004;**70**:153-226

[122] Cleveland CC, Townsend AR, Schimel DS, Fisher H, Howarth RW, Hedin LO, et al. Global patterns of terrestrial biological nitrogen (N_2) fixation in natural ecosystems. Global Biogeochemical Cycles. 1999;**13**:623-645

[123] Houlton BZ, Wang YP, Vitousek PM, Field CB. A unifying framework for dinitrogen fixation in the terrestrial biosphere. Nature. 2008;**454**:327-330

[124] Shiferaw B, Bantilan MCS, Serraj R. Harnessing the potential of BNF for poor farmers: Technological policy and institutional constraints and research need. In: Serraj R, editor. Symbiotic Nitrogen Fixation; Prospects for Enhanced Application in Tropical Agriculture. New Delhi: Oxford & IBH; 2004. p. 3

[125] Sprent JI. Nodulation in Legumes. Kew, London: Royal Botanic Gardens; 2001

[126] Jaftha JB, Strijdom BW, Steyn PL. Characterization of pigmented methylotrophic bacteria which nodulate *Lotononis bainesii*. Systematic and Applied Microbiology. 2002;**25**:440-449

[127] van Berkum P, Eardly BD. The aquatic budding bacterium *Blastobacter denitrificans* is a nitrogen-fixing symbiont of *Aeschynomene indica*. In: Finan TM, O'Brian MR, Layzell DB, Vessey JK, Newton WE, editors. Nitrogen Fixation: Global Perspectives. New York: CAB International; 2002. p. 520

[128] Rivas R, Velázquez E, Willems A, Vizcaíno N, Subba-Rao NS, Mateos PF, et al. Martínez-Molina E. a new species of Devosia that forms a unique nitrogen-fixing root nodule symbiosis with the aquatic legume *Neptunia natans*. Applied and Environmental Microbiology. 2002;**68**:5217-5222

[129] Chen W-M, Laevens S, Lee TM, Coenye T, de Vos P, Mergeay M, et al. *Ralstonia taiwanensis sp. nov*., isolated from root nodules of Mimosa species and sputum of a cystic fibrosis patient. International Journal of Systematic and Evolutionary Microbiology. 2001;**51**:1729-1735

[130] Moulin L, Munive A, Dreyfus B, Boivin-Masson C. Nodulation

of legumes by members of the β-subclass of Proteobacteria. Nature. 2001;**411**:948-950

[131] Bhardwaj D, Ansari MW, Sahoo RK, Tuteja N. Biofertilizers function as key player in sustainable agriculture by improving soil fertility, plant tolerance and crop productivity. Microbial Cell Factories. 2014;**13**:66

[132] Yadav SK, Dave A, Sarkar A, Singh HB, Sarma BK. Co-inoculated biopriming with *Trichoderma, Pseudomonas* and *Rhizobium* improves crop growth in *Cicer arietinum* and *Phaseolus vulgaris*. International Journal of Agriculture Environment & Biotechnology. 2013;**6**:255-259

[133] Herridge D, Peoples M, Boddey R. Global inputs of biological nitrogen fixation in agricultural systems. Plant and Soil. 2008;**311**:1-18

[134] Duca D, Lorv J, Patten CL, Rose D, Glick BR. Polygalacturonases, polygalacturonase-inhibiting proteins and pectic oligomers in plant-pathogen interactions. Biochimica et Biophysica Acta. 2014;**1696**:237-244

[135] Bhattacharyya PN, Jha DK. Plant growth-promoting rhizobacteria (PGPR): Emergence in agriculture. World Journal of Microbiological Biotechnology. 2012;**28**:1327-1350

[136] Yang J, Kloepper JW, Ryu CM. Rhizosphere bacteria help plant tolerate abiotic stress. Trends in Plant Science. 2009;**14**:1-4

[137] Snapp SS, Aggarwal VD, Chirwa RM. Note on phosphorus and genotype enhancement of biological nitrogen fixation and productivity of maize/bean intercrops in Malawi. Field Crops Research. 1998;**58**:205-212

[138] Hayat R, Ali S, Siddique MT, Chatha TH. Biological nitrogen fixation of summer legumes and their residual effects on subsequent rainfed wheat yield. Pakistan Journal of Botany. 2008a;**40**(2):711-722

[139] Hayat R, Ali S. Potential of summer legumes to fix nitrogen and benefit wheat crop under rainfed condition. Journal of Agronomy. 2004;**3**:273-281

[140] Arshad M, Frankenberger WT Jr. Plant growth regulating substances in the rhizosphere. Microbial production and function. Advances in Agronomy. 1998;**62**:46-51

[141] Jadhav RS, Thaker NV, Desai A. Involvement of the siderophore of cowpea rhizobium in the iron nutrition of the peanut. World Journal of Microbiology and Biotechnology. 1994;**10**:360-361

[142] Chabot R, Antoun H, Cescas MP. Growth promotion of maize and lettuce by phosphate-solubilizing *Rhizobium leguminosarumbio var. phaseoli*. Plant and Soil. 1996a;**184**:311-321

[143] Ehteshamul-Haque S, Ghaffar A. Use of rhizobia in the control of root diseases of sunflower, okra, soybean and mungbean. Journal of Phytopathology. 1993;**138**:157-163

[144] Kennedy IR, Choudhury AIMA, KecSkes ML. Non-Symbiotic bacterial diazotrophs in crop-farming systems: can their potential for plant growth promotion be better exploited? Soil Biology and Biochemistry. 2004;**6**(8):1229-1244

[145] BarassiCA, CreusCM, CasanovasEM, Sueldo RJ. Could Azospirillum Mitigate Abiotic Stress Effects in Plants? Auburn University; 2000

[146] Anjum MA, Sajjad MR, Akhtar N, Qureshi MA, Iqbal A, Jami AR, et al. Response of cotton to plant growth promoting rhizobacteria (PGPR) inoculation under different levels

of nitrogen. Journal of Agricultural Research. 2007;**45**(2):135-143

[147] Tran Vân V, Berge O, Ke SN, Balandreau J, Heulin T. Repeated beneficial effects of rice inoculation with a strain of *Burkholderia vietnamiensison* early and late yield components in low fertility sulphate acid soils of Vietnam. Plant and Soil. 2000;**218**:273-284

[148] Meyer SLF, Massoud SI, Chitwood DJ, Roberts DP. Evaluation of *Trichoderma virens* and *Burkholderia cepacia* for antagonistic activity against root-knot nematode, *Meloidogyne incognita*. Nematology. 2000;**2**:871-879

[149] Rodríguez H, Fraga R. Phosphate solubilizing bacteria and their role in plant growth promotion. Biotechnology Advances. 1999;**17**:319-339

[150] Rodríguez H, Fraga R, Gonzalez T, Bashan T. Genetics of phosphate solubilization and its potential applications for improving plant growth-promoting bacteria. Plant and Soil. 2006;**287**:15-21

[151] Banerjee MR, Yesmin L, Vessey JK. Plant growth promoting rhizobacteria as biofertilizers and biopesticides. In: Rai MK, editor. Handbook of Microbial Biofertilizers. New York: Haworth Press; 2006

Section 3

Relation to Human Health

Chapter 5

Hemolysin of *Vibrio* Species

Tamaki Mizuno, Anusuya Debnath and Shin-ichi Miyoshi

Abstract

Hemolysin is one of the major pathogenic factors among *Vibrio* species, which shows hemolytic activity against erythrocytes. It is associated with different *Vibrio* spp. that manifest either wound infection or intestinal infection as their clinical symptom. *V. vulnificus* and *V. alginolyticus* are well-known causative organisms for wound infection, whereas the gastrointestinal infection is caused by *V. cholerae*, *V. mimicus*, and *V. parahaemolyticus*. There are two major groups of hemolysins in *Vibrio* spp.: the thermostable direct hemolysin (TDH) from *V. parahaemolyticus* and the HlyA (El Tor hemolysin) from *V. cholerae*. These hemolysins have homology in certain degrees; however, the essential amino acids for the activity are variable depending on the species. This chapter summarizes the functions and features of hemolysins from *Vibrio* species, which has been reported so far.

Keywords: thermostable direct hemolysin, El Tor hemolysin, *Vibrio parahaemolyticus*, *vibrio cholerae*, *vibrio mimicus*, *vibrio vulnificus*

1. Introduction

The genus *Vibrio* is comprised of facultative, anaerobic, Gram-negative, curved-rod bacteria that are widely found in natural aquatic environments such as marine, estuarine, and freshwater [1]. More than 100 species have been currently described in this genus, and at least 12 species represented by *V. cholerae*, *V. parahaemolyticus*, and *V. vulnificus* cause a variety of clinical symptoms in human (Table) [1–5]. In addition, species such as *V. metoecus* and *V. navarrensis* are among the newly isolated species from human, and it is strongly suggested that they are human pathogens of *Vibrio* spp. [6, 7]. On the other hand, the major pathogenic *Vibrio* for aquatic vertebrates or invertebrates are *V. anguillarum*, *V. harveyi*, *V. ordalii* etc., responsible for fatal hemorrhagic septicemic disease called vibriosis in marine animals [5, 8–10]. *Vibrio* spp. prefer warm water temperature (15–35°C), so they are likely to flourish more with rising environmental water temperature due to global warming and thus the probability of infections caused by them.

Human diseases caused by pathogenic *Vibrio* spp. can be divided into two major types based on symptoms: intestinal infection and non-intestinal infection [1, 3]. The intestinal infection includes gastroenteritis and cholera, whereas non-intestinal infection includes septicemia and wound infection (**Table 1**). Cholera is caused by ingestion of food and drinking water contaminated with *V. cholerae* O1/O139 that produces cholera toxin (CT) as a major virulence determinant and characterized by severe diarrhea that rapidly leads to dehydration [11, 12]. Till date it remains a major public health disease with estimated 2.9 million cases and 95,000 deaths annually worldwide [13]. There are many clinical cases of gastroenteritis by *V. parahaemolyticus* due to ingestion of raw fish and shellfish [14, 15].

Species	Diseases	Hemolysin family
V. alginolyticus	Wound infection, otitis media	TDH
V. carchariae	Wound infection	
V. cincinnatiensis	Meningitis	
V. cholerae		
O1/O139	Cholera	HlyA
non-O1/non-O139	Gastroenteritis, wound infection	HlyA, TDH
V. damsela	Wound infection	HlyA
V. fluvialis	Gastroenteritis	HlyA
V. furnissii	Gastroenteritis	
V. hollisae	Gastroenteritis, septicemia	TDH
V. metschnikovii	Cholecystitis	
V. mimicus	Gastroenteritis	HlyA, TDH
V. parahaemolyticus	Gastroenteritis, wound infection	TDH
V. vulnificus	Septicemia, wound infection	HlyA

HlyA, El Tor hemolysin group; TDH, thermostable direct hemolysin family.

Table 1.
Pathogenic Vibrio species for human and hemolysins produced by them [1, 3].

The other species such as *V. cholerae* non-O1/non-O139, *V. mimicus*, and *V. fluvialis* are known as agents of foodborne illness [3, 16–19]. On the other hand, *V. vulnificus* is the most studied among *Vibrio* spp. as a causative bacterium of wound infections, though the clinical cases by *V. damsela* and *V. alginolyticus* are also reported [3, 20–24]. *V. vulnificus* is an opportunistic pathogen and poses a threat to individuals with compromised immunity because it can also cause septicemia, which leads to high lethal rates [24–26].

These pathogenic *Vibrio* have been reported to produce various virulence factors, including enterotoxin such as CT produced by *V. cholerae* O1/O139 [12, 27], hemolysin, and Type III secretion system (T3SS) in *V. parahaemolyticus* [28, 29] and extracellular protease in *V. vulnificus* [30]. This chapter has mainly summarized how hemolysins play an important role in the pathogenicity of Vibrio spp. based on studies till date.

2. Hemolysins

Hemolysin is a toxin that attacks membranes of mammalian erythrocyte and causes cell lysis called hemolysis. It is reported that hemolysins are produced by different species of bacteria like *Escherichia coli*, *Staphylococcus aureus*, and *Vibrio* [31–34]. In most cases, the evidence based either on in vivo experiments or clinical reports that suggests the involvement of hemolysins in the pathogenicity is reported [33, 35]. This toxin plays certainly an important role in the infection process initiated by *Vibrio* spp. Hemolysin from *Vibrio* spp. can be classified mainly into two groups, thermostable direct hemolysin (TDH) from *V. parahaemolyticus* and El Tor hemolysin (HlyA) from *V. cholerae*. Even though these toxins partially share the sequence homology, the essential amino acids for the activity, structural features, and function are different between TDH and HlyA.

2.1 Thermostable direct hemolysin (TDH), TRH, and others

2.1.1 Thermostable direct hemolysin (TDH)

V. parahaemolyticus was first isolated by Fujino et al. as a causative agent of food poisoning in Osaka, Japan [14]. The pathogenicity of *V. parahaemolyticus* is determined by multiple virulence factors including adhesins, thermostable direct hemolysin (Vp-TDH), TDH-related hemolysin (Vp-TRH), and two type III secretion systems (T3SS), T3SS1 and T3SS2 [28]. It has been reported that the clinical isolates of *V. parahaemolyticus* show β-hemolysis activity on Wagatsuma blood agar medium [36], whereas almost all non-clinical isolates are non-hemolytic. This hemolytic activity has been given a specific term known as Kanagawa phenomenon (KP), and it is due to Vp-TDH encoded by the *tdh* gene [37, 38]. Thus, Vp-TDH has been considered as an important virulence factor in gastroenteritis cases and KP reaction as a good marker for the identification of pathogenic strains. Thermostable direct hemolysin, Vp-TDH, was named so because of its characteristics. These characters include persistence of activity even after heating at 100°C for 10 minutes and the ability to act directly on erythrocytes with no enhancement in activity level even by the addition of lecithin [39]. This purified toxin has numerous biological activities such as hemolytic activity for erythrocytes of various species, cytotoxic activity for some mammalian cells, and enterotoxic activity measured by fluid accumulation (FA) in the rabbit ileal loop test [33, 40–42].

The mature form of Vp-TDH consists of 165 amino acids and is approximately of 19 kDa. It exists as a tetramer in solution, which is responsible for the membrane disruption [43, 44]. This is a pore-forming toxin, but it has no similarities with other bacterial pore formers except Vp-TDH homologs like Vp-TRH and TDH-like toxins from *V. cholerae* non-O1/non-O139, *V. mimicus,* and *V. hollisae* [45–49]. Vp-TDH forms pores of approximately 2 nm in diameter on erythrocyte membrane that results into colloidal osmotic lysis [50]; however, the exact mechanism of pore formation is not yet identified. The reactivity of Vp-TDH against erythrocytes from various animal species showed variability; for example, it causes hemolysis of erythrocytes from rat, human, rabbit, and sheep but not horse [51]. It is reported that the amino acid residues, Arg^{46}, Gly^{62}, Trp^{65}, and Gly^{90}, are critical for the hemolysis; in fact, the substitution of residue Arg^{46} by site-directed mutagenesis inhibits the formation of tetramer [33, 44, 52].

Enterotoxicity, which is another feature of Vp-TDH, has been evaluated by increase of FA in the rabbit ileal loop due to intestinal Cl^- secretion as a manifestation of diarrhea induced by *V. parahaemolyticus*. The Cl^- secretion from human colonic epithelial cells by Vp-TDH is caused by stimulation of Ca^{2+}-activated chloride ion channel not by pore formation on the cells [53]. Evidence suggests that Vp-TDH acts in three sequential steps: receptor binding on the epithelial cells, followed by increase in intracellular Ca^{2+} concentration due to protein kinase C activation, and finally, stimulation of Ca^{2+}-activated Cl^- channel. However, it is reported that the deletion of *tdh* only leads to partial decrease in enterotoxicity against rabbit intestinal cells, whereas cytotoxicity to Hela cells was not affected at all [54]. Moreover, a recent study provides a new evidence that Vp-TDH can also engage as an effector of T3SS and implicated to elevate FA in animal model [55]. Therefore, the reason behind pathogenicity of *V. parahaemolyticus* is perhaps not only because of Vp-TDH but also because it involves a synergistic action of multiple virulence factors including T3SS.

2.1.2 TDH-related hemolysin (Vp-TRH) and others

Vp-TRH is identified as a new hemolysin found in KP-negative strains from clinical samples, named TDH-related hemolysin (Vp-TRH) [45]. Vp-TRH protein has a conserved domain of Vp-TDH and immunologically similar to Vp-TDH. But unlike Vp-TDH, it is heat-labile and lost its activity when heated at 60°C for 10 minutes. It is reported that there are significant nucleotide differences that exist within the *trh* family of two subgroups (*trh1* and *trh2*), sharing 84% sequence identity, as opposed to the less diversity (<3.3%) of five *tdh* genes (*tdh1* to *tdh5*) [38, 56–58]. Vp-TRH also induces chloride ion secretion in human colonic epithelial cells like Vp-TDH; therefore, it is considered as one of the important virulence factors among KP-negative strains of *V. parahaemolyticus* [59].

TDH-like toxins have also been found in *V. cholerae* non-O1/non-O139, *V. mimicus*, and *V. hollisae* known as NAG-TDH, Vm-TDH, and Vh-TDH, respectively [47–49]. It is reported that all clinical isolates of *V. hollisae* possess *tdh* gene [60], whereas only some clinical strains of *V. cholerae* non-O1/non-O139 and *V. mimicus* contain *tdh* gene [46, 61]. The molecular weight of these toxins is similar to Vp-TDH and shows immunological cross-reactivity with Vp-TDH. Both NAG-TDH and Vm-TDH are stable on heating at 100°C for 10 minutes, and the hemolytic activity against erythrocytes of most animals is almost similar to Vp-TDH [47, 48]. On the other hand, Vh-TDH is a heat-labile toxin that gets inactivated by heating at 70°C for 10 minutes, unlike Vp-TDH [49]. Moreover, it is reported that *V. alginolyticus* also produce TDH-like toxin, and it shows toxicity for mouse and fish [62].

2.2 HlyA (El Tor hemolysin) and related toxins

2.2.1 HlyA of *V. cholerae*

V. cholerae O1/O139 is the causative agent of cholera, and its main virulence factors are cholera toxin (CT) and toxin-coregulated pilus (TcpA) [11, 12]. *V. cholerae* produces some other virulence factors such as hemolysin, hemagglutinin/protease (HA/protease), T3SS, etc., which can also serve as important elements for the pathogenesis, especially in the strains devoid of CT and TcpA [63–65]. The water-soluble cytolytic toxin produced by *V. cholerae* El Tor O1 and non-O1/non-O139 strains is known as El Tor hemolysin (HlyA)/*V. cholerae* cytolysin (VCC) [66, 67]. HlyA can facilitate lysis of erythrocytes from various animals and other mammalian cells [66, 68]. It can also exhibit potent enterotoxicity as measured by fluid accumulation in the rabbit ileal loop test. Thus, HlyA has been considered to play a crucial role in the pathogenesis of gastroenteritis caused by *V. cholerae* strains [63].

The HlyA encoded by *hlyA* gene is produced in the form of 82 kDa inactive precursor, termed pre-pro-HlyA [69, 70]. This 82 kDa precursor consists of 25 amino acid long signal peptide at the N-terminal, a pro-region of 14 kDa and mature region of 65 kDa at the C-terminal. The mature form of HlyA is generated via a two-step process [71]. In the first step, the 82 kDa precursor is converted to 79 kDa pro-HlyA by cleavage of the signal peptide during its translocation through the inner membrane and then secreted extracellularly in an inactive form. In the second step, the inactive pro-HlyA is converted to active HlyA through the proteolytic removal of pro-region, usually at the bond between Ala^{157} and Asn^{158} (**Figure 1**; Proteolytic cleavage site). It has been found that the pro-HlyA can be activated by extracellular metalloprotease (HA/protease), a major protease of *V. cholerae* and also by other exogenous or endogenous proteases. However, the exact proteolytic cleavage site depends on the specificity of the protease, which is different compared to native processing [72]. Moreover, it is reported that pro-HlyA can

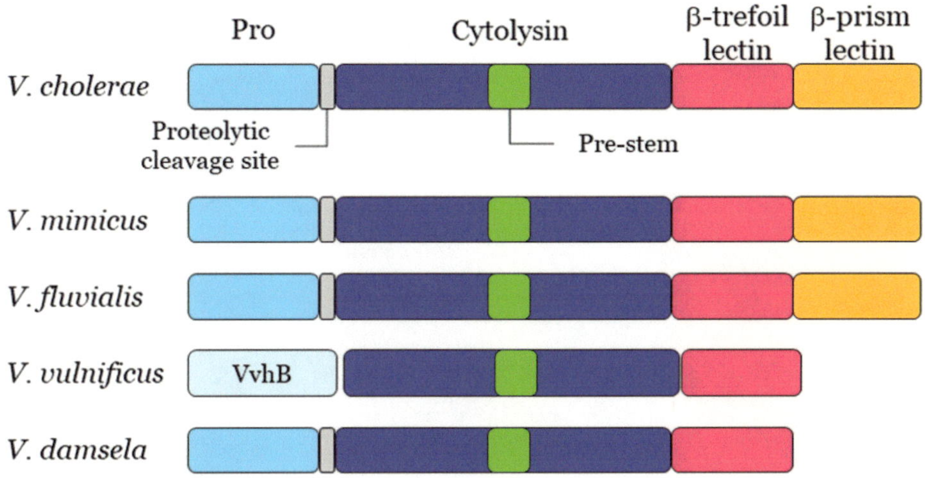

Figure 1.
Comparison between V. cholerae hemolysin (HlyA) and other Vibrio spp. hemolysins. HlyA of V. cholerae consists of pro-region (light blue), cytolysin domain (blue), β-trefoil lectin domain (pink), and β-prism lectin domain (yellow). There is a proteolytic cleavage site (gray) between pro-region and cytolysin domain for the conversion of pro-HlyA to mature HlyA. The hemolysins from V. mimicus and V. fluvialis have no significant differences in domain construction. V. vulnificus hemolysin lacks β-prism lectin domain and pro-region; instead, V. vulnificus produces VvhB that might act as chaperon-like pro-region. V. damsela produces HlyA-like hemolysin without β-prism lectin domain.

bind as a monomer to eukaryotic cell membrane, and then this bound pro-HlyA can be activated by exogenous, endogenous, extracellular, and even by cell-bound proteases [73]. It is well known that the pro-region can act as an intramolecular chaperone, an essential role of pro-region that governs the proper folding of HlyA pro-toxin [74].

HlyA belongs to bacterial β-barrel pore-forming toxins (β-PFTs) family that includes α-hemolysin of *Staphylococcus aureus* and aerolysin of *Aeromonas hydrophila* [75–77]. Consistent with generalized mode of action by β-PFT, the pore formation mechanism of HlyA has been proposed to follow three distinct steps (**Figure 2**); binding as a water-soluble monomer onto the target cell membrane, formation of pre-pore oligomeric intermediates by the self-assembly of toxin monomer, and finally insertion of the pore-forming stem-loop into the membrane, resulting into the formation transmembrane heptameric β-barrel pores on the cell membrane [78–80]. HlyA causes colloid osmotic lysis of mammalian cells by forming transmembrane pores on the target cell membranes [81, 82], which causes not only hemolysis but also potent cytotoxic effect such as vacuolation [83] and apoptosis [84, 85] of epithelial and immune cells.

The PFTs show affinity for a wide range of cell surface molecules such as cholesterol [86], glycosylphosphatidylinositol-anchored glycoproteins [87], and the human complement receptor CD59 [88]. In case of human erythrocyte membrane, glycophorin B has been reported to be a receptor for HlyA [89]. The hemolysis of rabbit erythrocytes by HlyA is competitively inhibited by asialofetuin and glycoproteins with multiple β_1-galactosyl residues [90]; this provides an evidence that cell surface carbohydrates are acting as functional receptors. *V. cholerae* cytolysin also shows strong preference for cholesterol- and sphingolipid-rich vesicles [91]. So, it can be said like other PFTs, HlyA also shows affinity for multiple cell surface receptor.

The mature HlyA is composed of three distinct domains: a central cytolysin domain and two lectin-like domains with β-trefoil and β-prism folds. The β-trefoil and β-prism domains exhibit structural homology to the carbohydrate-binding

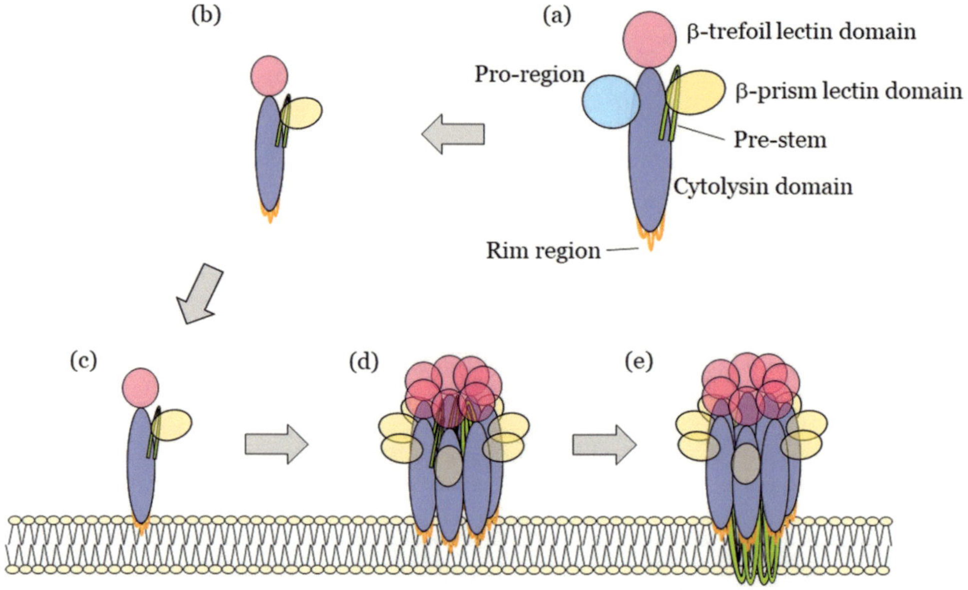

Figure 2.
Mechanism of transmembrane heptameric pore formation by HlyA. (a) Pro-HlyA structure. (b) Secreted pro-HlyA is activated through the removal of pro-region by protease. (c) HlyA monomer binds to the target membrane by using a rim region and/or β-prism lectin domain with membrane component such as cholesterol and carbohydrate receptor, respectively. (d) HlyA monomer assembles to heptameric pre-pore oligomeric intermediates. (e) The pre-stem of HlyA is inserted into the membrane, resulting into the formation of transmembrane heptameric β-barrel pores.

domain of the plant lectin ricin and jacalin, respectively [78]. In fact, the 15 kDa β-prism lectin domain has carbohydrate-binding activity [92], and the deletion of 15 kDa β-prism lectin domain generates a 50 kDa variant (HlyA50) with no effect on the global conformation of the monomer, but the hemolytic activity reduced by approximately 1000-fold [93, 94]. The β-prism domain has been shown to promote self-assembly of the toxin monomer in carbohydrate-independent manner, suggesting the hemolytic activity of HlyA50 is compromised due to reduction in pre-pore oligomeric intermediates [95]. Another study proposed the role of β-trefoil domain and showed that it is critical for the folding of cytolysin domain to its active conformation [96]. Recently, it is reported that the three loop sequences located in the bottom tip of the cytolysin domain play a critical role in the initial interaction with membrane lipid bilayer. This study showed that the replacement of the amino acid residues in the three loop sequences designated as "rim region" compromises the specific interaction of HlyA monomer with membrane lipid bilayer and blocks the pore formation process. Thus, it leads to repression in the lysis of human erythrocytes and reduced cytotoxic activity for HT-29 human colorectal adenocarcinoma cells [97]. In the next step that is pre-pore oligomerization, it has been shown that alteration of key amino acids affects not only the formation of oligomeric intermediates but also the subsequent formation of functional transmembrane pore [98]. Finally, pre-pore oligomeric intermediates lead to the formation of transmembrane β-barrel pore. Paul et al. confirmed that the transmembrane stem region plays a significant role in the functional pore formation. However, the deletion of "pre-stem" loop of cytolysin domain does not affect the membrane binding and pre-pore heptamer formation [99]. Therefore, it is considered that each step of HlyA pore formation mechanism plays an indispensable role in the generation of functional transmembrane pore in the target cell and thus enhances the virulence of *V. cholerae*.

2.2.2 Other related El Tor hemolysin of Vibrio species

Several studies have reported that other *Vibrio* species such as *V. mimicus*, *V. vulnificus*, and *V. fluvialis* also produce hemolysin that shares some common structural features with HlyA [100–102].

V. mimicus, a species closely related to *V. cholerae*, is a causative agent of human gastroenteritis [103]. Pathogenic strains of *V. mimicus* exhibit various clinical symptoms from watery to dysentery-like diarrhea [104]. This pathogen produces many kinds of virulence factors such as CT-like enterotoxin and heat-stable enterotoxin [105–108], with Vm-TDH as a causative factor in some clinical strains. However, most clinical strains lack the ability to produce any of these toxins. The heat-labile hemolysin/cytolysin (*V. mimicus* hemolysin; VMH) is thought to be the most common virulent enteropathogenic factor [109, 110]. In fact, VMH induces FA in a ligated rabbit ileal loop in dose-dependent manner, and the antibody against VMH apparently reduces enterotoxicity by *V. mimicus* in the living cells [100, 111]. These findings indicate that VMH is potently related to pathogenesis of this pathogen. The enterotoxic activity of VMH might be due to intestinal Cl^- secretion caused by the activation of both Ca^{2+}-dependent and cyclic AMP-dependent Cl^- secretion systems [111, 112]. Similar to HlyA, it has been indicated that VMH is also a pore-forming toxin. This toxin can disrupt various mammalian erythrocytes including bovine, rabbit, sheep, human, and mouse in colloid osmotic manner, and it shows the highest sensitivity for the horse erythrocytes [100].

VMH encoded by *vmhA* gene is predicted to be of 83 kDa with 82% similarity with *V. cholerae* HlyA. VMH is also secreted as 80 kDa precursor known as pro-VMH [113], which is then converted to 66 kDa mature toxin through the removal of N-terminal propeptide by trypsin-like protease of *V. mimicus* between the amino acid residues Arg^{151} and Ser^{152} [114, 115]. It has been assumed that VMH might be processed in a two-step reaction just like HlyA and pro-toxin can be activated by various proteases such as trypsin, chymotrypsin, and metalloprotease [115, 116]. Similar to 50 kDa variant of HlyA, mature VMH can be converted to 51 kDa of VMH (designated VMH51) through the removal of 15 kDa from C-terminal end by metalloprotease of *V. mimicus*. VMH51 almost showed no lytic activity toward horse erythrocytes because it lost the binding affinity toward erythrocyte membrane [116]. However, the VMH51 can associate with sheep erythrocyte membranes though the affinity is reduced as compared with intact VMH, suggesting that the truncated toxin interacts with other components in sheep erythrocyte membrane. It might be concluded that the 15 kDa C-terminal domain of VMH is functionally similar to β-prism lectin domain of HlyA.

V. fluvialis is one of the foodborne pathogens which can cause clinical symptoms similar to *V. cholerae* [117–119]. *V. fluvialis* secrets El Tor-like hemolysin, designed as *V. fluvialis* hemolysin (VFH), which can elicit lysis of erythrocytes from various animal. In addition to hemolytic activity, VFH can also trigger cytotoxicity toward Chinese hamster ovary (CHO) cells and induction of fluid accumulation in suckling mouse [102]. The purified VFH has molecular weight of 63 kDa, whose N-terminal amino acid sequence shares homology to HlyA from *V. cholerae* and VMH from *V. mimicus* [102]. It is suspected that VFH might play an important role in *V. fluvialis* pathogenicity.

V. vulnificus was first isolated from a leg ulcer, and it was wrongly reported as *V. parahaemolyticus* [120]. Later, it was found that some characters were different from *V. parahaemolyticus* such as positive lactose fermentation, so subsequently it was termed as *V. vulnificus* [20]. *V. vulnificus* can cause two types of illness, the primary septicemia and the wound infection [24]. The former is remarkable for its high fatality rate (over 50%). The primary septicemia is caused by the consumption of raw seafood, especially shellfish such as oyster contaminated by *V. vulnificus*, and it is reported that 95% of all seafood-related deaths are caused by *V. vulnificus* in the United States [121, 122]. Because most septicemia patients have an underlying disease such as hepatic cirrhosis, hepatitis, or diabetes, the septicemia by *V. vulnificus* is considered as an opportunistic infection [24]. Wound infections characterized clinical symptoms are edema, erythema, or necrosis and occurred after exposure to contaminated seawater or marine products. However, gastrointestinal symptom like diarrhea is very rare due to *V. vulnificus* infection [25, 26]. *V. vulnificus* produces various extracellular virulence factors such as hemolysin or protease [123, 124]. Hemolysin secreted by *V. vulnificus* called as *V. vulnificus* hemolysin (VVH) is also a toxin that can form pore on the target membranes of various mammalian cells. Purified VVH exhibits lytic activity against erythrocytes of various mammals and cultured cells such as CHO, mast, and pulmonary endothelial cells [101, 125–127]. In addition, it is reported that the sublytic doses of hemolysin can trigger apoptotic signaling pathway in human vascular endothelial cell line, ECV304 cells [128], and oligomerization of VVH is essential for the apoptotic activity in CHO cells [129].

VVH (VvhA) precursor has molecular weight of 51 kDa encoded by the structure gene *vvhA*, which constitutes an operon with *vvhB* gene. The *vvhB* gene is present upstream of *vvhA* and encodes 18 kDa protein VvhB. The VvhA precursor is composed of a signal peptide (20 amino acid residues) and cytolysin domain ($Gln^{1}$to Arg^{318}) including a putative pre-stem and β-trefoil lectin-like domain (His^{319} to Leu^{451}) (**Figure 1**); the pro-region and β-prism lectin domain are absent as compared with HlyA precursor [78, 130]. Although the function of VvhB is unknown, it might act as a chaperon in the absence of the pro-region like HlyA. This speculation is supported by the fact that even though VvhA is expressed in the absence of *vvhB* in vitro, the hemolytic activity cannot be detected [131]. Although VVH lacks β-prism lectin domain, the β-trefoil lectin domain has displayed binding capability for glycerol, N-acetyl-D-galactosamine, and N-acetyl-D-lactosamine unlike HlyA [92, 130]. In fact, VVH exhibits decreased ability to bind CHO cells when preincubated with methyl-beta-cyclodextrin, an oligosaccharide, and, thus, inhibition of its cytotoxic effect [132]. Similar to HlyA, it is believed that the VVH monomer binds to the cell membrane and forms oligomers [101, 133, 134] and the crystal structure of β-trefoil lectin domain of VVH reveals a heptameric ring arrangement [130]. It is strongly suggested that cholesterol is the receptor for VVH and facilitates conversion of monomer to oligomer [133, 135]. In addition, it is reported that Thr^{438} in the β-trefoil lectin domain is responsible for binding to cholesterol [131]. On the other hand, Phe^{334} in cytolysin domain that is located near the joint of two domains is essential for oligomerization of toxin monomer [136]. Moreover, it is shown that the mutation of Leu^{451} causes inhibition of hemolytic activity without reducing the membrane binding ability; this suggests that the Leu^{451} is essential for the oligomer formation [137]. Recently, a study showed that properties such as polarity and indole ring of amino acid Trp^{246} are essential for the binding of toxin to the target membrane [138]. It is assumed that hemolytic process of VVH is almost similar to HlyA though there are some differences in the function and structure of VVH.

It has been reported that a heat-labile hemolysin purified from *V. tubiashii*, a pathogen of juvenile bivalve, is similar to VVH [139]. Like VVH, this toxin has showed competitive inhibition by cholesterol and can lyse erythrocytes. In addition, the toxin exhibits cytotoxicity to CHO, Caco-2, and Atlantic menhaden liver cells in tissue culture.

V. damsela has been reported to cause wound infection by handling of fish, exposure to seawater and marine animals, and ingestion of raw seafood [21, 140–143]. It has been considered that there is not any other hemolysin in this bacterium, except of a hemolysin with phospholipase D activity known as damselysin [144].

Recently, it is reported that this bacterium possesses HlyA-like hemolysin encoded within a new virulence plasmid pPHDD1. The characteristics of this new HlyA-like hemolysin from *V. damsela* are not yet identified, but the predicted amino acid sequences show 69% similarity with HlyA of *V. cholerae*, missing the β-prism lectin-like domain (**Figure 1**) [145].

3. Conclusion

This chapter is focused on the hemolysins produced by *Vibrio* species, especially the human pathogens. Hemolysins are classified into two groups, namely, thermo-stable direct hemolysin (TDH) and El Tor hemolysin (HlyA). This chapter pays attention to Vp-TDH (*V. parahaemolyticus*), HlyA (*V. cholerae*), VMH (*V. mimicus*), and VVH (*V. vulnificus*) because these are well studied in terms of the toxin structure and their relation with the pathogenesis. The mechanism of action by HlyA and the essential amino acid residues have been clarified through the crystal structure of HlyA pro-toxin and the transmembrane heptameric oligomer over the past decade. Although the crystal structure has revealed the structural information about Vp-TDH and VVH, the exact mechanism of pore formation in the target membrane is yet to be studied. Several studies have indicated the involvement of novel virulence factors in pathogenesis like T3SS, but still Vp-TDH and Vp-TRH are considered to be the major virulence factors of *V. parahaemolyticus*, one of the important food poisoning bacteria in Japan and other eastern and Southeast Asian countries. HlyA is thought to be a major factor in CT-negative strains (e.g., *V. cholerae* non-O1/non-O139) that can cause diarrhea because it can induce enterotoxicity as well as apoptosis. *V. mimicus* hemolysin, VMH, is just one of the many enterotoxic factors. Even though there is detailed information about structural composition and mode of action of some of the hemolysin such as Vp-TDH and HlyA, still there is a lack of information about other hemolysins. Therefore, it is necessary to further enhance our knowledge regarding these toxins in order to thoroughly understand the mechanism of pathogenesis for the prevention of endemic infectious diseases associated with these pathogens.

Acknowledgements

This investigation was supported by a grant from the Program of the Japan Initiative for Global Research Network on Infectious Diseases (J-GRID), JP19fm0108002, from Ministry of Education, Culture, Sport, Science and Technology in Japan (MEXT), and from the Japan Agency for Medical Research and Development (AMED).

Author details

Tamaki Mizuno*, Anusuya Debnath and Shin-ichi Miyoshi
Graduate School of Medicine, Dentistry and Pharmaceutical Sciences, Okayama University, Okayama, Japan

*Address all correspondence to: mizuno-t@cc.okayama-u.ac.jp

References

[1] Farmer JJ III, Hickman-Brenner FW. The genera *Vibrio* and *Photobacterium*. In: Balows A, Trüper HG, Dworkin M, Harder W, Schleifer KH, editors. The Prokaryotes. A Handbook on the Biology of Bacteria: Ecophysiology, Isolation, Identification, and Applications. 2nd ed. Berlin (Germany): Springer; 2006. pp. 2952-3011. DOI: 10.1007/0-387-30746-x_18

[2] LPSN-List of prokaryotic names with standing in nomenclature (bacterio. net). 2015. Available from: http://www.bacterio.net/index.html

[3] Janda JM, Powers C, Bryant RG, Abbott SL. Current perspectives on the epidemiology and pathogenesis of clinically significant *Vibrio* spp. Clinical Microbiology Reviews. 1988;**1**(3): 245-267. DOI: 10.1128/cmr.1.3.245

[4] Romalde JL, Dieguez AL, Lasa A, Balboa S. New *Vibrio* species associated to molluscan microbiota: A review. Frontiers in Microbiology. 2013;**4**:413. DOI: 10.3389/fmicb.2013.00413

[5] Baker-Austin C, Oliver JD, Alam M, Ali A, Waldor MK, Qadri F, et al. *Vibrio* spp. infections. Nature Reviews. Disease Primers. 2018;**4**:8. DOI: 10.1038/s41572-018-0005-8

[6] Kirchberger PC, Turnsek M, Hunt DE, Haley BJ, Colwell RR, Polz MF, et al. *Vibrio metoecus* sp. nov., a close relative of *Vibrio cholerae* isolated from coastal brackish ponds and clinical specimens. International Journal of Systematic and Evolutionary Microbiology. 2014;**64**:3208-3214. DOI: 10.1099/ijs.0.060145-0

[7] Gladney LM, Tarr CL. Molecular and phenotypic characterization of *Vibrio navarrensis* isolates associated with human illness. Journal of Clinical Microbiology. 2014;**52**:4070-4074. DOI: 10.1128/JCM.01544-14

[8] Frans I, Michiels CW, Bossier P, Willems KA, Lievens B, Rediers H. *Vibrio anguillarum* as a fish pathogen: Virulence factors, diagnosis and prevention. Journal of Fish Diseases. 2011;**34**:643-661. DOI: 10.1111/j.1365-2761.2011.01379.x

[9] Austin B, Zhang XH. *Vibrio harveyi*: A significant pathogen of marine vertebrates and invertebrates. Letters in Applied Microbiology. 2006;**43**:119-124. DOI: 10.1111/j.1472-765X.2006.01989.x

[10] Schiewe MH, Trust TJ, Crosa JH. *Vibrio ordalii* sp. nov.: A causative agent of vibriosis in fish. Current Microbiology. 1981;**6**:343-348. DOI: 10.1007/BF01567009

[11] Howard-Jones N. Robert Koch and the cholera vibrio: A centenary. British Medical Journal. 1984;**288**:379-381

[12] Kaper JB, Morris JG, Levine MM. Cholera. Clinical Microbiology Reviews. 1995;**8**(1):48-86

[13] Ali M, Nelson AR, Lopez AL, Sack DA. Updated global burden of cholera in endemic countries. PLoS Neglected Tropical Diseases. 2015;**9**:e0003832. DOI: 10.1371/journal.pntd.000383

[14] Fujino T, Okuno Y, Nakada D, Aoyama A, Fukai K, Mukai T, et al. On the bacteriological examination of shirasu-food poisoning. Medical Journal of Osaka University. 1953;**4**:299-304

[15] Joseph SW, colwell RR, Kaper JB. *Vibrio parahaemolyticus* and related halophilic vibrios. Critical Reviews in Microbiology. 1982;**10**:77-124. DOI: 10.3109/10408418209113506

[16] Morris JG, Wilson R, Davis BR, Wachsmuth IK, Riddle CF, Wathen HG, et al. Non-O group 1 *Vibrio cholerae* gastroenteritis in the United States.

Annals of Internal Medicine. 1981;**94**:656-658. DOI: 10.7326/0003-4819-94-5-656

[17] Shandera WX, Jhonston JM, Davis BR, Blake PA. Disease from infection with *Vibrio mimicus*, a newly recognized *Vibrio* species. Annals of Internal Medicine. 1983;**99**:169-171. DOI: 10.7326/0003-4819-99-2-169

[18] Huq MI, Alam AKMJ, Brenner DJ, Morris GK. Isolation of *Vibrio*-like group, EF-6from patients with diarrhea. Journal of Clinical Microbiology. 1980;**11**:621-624

[19] Lee JV, Shread P, Furniss AL, Bryant TN. Taxonomy and description of *Vibrio fluvialis* sp. nov. (synonym group F *Vibrios*, group EF6). The Journal of Applied Bacteriology. 1981;**50**:73-94. DOI: 10.1111/j.1365-2672.1981.tb00873.x

[20] Farmer JJ. *Vibrio* ("*Beneckea*") *vulnificus*: The bacterium associated with sepsis, septicemia, and the sea. Lancet. 1979;**2**:903. DOI: 10.1016/s0140-6736(79)92715-6

[21] Morris JG Jr, Miller HG, Wilson R, Tacket CO, Hollis DG, Hickman FW, et al. Illness caused by *Vibrio damsela* and *Vibrio hollisae*. Lancet. 1982;**319**:1294-1297. DOI: 10.1016/s0140-6736(82)92853-7

[22] Pien F, Lee K, Higa H. *Vibrio alginolyticus* infections in Hawaii. Journal of Clinical Microbiology. 1977;**5**:670-672

[23] Rubin SJ, Tilton RC. Isolation of *Vibrio alginolyticus* from wound infections. Journal of Clinical Microbiology. 1975;**2**:556-558

[24] Blake PA, Weaver RE, Hollis DG. Disease of humans other than cholera caused by vibrios. Annual Review of Microbiology. 1980;**34**:341-367. DOI: 10.1146/annurev.mi.34.100180.002013

[25] Linkous DA, Oliver JD. Pathogenesis of *Vibrio vulnificus*. FEMS Microbiology Letters. 1999;**174**:207-214. DOI: 10.1111/j.1574-6968.1999.tb13570.x

[26] Strom MS, Paranjpye RN. Epidemiology and pathogenesis of *Vibrio vulnificus*. Microbes and Infection. 2000;**2**:177-188. DOI: 10.1016/S1286-4579(00)00270-7

[27] Mekalanos JJ. Cholera toxin: Genetic analysis, regulation, and role in pathogenesis. Current Topics in Microbiology and Immunology. 1985;**118**:97-118

[28] Makino K, Oshima K, Kurokawa K, Yokoyama K, Uda T, Tagomori K, et al. Genome sequence of *Vibrio parahaemolyticus*: A pathogenic mechanism distinct from that of *V. cholerae*. Lancet. 2003;**361**:743-749

[29] Park KS, Ono T, Rokuda M, Jang MH, Okada K, Iida T, et al. Functional characterization of two type III secretion systems of *Vibrio parahaemolyticus*. Infection and Immunity. 2004;**72**:6659-6665. DOI: 10.1128/IAI.72.11.6659-6665.2004

[30] Miyoshi S, Narukawa H, Tomochika K, Shinoda S. Actions of *Vibrio vulnificus* metalloprotease on human plasma proteinase-proteinase inhibitor systems: A comparative study of native protease with its derivative modified by polyethylene glycol. Microbiology and Immunology. 1995;**39**:959-966. DOI: 10.1111/j.1348-0421.1995.tb03299.x

[31] Ostolaza H, Bartolomé B, Serra JL, de la Cruz F, Goñi FM. Alpha-haemolysin from *E. coli*. Purification and self-aggregation properties. FEBS Letters. 1991;**280**:195-198. DOI: 10.1016/0014-5793(91)80291-A

[32] Bhakdi S, Tranum-Jensen J. Alpha-toxin of *Staphylococcus aureus*. Microbiological Reviews. 1991;**55**:733-751

[33] Iida T, Honda T. Hemolysins produced by Vibrios. Journal of Toxicology - Toxin Reviews. 1997;**16**:215-227. DOI: 10.3109/15569549709016457

[34] Zhang XH, Austin B. Haemolysins in *Vibrio* species. Journal of Applied Microbiology. 2005;**98**:1011-1019. DOI: 10.1111/j.1365-2672.2005.02583.x

[35] Shinoda S. Protein toxins produced by pathogenic vibrios. Journal of Natural Toxins. 1999;**8**:259-269

[36] Wagatsuma S. On a medium for hemolytic reaction (in Japanese). Media Circle. 1968;**13**:159-162

[37] Kaper JB, Campen KR, Seidler RJ, Baldini MH, Falkow S. Cloning of the thermostable direct or Kanagawa phenomenon associated haemolysin of *Vibrio parahaemolyticus*. Infection and Immunity. 1984;**45**:290-292

[38] Nishibuchi M, Kaper JB. Nucleotide sequence of the thermostable direct hemolysin gene of *Vibrio parahaemolyticus*. Journal of Bacteriology. 1985;**162**:558-564

[39] Sakurai J, Matsuzaki A, MiwataniI T. Purification and characterization of thermostable direct hemolysin of *Vibrio parahaemolyticus*. Infection and Immunity. 1973;**8**:775-780

[40] Ljungh A, Wadström T. Toxins of *Vibrio parahaemolyticus* and *Aeromonas hydrophila*. Journal of Toxicology - Toxin Reviews. 1982;**1**:257-307. DOI: 10.3109/15569548209019477

[41] Takeda Y. Thermostable direct hemolysin of *Vibrio parahaemolyticus*. Pharmacology and Therapeutics. 1983;**19**:123-146

[42] Honda T, Iida T. The pathogenicity of *Vibrio parahaemolyticus* and the role of the thermostable direct hemolysin and related haemolysins. Reviews in Medical Microbiology. 1993;**4**:106-113

[43] Hamada D, Higurashi T, Mayanagi K, Miyata T, Fukui T, Iida T, et al. Tetrameric structure of thermostable direct hemolysin from *Vibrio parahaemolyticus* revealed by ultracentrifugation, small-angle X-ray scattering and electron microscopy. Journal of Molecular Biology. 2007;**365**:187-195. DOI: 10.1016/j.jmb.2006.09.070

[44] Yanagihara I, Nakahira K, Yamane T, Kaieda S, Mayanagi K, Hamada D, et al. Structure and functional characterization of *Vibrio parahaemolyticus* thermostable direct hemolysin. The Journal of Biological Chemistry. 2010;**285**:16267-16274. DOI: 10.1074/jbc.M109.074526

[45] Honda T, Ni Y, Miwatani T. Purification and characterization of a hemolysin produced by a clinical isolate of Kanagawa phenomenon-negative *Vibrio parahaemolyticus* and related to the thermostable direct hemolysin. Infection and Immunity. 1988;**56**:961-965

[46] Nishibuchi M, Khaeomanee-iam V, Honda T, Kaper JB, Miwatani T. Comparative analysis of the hemolysin genes of *Vibrio cholerae* non-O1, *V. mimicus*, and *V. hollisae* that are similar to the *tdh* gene of *V. parahaemolyticus*. FEMS Microbiology Letters. 1990;**67**:251-256. DOI: 10.1016/0378-1097(90)90004-a

[47] Yoh M, Honda T, Miwatani T. Purification and partial characterization of a non-O1 *Vibrio cholerae* hemolysin that cross-reacts with thermostable direct hemolysin of *Vibrio parahaemolyticus*. Infection and Immunity. 1986;**52**:319-322

[48] Yoshida H, Honda T, Miwatani T. Purification and characterization of a hemolysin of *Vibrio mimicus* that relates to the thermostable direct hemolysin of *Vibrio parahaemolyticus*. FEMS Microbiology Letters;**84**:249-254. DOI: 10.1016/0378-1097(91)90364-G

[49] Yoh M, Honda T, Miwatani T. Purification and partial characterization of a *Vibrio hollisae* hemolysin that relates to the thermostable direct hemolysin of *Vibrio parahaemolyticus*. Canadian Journal of Microbiology. 1986;**32**: 632-636. DOI: 10.1139/m86-118

[50] Honda T, Ni Y, Miwatani T, Adachi T, Kim J. The thermostable direct haemolysin of *Vibrio parahaemolyticus* is a pore-forming toxin. Canadian Journal of Microbiology. 1992;**38**:1175-1180. DOI: 10.1139/m92-192

[51] Zen-Yoji H, Hitokoto H, Morozumi S, Le Clair RA. Purification and characterization of a hemolysin produced by *Vibrio parahaemolyticus*. The Journal of Infectious Diseases. 1971;**123**:665-667. DOI: 10.1093/infdis/123.6.665

[52] Baba K, Yamasaki S, Nishibuchi M, Takeda Y. Examination by site-directed mutagenesis of the amino acid residues of the thermostable direct hemolysin of *Vibrio parahaemolyticus* required for its hemolytic activity. Microbial Pathogenesis. 1992;**12**:279-287. DOI: 10.1016/0882-4010(92)90046-Q

[53] Takahashi A, Sato Y, Shiomi Y, Cantarelli VV, Iida T, Lee M, et al. Mechanisms of chloride secretion induced by thermostable direct haemolysin of *Vibrio parahaemolyticus* in human colonic tissue and a human intestinal epithelial cell line. Journal of Medical Microbiology. 2000;**49**:801-810. DOI: 10.1099/0022-1317-49-9-801

[54] Park KS, Ono T, Rokuda M, Jang MH, Lida T, Honda T. Cytotoxicity and enterotoxicity of the thermostable direct hemolysin-deletion mutants of *Vibrio parahaemolyticus*. Microbiology and Immunology. 2004;**48**:313-318. DOI: 10.1111/j.1348-0421.2004.tb03512.x

[55] Matsuda S, Okada R, Tandhavanat S, Hiyoshi H, Gotoh K, Iida T, et al. Export of a *Vibrio parahaemolyticus* toxin by the sec and type III secretion machineries in tandem. Nature Microbiology. 2019;**4**:781-788. DOI: 10.1038/s41564-019-0368-y

[56] Kishishita M, Matsuoka N, Kumagai K, Yamasaki S, Takeda Y, Nishibuchi M. Sequence variation in the thermostable direct hemolysin-related hemolysin (*trh*) gene of *Vibrio parahaemolyticus*. Applied and Environmental Microbiology. 1992;**58**:2449-2457

[57] Baba K, Shirai H, Terai A, Takeda Y, Nishibuchi M. Analysis of the *tdh* gene cloned from a *tdh* gene- and *trh* gene-positive strain of *Vibrio parahaemolyticus*. Microbiology and Immunology. 1991;**35**:253-258. DOI: 10.1111/j.1348-0421.1991.tb01554.x

[58] Nishibuchi M, Kaper JB. Duplication and variation of the thermostable direct haemolysin (*tdh*) gene in *Vibrio parahaemolyticus*. Molecular Microbiology. 1990;**4**:87-99. DOI: 10.1111/j.1365-2958.1990.tb02017.x

[59] Takahashi A, Kenjyo N, Imura K, Myonsun Y, Honda T. Cl^- secretion in colonic epithelial cells induced by the *Vibrio parahaemolyticus* hemolytic toxin related to thermostable direct hemolysin. Infection and Immunity. 2000;**68**:5435-5438. DOI: 10.1128/iai.68.9.5435-5438.2000

[60] Nishibuchi M, Ishibashi M, Takeda Y, Kaper JB. Detection of the thermostable direct hemolysin gene and related DNA sequences in *Vibrio parahaemolyticus* and other *Vibrio* species by the DNA colony hybridization test. Infection and immunity. 1985;**49**:481-486

[61] Honda T, Nishibuchi M, Miwatani T, Kaper JB. Demonstration of a plasmid-borne gene encoding a thermostable direct hemolysin in *Vibrio cholerae* non-O1 strains. Applied and Environmental Microbiology. 1986;**52**:1218-1220

[62] Cai SH, Wu ZH, Jian JC, Lu YS. Cloning and expression of gene encoding the thermostable direct hemolysin from *Vibrio alginolyticus* strain HY9901, the causative agent of vibriosis of crimson snapper (*Lutjanus erythopterus*). Journal of Applied Microbiology. 2007;**103**:289-296. DOI: 10.1111/j.1365-2672.2006.03250.x

[63] Ichinose Y, Yamamoto K, Nakasone N, Tanabe MJ, Takeda T, Miwatani T, et al. Enterotoxicity of El tor-like hemolysin of non-O1 *Vibrio cholerae*. Infection and Immunity. 1987;**55**:1090-1093

[64] Ghosh A, Saha DR, Hoque KM, Asakuna M, Yamasaki S, Koley H, et al. Enterotoxigenicity of mature 45-kilodalton and processed 35-kilodalton forms of hemagglutinin protease purified from a cholera toxin gene-negative *Vibrio cholerae* non-O1, non-O139 strain. Infection and Immunity. 2006;**74**:2937-2946. DOI: 10.1128/IAI.74.5.2937-2946.2006

[65] Shin OS, Tam VC, Suzuki M, Ritchie JM, Bronson RT, Waldor MK, et al. Type III secretion is essential for the rapidly fatal diarrheal disease caused by non-O1, non-O139 *Vibrio cholerae*. MBio. 2011;**2**:e00106-e00111. DOI: 10.1128/mBio.00106-11

[66] Yamamoto K, Al-Omani M, Honda T, Takeda Y, Miwatani T. Non-O1 *Vibrio cholerae hemolysin*: Purification, partial characterization, and immunological relatedness to El tor hemolysin. Infection and Immunity. 1984;**45**:192-196

[67] Yamamoto K, Ichinose Y, Nakasone N, Tanabe M, Nagahama M, Sakurai J, et al. Identity of hemolysins produced by *Vibrio cholerae* non-O1 and *V. cholerae* O1, biotype El tor. Infection and immunity. 1986;**51**:927-931

[68] Zitzer A, Wassenaar TM, Walev I, Bhakdi S. Potent membrane-permeabilizing and cytocidal action of *Vibrio cholerae* cytolysin on human intestinal cells. Infection and Immunity. 1997;**65**:1293-1298

[69] Alm RA, Stroeher UH, Manning PA. Extracellular proteins of *Vibrio cholerae*: Nucleotide sequence of the structural gene (*hlyA*) for the haemolysin of the haemolytic El tor strain O17 and characterization of the *hlyA* mutation in the non-haemolytic classical strain 569B. Molecular Microbiology. 1988;**2**:481-488

[70] Rader AE, Murphy JR. Nucleotide sequences and comparison of the hemolysin determinants of *Vibrio cholerae* El tor RV79(Hly+) and RV79(Hly-) and classical 569B(Hly-). Infection and Immunity. 1988;**56**:1414-1419

[71] Yamamoto K, Ichinose Y, Shinagawa H, Makino K, Nakata A, Iwanaga M, et al. Two-step processing for activation of the cytolysin/hemolysin of *Vibrio cholerae* O1 biotype El tor: Nucleotide sequence of the structural gene (*hlyA*) and characterization of the processed products. Infection and Immunity. 1990;**58**:4106-4116

[72] Nagamune K, Yamamoto K, Naka A, Matsuyama J, Miwatani T, Honda T. In vitro proteolytic processing and activation of the recombinant precursor of El tor cytolysin/hemolysin (pro-HlyA) of *Vibrio cholerae* by soluble hemagglutinin/protease of *V. cholerae*, trypsin, and other proteases. Infection and Immunity. 1996;**64**:4655-4658

[73] Valeva A, Walev I, Weis S, Boukhallouk F, Wassenaar TM, Endres K, et al. A cellular metalloproteinase activates *Vibrio cholerae* pro-cytolysin. The Journal of Biological Chemistry. 2004;**279**:25143-25148. DOI: 10.1074/jbc.M313913200

[74] Nagamune K, Yamamoto K, Honda T. Intramolecular chaperone

activity of the pro-region of *Vibrio cholerae* El tor cytolysin. The Journal of Biological Chemistry. 1997;**272**: 1338-1343. DOI: 10.1074/jbc.272.2.1338

[75] Song L, Hobaugh MR, Shustak C, Cheley S, Bayley H, Gouaux JE. Structure of staphylococcal alpha-hemolysin, a heptameric transmembrane pore. Science. 1996;**274**:1859-1866. DOI: 10.1126/science.274.5294.1859

[76] Parker MW, Buckley JT, Postma JP, Tucker AD, Leonard K, Pattus F, et al. Structure of the *Aeromonas* toxin proaerolysin in its water-soluble and membrane-channel states. Nature. 1994;**367**:292-295. DOI: 10.1038/367292a0

[77] Los FC, Randis TM, Aroian RV, Ratner AJ. Role of pore-forming toxins in bacterial infectious diseases. Microbiology and Molecular Biology Reviews. 2013;**77**:173-207. DOI: 10.1128/MMBR.00052-12

[78] Olson R, Gouaux E. Crystal structure of the *Vibrio cholerae* cytolysin (VCC) pro-toxin and its assembly into a heptameric transmembrane pore. Journal of Molecular Biology. 2005;**350**:997-1016. DOI: 10.1016/j.jmb.2005.05.045

[79] He Y, Olson R. Three-dimensional structure of the detergent-solubilized *Vibrio cholerae* cytolysin (VCC) heptamer by electron cryomicroscopy. Journal of Structural Biology. 2010;**169**:6-13. DOI: 10.1016/j.jsb.2009.07.015

[80] De S, Olson R. Crystal structure of the *Vibrio cholerae* cytolysin heptamer reveals common features among disparate pore-forming toxins. Proceedings of the National Academy of Sciences of the United States of America. 2011;**108**:7385-7390. DOI: 10.1073/pnas.1017442108

[81] Löhner S, Walev I, Boukhallouk F, Palmer M, Bhakdi S, Valeva A. Pore formation by *Vibrio cholerae* cytolysin follows the same archetypical mode as beta-barreltoxins from gram-positive organisms. The FASEB Journal. 2009;**23**:2521-2528. DOI: 10.1096/fj.08-127688

[82] Krasilnikov OV, Muratkhodjaev JN, Zitzer AO. The mode of action of *Vibrio cholerae* cytolysin. The influences on both erythrocytes and planar lipid bilayers. Biochimica et Biophysica Acta. 1992;**1111**:7-16. DOI: 10.1016/0005-2736(92)90268-Q

[83] Figueroa-Arredondo P[1], Heuser JE, Akopyants NS, Morisaki JH, Giono-Cerezo S, Enríquez-Rincón F, Berg DE. Cell vacuolation caused by *Vibrio cholerae* hemolysin. Infection and Immunity 2001;69:1613-1624. DOI:10.1128/IAI.69.3.1613-1624.2001

[84] Saka HA, Bidinost C, Sola C, Carranza P, Collino C, Ortiz S, et al. *Vibrio cholerae* cytolysin is essential for high enterotoxicity and apoptosis induction produced by a cholera toxin gene-negative *V. cholerae* non-O1, non-O139 strain. Microbial Pathogenesis. 2008;**44**:118-128

[85] Mukherjee G, Biswas A, Banerjee KK, Biswas T. *Vibrio cholerae* hemolysin is apoptogenic to peritoneal B-1a cells but its oligomer shepherd the cells for IgA response. Molecular Immunology. 2008;**45**:266-270. DOI: 10.1016/j.molimm.2007.04.024

[86] Tweten RK. Cholesterol-dependent cytolysins, a family of versatile pore-forming toxins. Infection and Immunity. 2005;**73**:6199-6209. DOI: 10.1128/IAI.73.10.6199-6209.2005

[87] Fivaz M, Vilbois F, Thurnheer S, Pasquali C, Abrami L, Bickel PE, Parton RG, van der Goot FG. Differential sorting and fate of endocytosed GPI-anchored proteins. The EMBO Journal 2002;21:3989-4000. DOI: 10.1093/emboj/cdf398

[88] Giddings KS, Zhao J, Sims PJ, Tweten RK. Human CD59 is a receptor for the cholesterol-dependent cytolysin intermedilysin. Nature Structural and Molecular Biology. 2004;**1**:1173-1178. DOI: 10.1038/nsmb862

[89] Zhang D, Takahashi J, Seno T, Tani Y, Honda T. Analysis of receptor for *Vibrio cholerae* El tor hemolysin with a monoclonal antibody that recognizes glycophorin B of human erythrocyte membrane. Infection and Immunity. 1999;**67**:5332-5337

[90] Saha N, Banerjee KK. Carbohydrate-mediated regulation of interaction of *Vibrio cholerae* hemolysin with erythrocyte and phospholipids vesicle. The Journal of Biological Chemistry. 1997;**272**:162-167. DOI: 10.1074/jbc.272.1.162

[91] Zitzer A, Zitzer O, Bhakdi S, Palmer M. Oligomerization of *Vibrio cholerae* cytolysin yields a pentameric pore and has a dual specificity for cholesterol and sphingolipids in the target membrane. The Journal of Biological Chemistry. 1999;**274**: 1375-1380. DOI: 10.1074/jbc.274.3.137

[92] Levan S, De S, Olson R. *Vibrio cholerae* cytolysin recognizes the heptasaccharide core of complex N-glycans with nanomolar affinity. Journal of Molecular Biology. 2013;**425**:944-957. DOI: 10.1016/j.jmb.2012.12.016

[93] Dutta S, Mazumdar B, Banerjee KK, Ghosh AN. Three-dimensional structure of different functional forms of the *Vibrio cholerae* hemolysin oligomer: A cryo-electron microscopic study. Journal of Bacteriology. 2010;**192**:169-178. DOI: 10.1128/JB.00930-09

[94] Mazumdar B, Ganguly S, Ghosh AN, Banerjee KK. The role of C-terminus carbohydrate-binding domain of *Vibrio cholerae* haemolysin/cytolysin in the conversion of the pre-pore β-barrel oligomer to a functional diffusion channel. The Indian Journal of Medical Research. 2011;**133**:131-137

[95] Ganguly S, Mukherjee A, Mazumdar B, Ghosh AN, Banerjee KK. The β-prism lectin domain of *Vibrio cholerae* hemolysin promotes self-assembly of the β-pore-forming toxin by a carbohydrate-independent mechanism. The Journal of Biological Chemistry. 2014;**289**:4001-4008. DOI: 10.1074/jbc.M113.522284

[96] Mukherjee A, Ganguly S, Chatterjee NS, Banerjee KK. *Vibrio cholerae* hemolysin: The β-trefoil domain is required for folding to the native conformation. Biochemistry and Biophysics Reports. 2016;**8**:242-248. DOI: 10.1016/j.bbrep.2016.09.009

[97] Rai AK, Chattopadhyay K. Revisiting the membrane interaction mechanism of a membrane-damaging β-barrel pore-forming toxin *Vibrio cholerae* cytolysin. Molecular Microbiology. 2015;**97**:1051-1062. DOI: 10.1111/mmi.13084

[98] Rai AK, Chattopadhyay K. Trapping of *Vibrio cholerae* cytolysin in the membrane-bound monomeric state blocks membrane insertion and functional pore formation by the toxin. The Journal of Biological Chemistry. 2014;**289**:16978-16987. DOI: 10.1074/jbc.M114.567099

[99] Paul K, Chattopadhyay K. Pre-pore oligomer formation by *Vibrio cholerae*: Insights from a truncated variant lacking the pore-forming pre-stem loop. Biochemical and Biophysical Research Communications. 2014;**443**:189-193. DOI: 10.1016/j.bbrc.2013.11.078

[100] Miyoshi S, Sasahara K, Akamatsu S, Rahman MM, Katsu T, Tomochika K, et al. Purification and characterization of a hemolysin produced by *Vibrio mimicus*. Infection and Immunity. 1997;**65**:1830-1835

[101] Yamanaka H, Satoh T, Katsu T, Shinoda S. Mechanism of haemolysis by *Vibrio vulnificus* haemolysin. Journal of General Microbiology. 1987;**133**:2859-2864. DOI: 10.1099/00221287-133-10-2859

[102] Kothary MH, Lowman H, McCardell BA, Tall BD. Purification and characterization of enterotoxigenic El tor-like hemolysin produced by *Vibrio fluvialis*. Infection and Immunity. 2003;**71**:3213-3220. DOI: 10.1128/IAI.71.6.3213-3220.2003

[103] Davis BR, Fanning GR, Madden JM, Steigerwalt AG, Bradford HB Jr, Smith HL, et al. Characterization of biochemically atypical *Vibrio cholerae* strains and designation of a new pathogenic species, *Vibrio mimicus*. Journal of Clinical Microbiology. 1981;**14**:631-639

[104] Hoge CW, Watsky D, Pealer RN, Libonati JP, Israel E, Morris JG Jr. Epidemiology and spectrum of *Vibrio* infections in a Chesapeake Bay community. The Journal of Infectious Diseases. 1989;**160**:985-993. DOI: 10.1093/infdis/160.6.985

[105] Ramamurthy T, Albert MJ, Huq A, Colwell RR, Takeda Y, Takeda T, et al. *Vibrio mimicus* with multiple toxin types isolated from human and environmental sources. Journal of Medical Microbiology. 1994;**40**:194-196. DOI: 10.1099/00222615-40-3-194

[106] Shinoda S, Miyoshi S. Enteropathogenic factors produced by vibrios other than cholera toxin. Journal of Natural Toxins. 2000;**9**:231-249

[107] Spira WM, Fedorka-Cray PJ. Purification of enterotoxin from *Vibrio mimicus* that appear to be identical to cholera toxin. Infection and Immunity. 1984;**45**:679-684

[108] Chowdhury MAR, Aziz KMS, Kay BA, Rahim Z. Toxin production by *Vibrio mimicus* strains isolated from human and environmental sources in Bangladesh. Journal of Clinical Microbiology. 1987;**25**:2200-2203

[109] Shi L, Miyoshi S, Bi K, Nakamura M, Hiura M, Tomochika K, et al. Presence of hemolysin genes (*vmh*, *tdh* and *hlx*) in isolates of *Vibrio mimicus* determined by polymerase chain reaction. The Journal of Health Science. 2000;**46**:63-65

[110] Shinoda S, Nakagawa T, Shi L, Bi K, Kanoh Y, Tomochika K, et al. Distribution of virulence-associated genes in *Vibrio mimicus* isolates from clinical and environmental origins. Microbiology and Immunology. 2004;**48**:547-551. DOI: 10.1111/j.1348-0421.2004.tb03551.x

[111] Li Y, Okamoto K, Takahashi E, Miyoshi S, Shinoda S, Tsuji T, et al. A hemolysin of *Vibrio mimicus* (VMH) stimulates cells to produce ATP and cyclic AMP which appear to be secretory mediators. Microbiology and Immunology. 2005;**49**:73-78. DOI: 10.1111/j.1348-0421.2005.tb03631.x

[112] Takahashi A, Miyoshi S, Takata N, Nakano M, Hamamoto A, Mawatari K, et al. Haemolysin produced by *Vibrio mimicus* activates two Cl^- secretory pathways in cultured intestinal-like Caco-2 cells. Cellular Microbiology. 2007;**9**:583-595. DOI: 10.1111/j.1462-5822.2006.00809.x

[113] Sultan Z, Mizuno T, Sakurai A, Takata N, Okamoto K, Miyoshi S. Growth phase dependant activation of the precursor of *Vibrio mimicus* hemolysin (pro-VMH). The Journal of Health Science. 2007;**53**:430-434. DOI: 10.1248/jhs.53.430

[114] Rahman MM, Miyoshi S, Tomochika K, Wakae H, Shinoda S.

Analysis of the structural gene encoding a hemolysin in *Vibrio mimicus*. Microbiology and Immunology. 1997;**41**:169-173. DOI: 10.1111/j.1348-0421.1997.tb01183.x

[115] Mizuno T, Nanko A, Maehara Y, Shinoda S, Miyoshi S. A novel extracellular protease of *Vibrio mimicus* that mediates maturation of an endogenous hemolysin. Microbiology and Immunology. 2014;**58**:503-512. DOI: 10.1111/1348-0421.12177

[116] Mizuno T, Sultan SZ, Kaneko Y, Yoshimura T, Maehara Y, Nakao H, et al. Modulation of *Vibrio mimicus* hemolysin through limited proteolysis by an endogenous metalloprotease. The FEBS Journal. 2009;**276**:825-834. DOI: 10.1111/j.1742-4658.2008.06827.x

[117] Chowdhury G, Pazhani GP, Dutta D, Guin S, Dutta S, Ghosh S, et al. *Vibrio fluvialis* in patients with diarrhea, Kolkata, India. Emerging Infectious Diseases. 2012;**18**:1868-1871. DOI: 10.3201/eid1811.120520

[118] Liang P, Cui XY, Du XL, Kan B, Liang WL. The virulence phenotypes and molecular epidemiological characteristics of *Vibrio fluvialis* in China. Gut Pathogen. 2013;**5**:6. DOI: 10.1186/1757-4749-5-6

[119] Ramamurthy T, Chowdhury G, Pazhani GP, Shinoda S. *Vibrio fluvialis*: An emerging human pathogen. Frontiers in Microbiology. 2014;**5**:91. DOI: 10.3389/fmicb.2014.00091

[120] Roland FP. Leg gangrene and endotoxin shock due to *Vibrio parahaemolyticus*: An infection acquired in New England coastal waters. The New England Journal of Medicine. 1970;**282**:1306. DOI: 10.1056/NEJM197006042822306

[121] Klontz KC, Lieb S, Schreiber M, Janowski HT, Baldy LM, Gunn RA. Syndromes of *Vibrio vulnificus* infections: Clinical and epidemiological features in Florida cases, 1981-1987. Annals of Internal Medicine. 1988;**109**:318-323. DOI: 10.7326/0003-4819-109-4-318

[122] Oliver JD. The viable but non-culturable state in the human pathogen *Vibrio vulnificus*. FEMS Microbiology Letters. 1995;**133**: 203-208. DOI: 10.1111/j.1574-6968.1995.tb07885.x

[123] Miyoshi S, Oh EG, Hirata K, Shinoda S. Exocellular toxic factors produced by *Vibrio vulnificus*. Journal of Toxicology - Toxin Reviews. 1993;**12**:253-288. DOI: 10.3109/15569549309014409

[124] Miyoshi S, Shinoda S. Bacterial metalloprotease as the toxic factor in infection. Journal of Toxicology - Toxin Reviews. 1997;**16**:177-194. DOI: 10.3109/15569549709016455

[125] Kreger A, Lockwood D. Detection of extracellular toxin(s) produced by *Vibrio vulnificus*. Infection and Immunity. 1981;**33**:583-590

[126] Yamanaka H, Sugiyama K, Furuta H, Miyoshi S, Shinoda S. Cytolytic action of *Vibrio vulnificus* haemolysin on mast cells from rat peritoneal cavity. Journal of Medical Microbiology. 1990;**32**:39-43. DOI: 10.1099/00222615-32-1-39

[127] Kim JS. Cytotoxicity of *Vibrio vulnificus* cytolysin on pulmonary endothelial cells. Experimental and Molecular Medicine. 1997;**29**:117-121. DOI: 10.1038/emm.1997.17

[128] Kwon KB, Yang JY, Ryu DG, Rho HW, Kim JS, Park JW, et al. *Vibrio vulnificus* cytolysin induces superoxide anion-initiated apoptotic signaling pathway in human ECV304 cells. The Journal of Biological Chemistry.

2001;**276**:47518-47523. DOI: 10.1074/jbc.M108645200

[129] Kashimoto T, Ueno S, Ehara H, Fukudome S, Komai M, Susa N. Oligomerization is essential for apoptotic activity of *Vibrio vulnificus* hemolysin. The Journal of Veterinary Medical Science. 2009;**71**:1403-1406. DOI: 10.1292/jvms.001403

[130] Kaus K, Lary JW, Cole JL, Olson R. Glycan specificity of the *Vibrio vulnificus* hemolysin lectin outlines evolutionary history of membrane targeting by a toxin family. Journal of Molecular Biology. 2014;**426**:2800-2812. DOI: 10.1016/j.jmb.2014.05.021

[131] Senoh M, Okita Y, Shinoda S, Miyoshi S. The crucial amino acid residue related to inactivation of *Vibrio vulnificus* hemolysin. Microbial Pathogenesis. 2008;**44**:78-83. DOI: 10.1016/j.micpath.2007.07.002

[132] Sugiyama H, Kashimoto T, Ueno S, Susa N. Inhibition of binding of *Vibrio vulnificus* hemolysin (VVH) by MβCD. The Journal of Veterinary Medical Science. 2013;**75**:649-452. DOI: 10.1292/jvms.12-0387

[133] Shinoda S, Miyoshi S, Yamanaka H, Miyoshi-Nakahara N. Some properties of *Vibrio vulnificus* hemolysin. Microbiology and Immunology. 1985;**29**:583-590

[134] Kim HR, Rho HW, Jeong MH, Park JW, Kim JS, Park BH, et al. Hemolytic mechanism of cytolysin produced from *V. vulnificus*. Life Sciences. 1993;**53**:571-577. DOI: 10.1016/0024-3205(93)90714-e

[135] Kim BS, Kim JS. Cholesterol induce oligomerization of *Vibrio vulnificus* cytolysin specifically. Experimental and Molecular Medicine. 2002;**34**:239-242. DOI: 10.1038/emm.2002.33

[136] Kashimoto T, Ueno S, Koga T, Fukudome S, Ehara H, Komai M, et al. The aromatic ring of phenylalanine 334 is essential for oligomerization of *Vibrio vulnificus* hemolysin. Journal of Bacteriology. 2010;**192**:568-574. DOI: 10.1128/JB.01049-09

[137] Miyoshi S, Abe Y, Senoh M, Mizuno T, Maehara Y, Nakao H. Inactivation of *Vibrio vulnificus* hemolysin through mutation of the N- or C-terminus of the lectin-like domain. Toxicon. 2011;**57**:904-908. DOI: 10.1016/j.toxicon.2011.03.013

[138] Kashimoto T, Akita T, Kado T, Yamazaki K, Ueno S. Both polarity and aromatic ring in the side chain of tryptophan 246 are involved in binding activity of *Vibrio vulnificus* hemolysin to target cells. Microbial Pathogenesis. 2017;**109**:71-77. DOI: 10.1016/j.micpath.2017.05.029

[139] Kothary MH, Delston RB, Curtis SK, McCardell BA, Tall BD. Purification and characterization of a vulnificolysin-like cytolysin produced by *Vibrio tubiashii*. Applied and Environmental Microbiology. 2001;**67**:3707-3711. DOI: 10.1128/AEM.67.8.3707-3711.2001

[140] Alvarez JR, Lamba S, Dyer KY, Apuzzio JJ. An unusual case of urinary tract infection in a pregnant woman with *Photobacterium damsela*. Infectious Diseases in Obstetrics and Gynecology. 2006;**80682**:1-3. DOI: 10.1155/IDOG/2006/80682

[141] Asato J, Kanaya F. Fatal infection of the hand due to *Photobacterium damsela*: A case report. Clinical Infectious Diseases. 2004;**38**:e100-e101. DOI: 10.1086/383468

[142] Kim HR, Kim JW, Lee MK, Kim JG. Septicemia progressing to fatal hepatic dysfunction in a cirrhotic patient after oral ingestion of *Photobacterium damsela*: A case

report. Infection. 2009;**37**:555-556. DOI: 10.1007/s15010-009-9049-8

[143] Yamane K, Asato J, Kawade N, Takahashi H, Kimura B, Arakawa Y. Two cases of fatal necrotizing fasciitis caused by *Photobacterium damsela* in Japan. Journal of Clinical Microbiology. 2004;**42**:1370-1372. DOI: 10.1128/jcm.42.3.1370-1372.2004

[144] Kreger AS, Bernheimer AW, Etkin LA, Daniel LW. Phospholipase D activity of *Vibrio damsela* cytolysin and its interaction with sheep erythrocytes. Infection and Immunity. 1987;**55**:3209-3212

[145] Rivas AJ, Balado M, Lemos ML, Osorio CR. The *Photobacterium damselae* subsp. damselae hemolysins damselysin and HlyA are encoded within a new virulence plasmid. Infection and Immunity. 2011;**79**:4617-4627. DOI: 10.1128/IAI.05436-11

Chapter 6

The Genus *Enterococcus* and Its Associated Virulent Factors

Hassan Bin-Asif and Syed Abid Ali

Abstract

Enterococci, the Gram-positive, catalase negative, non-spore forming and aero-tolerant fermentative organisms form the second largest group of bacteria studied with reference to microbial source tracking in view of their ability to survive adverse environmental conditions and adaptable nature to revolutionize from low number commensals to a predominant population of host microbiota thus creating a consequence for pathogenesis. Despite being a member of normal human intestinal flora, they are not regarded anymore as generally recognized as safe (GRAS) organisms and some of its species may turned out to be a major cause of nosocomial infections. Ecological and epidemiological studies showed that these bacteria enter in the environment via feces and colonize because of their high adaptability. The main contributors in pathogenesis of enterococci are the presence of various virulence factors and antibiotic resistance genes. This chapter aims to highlight the infections caused by enterococci and their respective virulent determinants.

Keywords: *enterococcus*, virulence, resistance, hemolysis, lactic acid bacteria, nosocomial infections

1. Introduction

Enterococci (ENT), the Gram-positive (G +ve), catalase negative, benzidine negative, non-spore forming and aero-tolerant fermentative organisms form the second largest group of bacteria studied with reference to microbial source tracking (MST) [1, 2]. It is a non-filamentous microorganism but some species like *E. casseliflavus* and *E. gallinarum* exhibit motility by scanty flagella. They produce lactic acid [L (+)- lactic acid enantiomer in case of glucose fermentation] by homofermentative Embden-Meyerhof-Parnas pathway, hence called Lactic Acid Bacteria (LAB). All the species except *E. faecalis* [(*E. fl*) (which contains lysine alanine 2–3 type)] contains lysine-D-asparagine linkages with D-isoasparagine as cross bridge in peptidoglycan. Their ability to survive in adverse environmental conditions and adaptable nature revolutionize them from low number commensals to a predominant population of host microbiota which ultimately results in creating a consequence for their pathogenesis [3]. Despite being a member of normal human intestinal flora, they are not regarded as GRAS (Generally Recognized As Safe) organisms anymore [4] as some of its species have turned out to be a major cause of nosocomial infections including hepatobiliary sepsis, urinary tract infections (UTI), surgical wound infections, endocarditis, bacteremia and neonatal sepsis [5]. From a medical perspective, ENT have been recognized as an important hospital acquired pathogen due to their ability to transfer or acquire resistance genes via

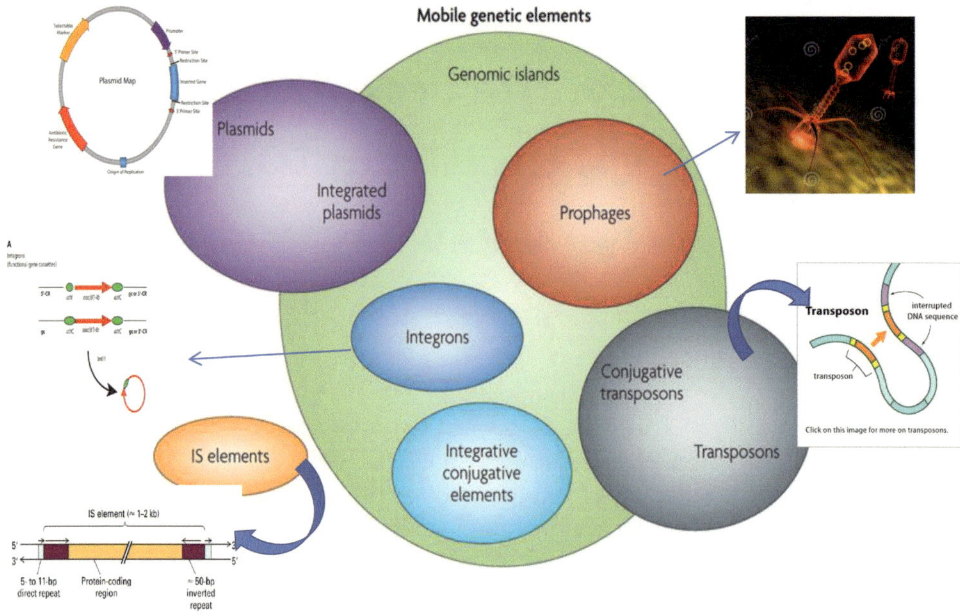

Figure 1.
Bacterial mobile genetic elements.

chromosomal exchange as well as plasmid or transposon (**Figure 1**). This can lead to increment in dangerous nosocomial infections, thus limiting therapeutic options [6]. This is the reason for exploitation of this genus as an important key indicator bacterium for humans and veterinary resistance surveillance system [7].

2. Enterococcal infections and their treatments

Over the past few decades, members of the genus *Enterococcus* have emerged as an important nosocomial pathogen causing different infections. Their transformation from gut commensal to pathogen is attributed by increasing antibiotic resistance especially resistance to vancomycin, high-level aminoglycosides (HLA), and penicillin is of interest. Moreover, resistance to new antimicrobial agents, like linezolid, quinupristin/dalfopristin, and daptomycin has also been emerged (**Figure 2**). Being more resistant than *E. fl*, *E. faecium* (*E. fm*) has come out to be the leading cause of multidrug resistant (MDR) infections in U.S. Because of its resistance to vancomycin, ampicillin and high-level aminoglycosides, infections caused by this species is very difficult to treat. According to National Healthcare Safety Network (NHSN) report, majority of device associated infections (for example, central lines infections, urinary drainage catheters infection and ventilator infections) were caused by 80% vancomycin and 90.4% ampicillin resistant *E. faecium* [8]. Other enterococcal species including *E. avium, E. casseliflavus, E. durans, E. hirae, E. raffinosus, E. gallinarum and E. mundtii* accounts for less human's infection [9]. Enterococci can cause variety of infections directly as sole cause of an infection or indirectly as a contributor in co-infection with other microorganisms [10] (**Figure 3**).

Enterococcal infections particularly those caused by vancomycin resistant enterococci (VRE) are associated with prolonged hospital stay and excess mortality. World Health Organization (WHO), in its report published in February 2017 placed Vancomycin Resistant *E. faecium* in the "HIGH PRIORITY category in global priority pathogens list (global PPL)" of antibiotic resistant bacteria to help in prioritizing

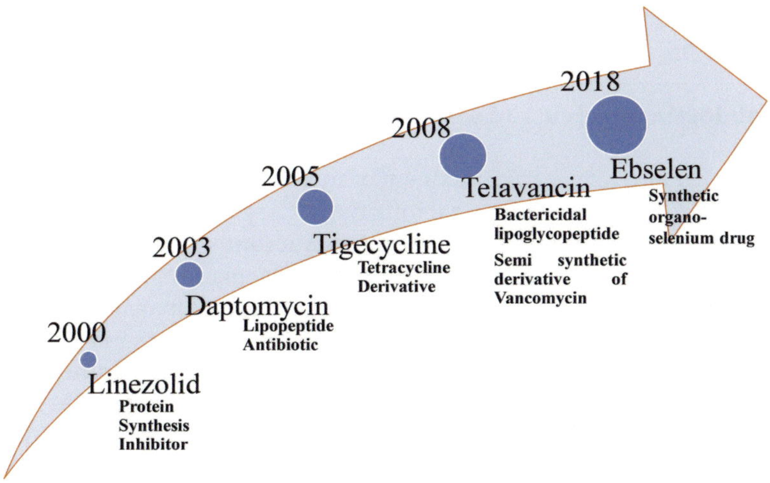

Figure 2.
Examples of recently approved drugs.

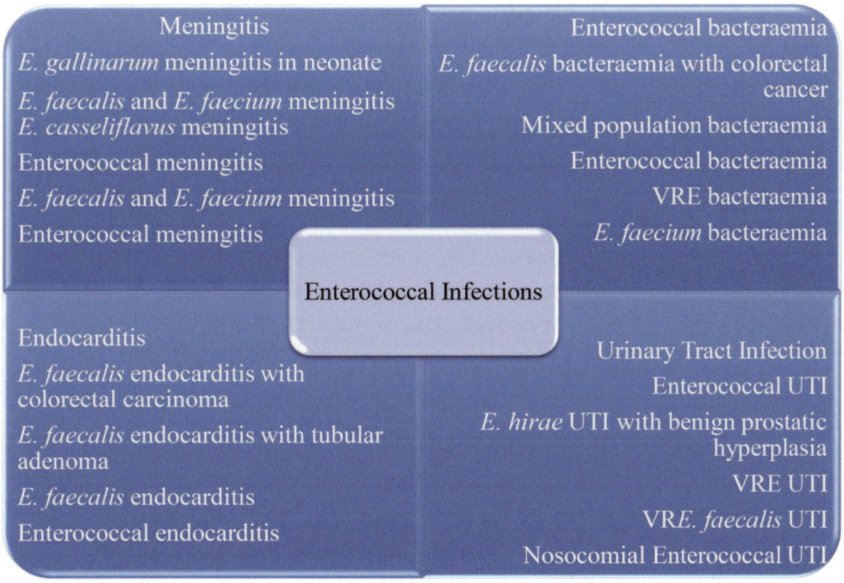

Figure 3.
Different infections caused by genus Enterococcus.

the research and development of new and effective antibiotic treatments [11, 12]. Earlier to this, VRE was also categorized as "microorganisms with a threat level of serious" with estimated 20,000 drugs resistant enterococcal infections, 1300 death tolls and 66,000 *Enterococcus* infections per year in United States [13].

2.1 Urinary tract infections (UTIs)

UTIs including prostatitis, epididymitis and cystitis are the most common types of infections caused by ENT. Majority of the patients includes older men as compared to young women. Upper UTIs which lead to bacteremia also occurred in young men [14]. According to a report presented to NHSN by center of disease control and prevention (CDC), *Enterococcus spp.* account for 14.9% of the total catheter associated UTIs between 2006 and 2007 [8]. Moreover, it is also reported that 15% of UTIs occur in ICU setting with VRE being the major health care associated pathogen [15].

2.2 Intra-abdominal, pelvic and soft tissue infections

ENT are often recovered as a component of mixed microbial flora from cultures of pelvic, soft tissues and intra-abdominal infections. They rarely cause monomicrobial infections at these sites. Enterococcal bacteremia is accompanied with intra-abdominal and pelvic abscesses and wounds; this is the reason why many clinicians prescribe antibiotic regimens for infections at these sites [14, 16, 17]. Moreover, ENT are frequently found in cultures from foot ulcers, decubiti and in diabetics in association with osteomyelitis [15]. Tigecycline, a semi synthetic, bacteriostatic in nature analogue of TET is active against many Gram negative (G –ve) and G +ve bacteria has been used use for the treatment of skin, intra-abdominal and soft tissue infections [18].

2.3 Bacteremia

Incidence of enterococcal blood stream infections are rising day by day [19]. Starting from 6th position in early 80's, ENT is now the 2nd most common cause of health care associated bacteremia [8]. Bacteremia is designated as a major cause of mortality with *Enterococcus* spp. being the third and fourth most common etiological agent of blood stream infections in U.S and Denmark, respectively [20–22]. Genitourinary tract, intra-abdominal, biliary sources, soft tissues infections and indwelling central lines are the common sources of bacteremia from which ENT are isolated as a polymicrobial component [17]. Although enterococcal bacteremia occurs in patients with underlying immunity and illnesses, it rarely affects distant organs or cause metastatic abscesses. Usage of inappropriate antibiotics or late treatment is associated with excess mortality [19]. However, some studies found no decrease in mortality with appropriate antibiotic treatment [23, 24], while some revealed a better outcome after using appropriate antibiotics both for vancomycin and high-level gentamicin resistant enterococci [25, 26].

2.4 Endocarditis

Endocarditis is one of the major enterococcal infections for which antibiotic treatment is difficult because of enterococci's intrinsic resistance to many antibiotics. First case report of endocarditis with details of clinical and pathological description of a strain called *Micrococcus zymogens* (*Enterococcus faecalis*) was published in 1899 [27]. Since then this species is responsible for 8–17% of all infective endocarditis (IE) cases affecting mainly elderly patients with prosthetic heart valve, degenerative heart valve diseases, urinogenital or GIT infections leading to bacteremia and becoming third most frequent etiologic agent of both native and prosthetic valve IE [28–30]. In certain cases, dual antibiotic therapy including aminoglycoside (preferably gentamicin) and cell-wall synthesis inhibitor (vancomycin or β-lactam) is required for IE therapy.

American Heart Association (AHA) and European Society of Cardiology (ESC) recommends 4 to 6 weeks of combined antibiotic treatment with success rate of 80%. Due to nephrotoxic effects of long-term aminoglycoside usage, Danish guidelines on endocarditis treatment endorsed aminoglycoside usage but for 2 weeks only [30]. In case of VRE and HLGR enterococcal IE, surgery remains the only option to remove the infected valve [15]. Among *Enterococcus* spp., *E. fl* was thought to be the most common causative agent of endocarditis infecting mostly older persons as compared to women [31, 32] but recently a more problematic MDR strain of

E. fm belonging to well characterized hospital-associated clade was also identified as a cause of IE. The strains of *E. fm* has high resistance against first line antibiotics (i.e., MIC >64 mg/L ampicillin and vancomycin) due to which their application in curing IE is obsolete [33, 34]. In response to this, AHA recommends Quinupristin-dalfopristin (Q/D; 30% Streptogramin B and 70% A) and linezolid as alternate to treat MDR *E. fm* IE [35]. In fact, many reports suggest better efficacy of Q/D (24 g/day) when use in combination with imipenem, levofloxacin, doxycycline, rifampicin, high-dose ampicillin [36, 37]. Two main and critical steps in the pathogenesis of IE are attachment to tissues and production of biofilm. Biofilm associated proteins which facilitates occurrence of IE includes aggregation substance protein, i.e., *Asc10* [38], microbial surface components recognizing adhesive matrix molecules (MSCRAMM) proteins *ace* for *E. fl* [39] and *acm* for *E. fm* [40], *esp* and its homolog in *E. fm*, esp_{fm} [41, 42], endocarditis and biofilm associated pili of *E. fl*, i.e., *ebp* [42–44]. The main complication of enterococcal IE is heart failure occurring in half of the patients. Moreover, MDR *E. fm* is also an important factor in increasing epidemiology of enterococcal IE because >90% of *E. fl* are susceptible to ampicillin and vancomycin [45].

3. Pathogenesis and virulence associated with enterococci

Virulence factors are potential traits that define the pathogenesis of most infections which involves a series of events namely, colonization, adhesion to the host's cells, tissue invading and resistance to non-specific defensive mechanisms. Researchers are encouraged to characterize the factors involved in etiology of infections caused by pathogenic ENT in immunocompromised or impaired immunity patients. Two major classes of virulent factors have been well characterized: (1) surface factors that promote colonization in host cells, and (2) protein and peptides secreted by ENT that damage the tissues [46].

3.1 Gelatinase (*gelE*), serine protease (*sprE*) and *fsr* regulator

Gelatinase is a zinc metalloprotease expressed extracellularly and hydrolyze gelatin, collagen and casein [47]. It is proved to be a full virulence factor expressed in mouse model of peritonitis, endocarditis [48, 49], endophthalmitis [50], in nematode [51] and in vitro translocation [52]. It is encoded by *gelE* and *sprE* operon and expressed in regulation by a quorum sensing system encoded by the *fsr* locus [53]. The *fsr* locus (*E. fl* regulator) is a well characterized locus containing *fsrA*, *fsrB*, *fsrC* and *fsrD* genes which is homologs to staphylococcal *agrBCA* loci [54]. A signaling peptide in *fsrB* liberates gelatinase biosynthesis activating pheromone (GBAP) peptide by auto-processing and a quorum sensing system. *gelE* and *sprE* genes are induced when GBAP accumulates from exponential to stationary phase. *Fsr* regulon is present above the *sprE* and *gelE* and encode a serine protease and gelatinase, respectively [55]. Possible molecular mechanism behind the expression of *gelE* and *sprE* is shown in **Figure 4** [56]. Epidemiological data suggests the involvement of *fsr* locus and gelatinase in virulence traits, like adhesion capacity (biofilm) established by processing of C-terminal gelatinase protein [57, 58].

3.2 Catalase (EC 1.11.1.6)

Catalase is a renowned enzyme present in all three domains of life. It catalyzes the decomposition of hydrogen peroxide (HP) to water and oxygen, protecting the cell from oxidative damage of HP. HP is a reactive oxygen species (ROS) in

fsrD encodes

↓

Peptide lactone accumulates ⟶ Exported to extracellular space by *fsrB* ⟶ Sense by *fsrC* histidine kinase ⟶ Activation of response regulator and *fsrA* by phosphorylation

↓

Phosphorylated *fsrA* activates expression of *gelE* and *sprE.*

Figure 4.
Flow diagram showing the possible mechanism of gelE *and* sprE *gene expression.*

biosphere. It is produced as a by-product in aerobic metabolism such as in oxygen activation, in photosynthetic and respiratory electron transport chain and as product of oxidases activity. First step of catalase reaction is the reduction of HP to water forming cationic heam radical and an oxoiron [compound 1 (Fe^{IV}=O ion)]. In the second step, dismutation is completed by the reaction of a second HP, resulting in the release of oxygen and water. The enzyme is regenerated in the resting Fe^{III} state. NADPH binding catalases prevent the build-up of an inactive partially oxidized dead-end form of the enzyme called compound II [59].

Catalases are of three types: Prokaryotic Mn-catalases (minor bacterial protein family), bifunctional catalase peroxidases (not found in plants and animals and exhibit both catalytic and peroxidative activities) and haem catalases (most abundant group found in Archaebacteria, Eubacteria, Fungi, Protista, Animalia and Plantae). Despite catalyzing the same reaction ($2H_2O_2 \rightarrow 2H_2O + O_2$), all three families differ in architecture of active site and mode of reaction [60]. Among G +ve lactic acid bacteria (LAB), *E. fl* are unable to make porphyrin compounds, including heam groups. It exhibits catalase activity but only when it is grown in heme containing medium [61]. *E. fl* catalase (*katA*) is a homo-tetrameric protein containing only one heme group (protoheme IX) and belongs to the group of heme containing mono functional catalases [62]. In the absence of heme, *E. fl* produces NADH peroxidase (*Npr*) that degrades HP to water. Factors involve in biogenesis of catalase was not known until Baureder and Hederstedt [62] carried out a research in which they used two different transposon systems to construct libraries of *E. fl* mutants and screened for clone defective in catalase activity by using colony zymogram staining procedure. They identified nine genes (in addition to *katA,* which codes for catalase enzyme protein) distributed over five chromosomal loci which are important for expression of catalase activity in *E. fl*. The proteins encoded by those genes have diverse functions such as NADH oxidation and HP detoxification (*npr*), global regulation of RNA turnover (*rnjA*, *srmB*), membrane transport (*oppBC*) and/or stress response (*etaR*) [62].

3.3 Hyaluronidase (EC 4.2.2.1)

These are the enzymes capable of degrading hyaluronate (Hyaluronic acid, hyaluronan) found in several body parts, like umbilical cord, synovial fluid, cartilage, brain, muscles and extracellular matrix (ECM) in connective tissues. Almost half of the total body hyaluronate is found in the skin. The viscous ground substance release by the connective tissues provides a barrier for the entry of bacteria or toxin into the body. However, ground substance contains hyaluronate as a major component which is degraded by hyaluronidases. Rooster's combs and certain bacteria like streptococci also produce hyaluronidases [63]. Many pathogenic bacteria release some extracellular products which helps them in damaging the tissue thus acting as a virulent factor and smoothen the progress of bacterial toxin into the tissues and are commonly named as "spreading factors." Bacterial hyaluronidases (BH) are among some of the spreading factors released by certain G +ve and G −ve bacteria. BH belongs to the third type of hyaluronidases commonly called as hyaluronate lyases. They eliminate β 1–4 linkage resulting in the production of unsaturated disaccharides by acting as endo-*N*-acetylhexosaminidases [63]. Different models of *E. fm* trans conjugant's virulence that harbors conjugative mega plasmid have been reported [64, 65] to carried *hyl* gene. According to some previous studies, the *hyl* gene was more prevalent in clinical isolates rather than community base isolates. According to a recent study, *hyl* gene is considered as a passive marker of virulence because deletion of this gene caused no effect on mouse peritonitis model [66, 67].

3.4 Cytolysin (Cyt)

Enterococcal Cyt is a broad range prokaryotic and eukaryotic lysin usually plasmid encoded. It is reported to enhance virulence of *E. fl* in animal models. It was originally described as lanthionine-containing bacteriocins of G +ve bacteria [68]. The Cyt operon is a part of *E. fl* PAI consisting of 6 genes related to toxin biosynthesis and two promoters namely P_L (involve in regulation of transcription of genes related to toxin structure and function) and P_{REG} (involve in transcription of regulatory genes) and present near *esp* gene [69]. Like gelatinase, expression of Cyt is quorum sensing dependent and regulated by two component systems [70]. The regulatory system of Cyt consists of two open reading frames (ORFs) namely *cylR1* and *cylR2* which encodes a transmembrane protein of unknown function (cylR1) and a helix-turn-helix DNA binding protein (cylR2) [71]. The Cyt operon is either present on conjugative pheromone responsive plasmid such as pAD1 [72] or encoded by chromosome within 150 kb PAI [73, 74]. Todd et al. [75] conducted the first comprehensive study on hemolysin molecule after the observation of hemolysis zones on blood agar plates produced by *E. fl*. Increased virulence due to Cyt in *E. fl* was first described in the study of Ike and colleagues [76] through dose dependent intraperitoneal injections of *E. fl* strains harboring plasmid pAD1 which encodes Cyt. Later, various researchers showed the lyses of mouse erythrocytes, macrophages, and PMNs or death of experimental animals/organism like mouse, rabbits and *C. elegans* with Cyt [58, 73, 77–80]. Self-lysis of Cyt producing cells is prevented by an unknown mechanism. However, immunity proteins or ABC transporters protects other lantibiotic producing bacteria from self-lysis [81, 82]. In *E. fl*, a zinc metalloprotease and transmembrane protein, CylI (immunity factor) is shown to protect from Cyt mediated bacterial cell death [83].

Despite having a virulence face, Cyt can also act as beneficiary trait for both *E. fl* and its host. Possible beneficial activities might include, acting as colonization factor, providing self-defense against something which is more harmful (probably an intestinal parasite), facilitating nutrient acquisition from prokaryotic or eukaryotic sources, function as signaling molecule to monitor bacterial population size and probe the environment for target cells and last but not the least, bacteriocin activity of Cyt allows *E. fl* to occupy a novel host niche which non-cytolytic bacteria cannot access [68, 84, 85].

3.5 Enterococcal surface protein (*esp*)

Esp, a putative virulent factor is found in both *E. fl* and *E. fm*. It is located on pathogenicity island (PAI) at the surface of the bacterium [56]. It was initially identified in a highly virulent gentamicin resistant strain of *E. fl* [69]. *Esp* shares global structural similarity with *Streptococcus agalactiae* Rib [86], *S. pyogenes* R28, C-alpha protein, and *S. aureus* Bap (biofilm associated protein) [87]. These similarities are restricted to a highly conserved region within the C repeat units of *Esp* proteins and group A and B of streptococcal proteins of streptococci and nonrepeat N-terminal region of Bap protein [71]. Bap protein from *S. aureus* is associated with biofilm formation and shares a sequence and structural similarity with *Esp*. *Esp* is also associated with *E. fl* biofilm formation on different surfaces, like polysterylene plates and hospital equipment like catheters, prosthetic heart valves, orthopedic appliances, artificial cardiac pace makers [47], ureteral stents [88], intravascular catheters [89], silicone gastrostomy devices [90], and biliary stents [91].

A variant of *Esp* is also reported in *E. fm* isolates [92]. *E. fm esp* is predominantly present in nosocomial settings in contrast to *E. fl esp* which is widely distributed among environmental strains [93, 94]. Expression of *esp* is affected by environmental conditions like temperature (maximum at 37°C) and availability of oxygen, i.e., under anaerobiosis [56]. Several research groups demonstrated the role of *E. fm esp* in pathogenesis of experimental endocarditis, UTIs, and bacteremia. While no specific role of *esp* was found in peritonitis, and colonization of GIT [95, 96]. Role of *esp* was also established by a genetic approach. In a study conducted by Tendolkar *et al.* [97], *esp*-lacking *E. fl* strains produced biofilm in large amounts after successful induction and expression of *esp* gene. In contrast, several studies suggest that *esp* is not necessary for biofilm formation [98, 99]. Study conducted by Kristich *et al.* [100] demonstrated that *E. fl* OG1RF produced biofilms not only in the absence of *esp* and entire PAI that harbors it. In other studies, conducted on clinical enterococcal isolates, majority of the *esp*-negative isolates produced biofilms and no correlation was found among *esp* gene and biofilm forming capacity [89, 101].

3.6 Aggregation substance (AS)

AS is a group of proteins encoded by pheromone-induced conjugative plasmids. AS directed bacteria to aggregate which results in close cell contact between donor and recipient. Several studies showed that AS mediated internalization of *E. fl* by cultured human intestinal epithelial cells and increased in vitro adhesion to cultured renal tubular cells [102]. Among the best studied AS proteins are Asa I, Asp I, and Acs 10 encoded by *asa1*, *aspI* and *prgB* genes of conjugated plasmids *pAD1* and *pCF10*, respectively, and shows >90% sequence identity [56, 66]. These proteins contain an N-terminal domain, a central domain, a variable region and two Arg-Gly-Asp (RGD) motifs which are also found in fibronectin and associated with integrin binding proteins [102, 103]. Apart from their function in conjugation transfer, these RGD motifs are also involve in eukaryotic cell binding and binding to renal epithelial cells [102]. It is demonstrated in a study that central domain and N-terminal domain are responsible for aggregation of Asc 10 [104]. Beside its role in conjugation, AS also serves as a virulence factor in *E. fl* by promoting cell-cell contact, adhesions to host cells and ECM proteins (including thrombospondin, fibronectin, vitronectin, and collagen type I), increased vegetation in experimental endocarditis, resistance to killing by polymorphonuclear leukocytes (PMNs) by inhibition of respiratory burst (production of ROS) in the macrophages, increased cell surface hydrophobicity [71]. All the proteins aid in the pathogenesis of AS in *E. fl,* like Asa I increases adherence to human macrophages and renal tubular cells, Asc 10 facilitates internalization and intracellular survival in PMNs [74, 103, 105]. Both

Virulent factors	Gene	Primer sequence (5′-3′)	Product length (bp)	Reference
Biofilm associated genes	*espTIM*	CTT-TGA-TTC-TTG-GTT-GTC-GGA-TAC TCC-AAC-TAC-CAC-GGT-TTG-TTT-ATC	475	[111]
	agg	AAG-AAA-AAG-AAG-TAG-ACC-AAC AAA-CGG-CAA-GAC-AAG-TAA-ATA	1553	[112]
	acm	GGC-CAG-AAA-CGT-AAC-CGA-TA CGC-TGG-GGA-AAT-CTT-GTA-AA	353	[113]
	efaAfm	AAC-AGA-TCC-GCA-TGA-ATA CAT-TTC-ATC-ATC-TGA-TAG-TA	735	[92]
	efaAfs	GAC-AGA-CCC-TCA-CGA-ATA AGT-TCA-TCA-TGC-TGT-AGT-A	705	
	asa	GCA-CGC-TAT-TAC-GAA-CTA-TGA TAA-GAA-AGA-ACA-TCA-CCA-CGA	375	[114]
	ace	AAA-GTA-GAA-TTA-GAT-CCA-CAC TCT-ATC-ACA-TTC-GGT-TGC-G	320	[115]
	ccf	GGG-AAT-TGA-GTA-GTG-AAG-AAG AGC-CGC-TAA-AAT-CGG-TAA-AAT	542	[112]
	cpd	TGG-TGG-GTT-ATT-TTT-CAA-TTC TAC-GGC-TCT-GGC-TTA-CTA	782	
	cob	AAC-ATT-CAG-CAA-ACA-AAG-C TTG-TCA-TAA-AGA-GTG-GTC-AT	1405	
	eep	GAG-CGG-GTA-TTT-TAGTTC-GT TAC-TCCAGCATTGGATGCT	937	
Gelatinase operon genes	*gelE*	ACC-CCG-TAT-CAT-TGG-TTT ACG-CAT-TGC-TTT-TCC-ATC	419	[116]
	sprE	TTG-AGC-TCC-GTT-CCT-GCC-GAA-AGT-CAT-TC TTG-GTA-CCG-ATT-GGG-GAA-CCA-GAT-TGA-CC	591	
	fsrA	ATG-AGT-GAA-CAA-ATG-GCT-ATT-TA CTA-AGT-AAG-AAA-TAG-TGC-CTT-GA	740	
	fsrB	GGG-AGC-TCT-GGA-CAA-AGT-ATT-ATC-TAA-CCG TTG-GTA-CCC-ACA-CCA-TCA-CTG-ACT-TTT-GC	566	
	fsrC	ATG-ATT-TTG-TCG-TTA-TTA-GCT-ACT CAT-CGT-TAA-CAA-CTT-TTT-TAC-TG	1343	
Cytolysin operon genes	*cylL$_L$*	GAT-GGA-GGG-TAA-GAA-TTA-TGG GCT-TCA-CCT-CAC-TAA-GTT-TTA-TAG	253	[117]
	cylL$_s$	GAA-GCA-CAG-TGC-TAA-ATA-AGG GTA-TAA-GAG-GGC-TAG-TTT-CAC	240	
	cylM	AAA-AGG-AGT-GCT-TAC-ATG-GAA-GAT CAT-AAC-CCA-CAC-CAC-TGA-TTC-C	2940	
	cylB	AAG-TAC-ACT-AGT-AGA-ACT-AAG-GGA ACA-GTG-AAC-GAT-ATA-ACT-CGC-TAT-T	2020	
	cylA	ACT-CGG-GGA-TTG-ATA-GGC GCT-GCT-AAA-GCT-GCG-CTT	688	

Virulent factors	Gene	Primer sequence (5′-3′)	Product length (bp)	Reference
Enterocin genes	*entA*	ATG-AAA-CAT-TTA-AAA-ATT-TTG-TCT-ATT-AAA-G TTA-GCA-CTT-CCC-TGG-AAT-TGC-TCC	1770	[118]
	entB	AGA-CCT-AAC-AAC-TTA-TCT-AAA-G GTT-GCA-TTT-AGA-GTA-TAC-ATT-TGC	126	
	entP	ATG-AGA-AAA-AAA-TTA-TTT-AGT-TTA-GCT-CTT-ATT-GG TTA-ATG-TCC-CAT-ACC-TGC-CAA-ACC-AG	216	
	Ef1097	GGC-GAT-GGC-ATT-ACT-AAT-GAC-ATT-AGG CTT-AGC-CCA-CAT-TGA-ACT-GCC-CAT-AAA-GC	408	
	enlA	CGA-TTT-CTG-TTG-TAG-GAA-CC GTA-CAT-CTC-CAT-ATA-CTT-TTC-C	1405	
Insertion sequence element gene	*IS16*	CATG-TTC-CAG-CAA-CCA-GAG TCA-AAA-AGT-GGG-CTT-GGC	547	[111]
Hyaluronidase gene	*hyl*	ACA-GAA-GAG-CTG-CAG-GAA-ATG GAC-TGA-CGT-CCA-AGT-TTC-CAA	276	[119]
Catalase gene	*kat*	ACC-CCG-TAT-CAT-TGG-TTT ACG-CAT-TGC-TTT-TCC-ATC	419	[110]
Lipase gene	*Lip-fm*	TTG-AGC-TCC-GTT-CCT-GCC-GAA-AGT-CAT-TC TTG-GTA-CCG-ATT-GGG-GAA-CCA-GAT-TGA-CC	591	[108]
	Lip-fl	ATG-AGT-GAA-CAA-ATG-GCT-ATT-TA CTA-AGT-AAG-AAA-TAG-TGC-CTT-GA	740	

Table 1.
List of primers reported for the genotypic assessment of major virulence factors.

	Thermal cycler programme	Gene name (Reference)
A	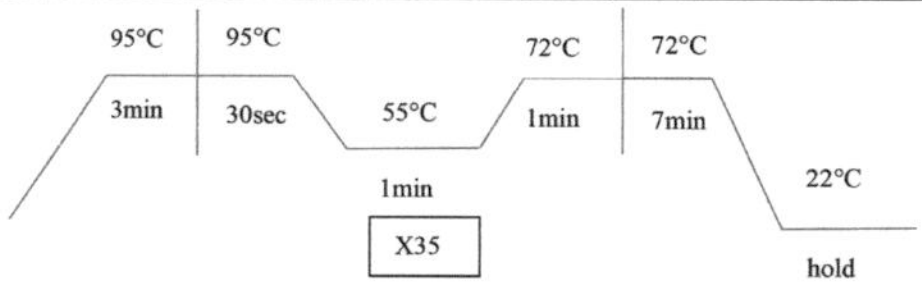	*gelE*, *sprE*, *fsrA*, *fsrB*, *fsrC* genes [116]
B	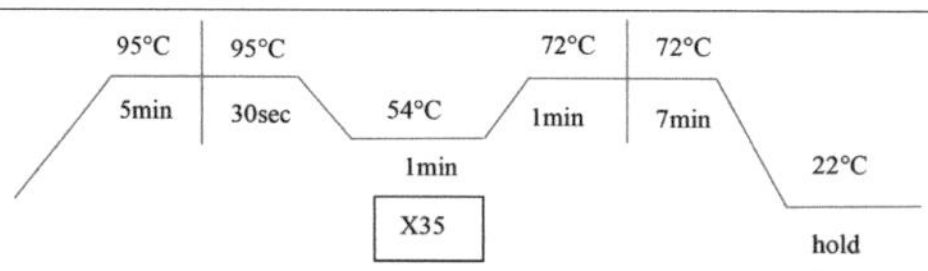	*lip-fm* and *lip-fl* genes [108]
C	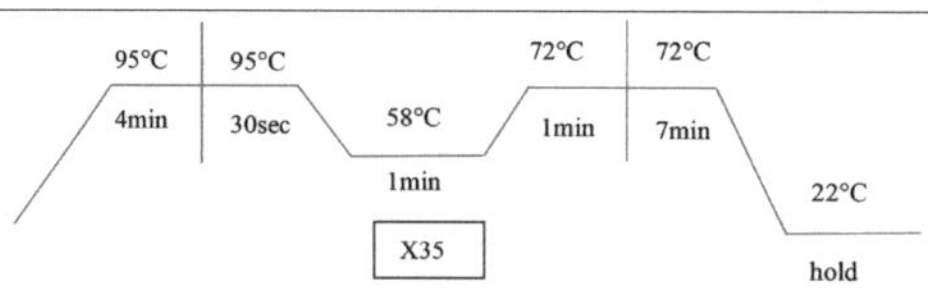	*kat* gene [110]
D	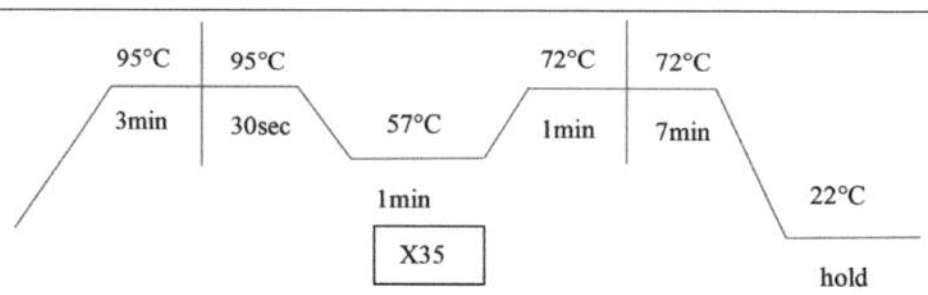	*cylL*$_L$, *cylM*, *cylB*, *cylA* [117]

	Thermal cycler programme	Gene name (Reference)
E	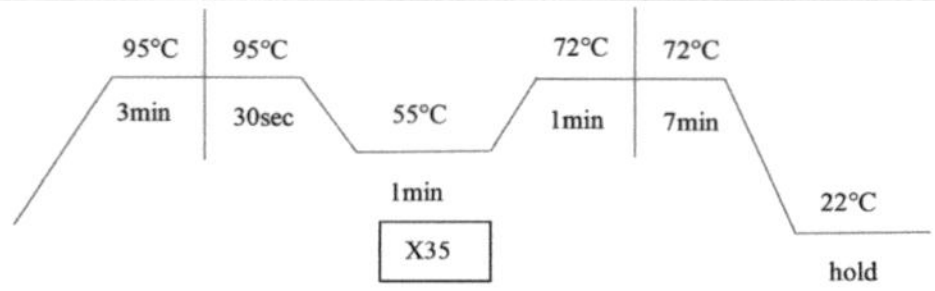	*efaAfs* and *efaAfm* [92]
F	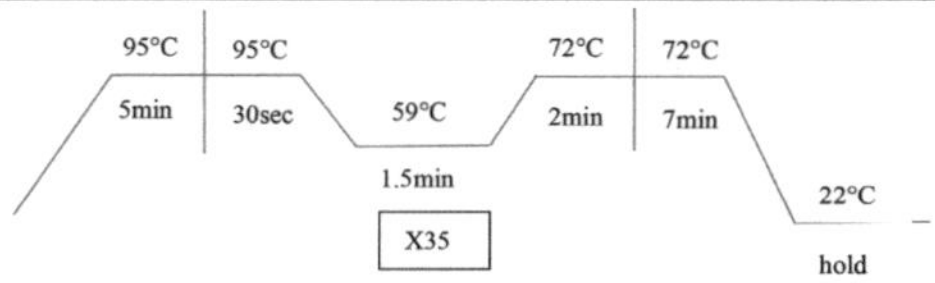	*espTIM* [111]
G	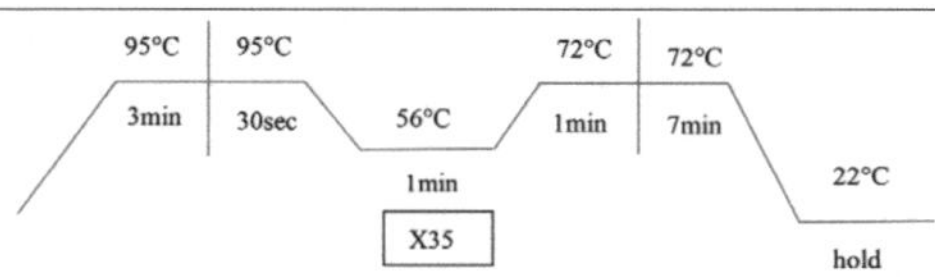	*acm* gene [113]
H	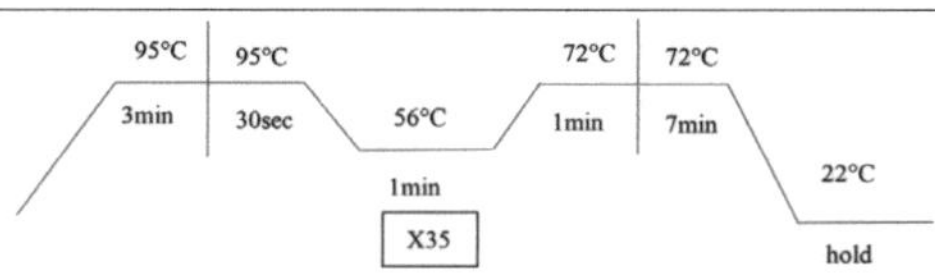	*ace* and *asa1* gene [114, 115]
I	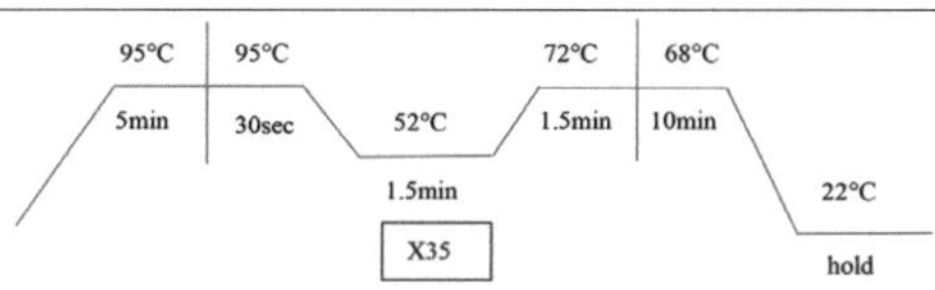	*agg* or *AP* gene [112]
J	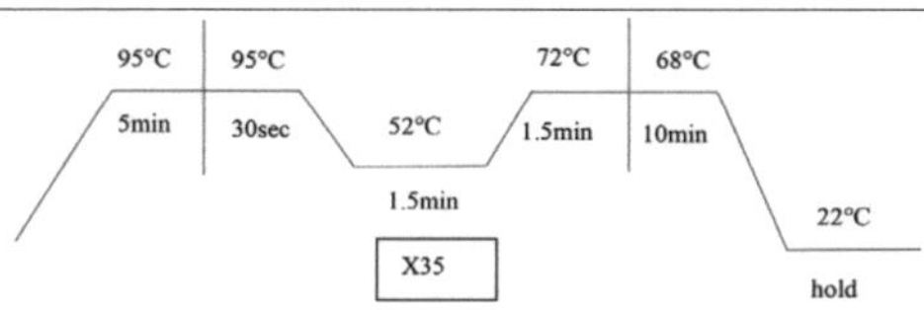	*ccf*, *cpd*, *eep*, *cob* genes [112]
K	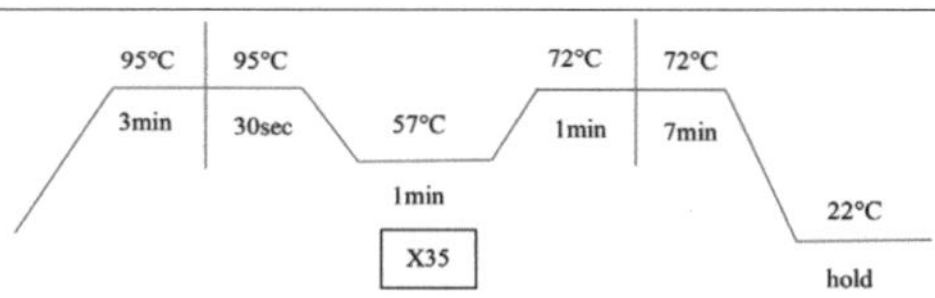	*entP, entA, entB, ef1097* gene [118]
L	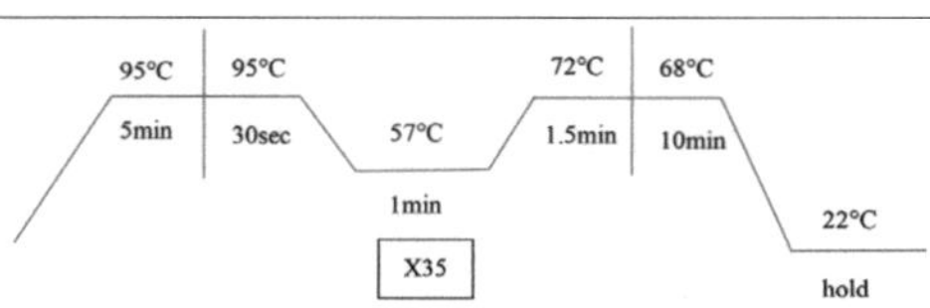	*enlA* gene [118]
M	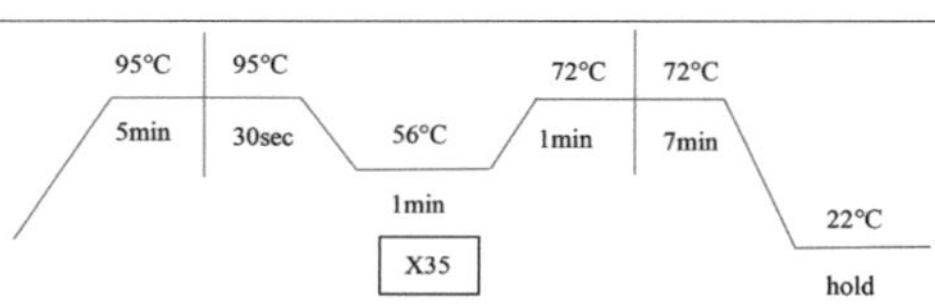	*IS16* gene [111]
N	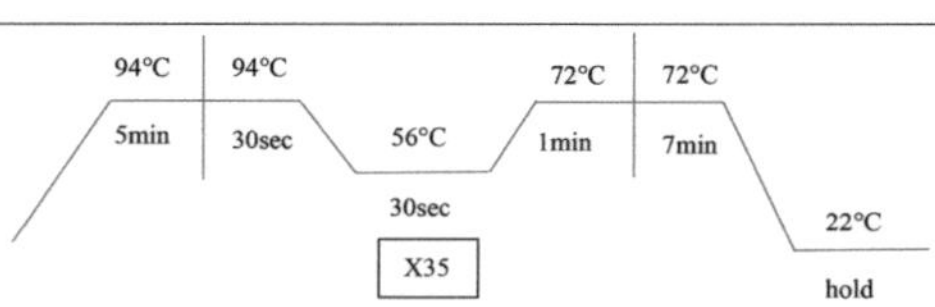	*Hyl gene* [119]

Table 2.
Illustrations for the PCR conditions for the amplification of various putative virulence genes.

Asa I and Asc 10 increase virulence of *E. fl* in rabbit endocarditis model by increasing adherence to certain ECM proteins [79, 106].

4. Conclusions

In conclusion, acquired resistance to certain antibiotics is an important feature of the genus *Enterococcus*. Persistent use of antibiotics in humans and animals for therapy and as growth promoters plus the presence of insertion sequences, transposons, integerons and plasmids make them large reservoirs of transferable antibiotic resistance and virulence genes in various ecosystems including soil, water, and food. Due to its rapid popularity, as resistant bacteria, ENT serves as an important key indicator in the surveillance of many humans and veterinary resistance profile. Adherence capability plus antibiotic resistance make them more problematic for effective therapeutic decisions. Till now only food consumption is considered as an option for the spread of antibiotic resistant bacteria to humans but the detection of resistant bacteria in soil opens a new route for the exposure of environmental antibiotic resistance to humans. Results of different studies from our lab concludes that soil, poultry, animals and birds carried high burdens of ENT which are fully armed with potential virulent and antibiotic resistance genes [107–110]. In Pakistan, there is paucity of information regarding prevalence, types and genetic characteristics of enterococci along with their resistance/virulence genes and clones especially from clinical and other environmental sources. In this respect, regular environmental monitoring using most advance molecular genotyping (**Tables 1** and **2**) as routine testing is recommended. Genes mirror the requirements of life. As our understanding of enterococcal genomics grows, bacterial genomics will become an important tool for providing new insights into the nature, biology and habitats of the enterococci. Presence of insertion sequence (IS16) gene in soil isolates verified the dissemination of hospital associated ENT into the environment via inappropriate handling of hospital wastes [108]. It is therefore also recommended to dispose clinical/hospital waste properly and appropriately.

Acknowledgements

The financial support from Higher Education Commission (HEC) Islamabad, Pakistan to SA Ali (HEC No. 20-1339/R&D/09) is greatly acknowledged.

Author details

Hassan Bin-Asif and Syed Abid Ali*
H.E.J. Research Institute of Chemistry, International Centre for Chemical and Biological Sciences (ICCBS), University of Karachi, Karachi, Pakistan

*Address all correspondence to: abid.ali@iccs.edu

References

[1] Jackson CR, Spicer LM, Barrett JB, Hiott LM. Application of multiplex PCR, pulsed-field gel electrophoresis (PFGE), and BOX-PCR for molecular analysis of enterococci. In: Gel Electrophoresis-Principles and Basics. Rijeka: IntechOpen; 2012

[2] Domig KJ, Mayer HK, Kneifel W. Methods used for the isolation, enumeration, characterisation and identification of *Enterococcus* spp.: 1. Media for isolation and enumeration. International Journal of Food Microbiology. 2003;**88**(2-3):147-164

[3] Staley C, Dunny GM, Sadowsky MJ. Environmental and animal-associated enterococci. In: Advances in Applied Microbiology. Jan 1, 2014;**87**:147-186

[4] Rathnayake IU, Hargreaves M, Huygens F. Antibiotic resistance and virulence traits in clinical and environmental *Enterococcus faecalis* and *Enterococcus faecium* isolates. Systematic and Applied Microbiology. 2012;**35**(5):326-333

[5] Elhani D, Klibi N, Dziri R, Hassan MB, Mohamed SA, Said LB, et al. vanA-containing *E. faecium* isolates of clonal complex CC17 in clinical and environmental samples in a Tunisian hospital. Diagnostic Microbiology and Infectious Disease. 2014;**79**(1):60-63

[6] Dadfarma N, Fooladi AA, Oskoui M, Hosseini HM. High level of gentamicin resistance (HLGR) among enterococcus strains isolated from clinical specimens. Journal of Infection and Public Health. 2013;**6**(3):202-208

[7] Borck Høg B, Korsgaard HB, Wolff Sönksen U, Bager F, Bortolaia V, Ellis-Iversen J, et al. DANMAP 2016 - Use of antimicrobial agents and occurrence of antimicrobial resistance inbacteria from food animals, food and humans in Denmark. Statens Serum Institut, National Veterinary Institute, Technical University of Denmark National Food Institute, Technical University of Denmark; 2017

[8] Hidron AI, Edwards JR, Patel J, Horan TC, Sievert DM, Pollock DA, et al. Antimicrobial-resistant pathogens associated with healthcare-associated infections: Annual summary of data reported to the National Healthcare Safety Network at the Centers for Disease Control and Prevention, 2006-2007. Infection Control and Hospital Epidemiology. 2008;**29**(11):996-1011

[9] Gordon S, Swenson JM, Hill BC, Pigott NE, Facklam RR, Cooksey RC, et al. Antimicrobial susceptibility patterns of common and unusual species of enterococci causing infections in the United States. Enterococcal Study Group. Journal of Clinical Microbiology. 1992;**30**(9):2373-2378

[10] Hoge CW, Adams J, Buchanan B, Sears SD. Enterococcal bacteremia: To treat or not to treat, a reappraisal. Reviews of Infectious Diseases. 1991;**13**(4):600-605

[11] WHO, Global Priority List of Antibiotic- Resistant Bacteria to Guide Research, Discovery, and Development of New Antibiotics. 2017. Available on-line at://www.who.int/medicines/publications/global-priority-list-antibiotic-resistant-bacteria/en/

[12] Remschmidt C, Behnke M, Kola A, Diaz LA, Rohde AM, Gastmeier P, et al. The effect of antibiotic use on prevalence of nosocomial vancomycin-resistant enterococci-an ecologic study. Antimicrobial Resistance and Infection Control. 2017;**6**(1):95

[13] CDC, Antibiotic Resistance Threats in the United States, 2013; Available at: http://www.cdc.gov/drugresistance/threat-report-2013/

[14] Graninger W, Ragette R. Nosocomial bacteremia due to *Enterococcus faecalis* without endocarditis. Clinical Infectious Diseases. 1992;**15**(1):49-57

[15] Higuita NI, Huycke MM. Enterococcal disease, epidemiology, and implications for treatment. In: Enterococci: From Commensals to Leading Causes of Drug Resistant Infection [Internet]. Massachusetts Eye and Ear Infirmary; 2014

[16] Noskin GA, Stosor V, Cooper I, Peterson LR. Recovery of vancomycin-resistant enterococci on fingertips and environmental surfaces. Infection Control and Hospital Epidemiology. 1995;**16**(10):577-581

[17] Patterson JE, Sweeney AH, Simms M, Carley N, Mangi R, Sabetta J, et al. An analysis of 110 serious enterococcal infections. Epidemiology, antibiotic susceptibility, and outcome. Medicine. 1995;**74**(4):191-200

[18] Rose WE, Rybak MJ. Tigecycline: First of a new class of antimicrobial agents. Pharmacotherapy: The Journal of Human Pharmacology and Drug Therapy. 2006;**26**(8):1099-1110

[19] Suppli M, Aabenhus R, Harboe ZB, Andersen LP, Tvede M, Jensen JU. Mortality in enterococcal bloodstream infections increases with inappropriate antimicrobial therapy. Clinical Microbiology and Infection. 2011;**17**(7):1078-1083

[20] National Nosocomial Infections Surveillance (NNIS). System report, data summary from January 1990-May 1999, issued June 1999. American Journal of Infection Control. 1999;**27**:520-532

[21] Wisplinghoff H, Bischoff T, Tallent SM, Seifert H, Wenzel RP, Edmond MB. Nosocomial bloodstream infections in US hospitals: Analysis of 24,179 cases from a prospective nationwide surveillance study. Clinical Infectious Diseases. 2004;**39**(3):309-317

[22] Schønheyder HC, Søgaard M. Hospital-acquired bacteraemia and fungaemia. A regional study with national implications. Ugeskrift for Laeger. 2007;**169**(48):4175-4179

[23] Bryan CS, Reynolds KL, Brown JJ. Mortality associated with enterococcal bacteremia. Surgery, Gynecology & Obstetrics. 1985;**160**(6):557-561

[24] Lautenbach E, Bilker WB, Brennan PJ. Enterococcal bacteremia: Risk factors for vancomycin resistance and predictors of mortality. Infection Control and Hospital Epidemiology. 1999;**20**(5):318-323

[25] Vergis EN, Hayden MK, Chow JW, Snydman DR, Zervos MJ, Linden PK, et al. Determinants of vancomycin resistance and mortality rates in enterococcal bacteremia: A prospective multicenter study. Annals of Internal Medicine. 2001;**135**(7):484-492

[26] Vergis EN, Shankar N, Chow JW, Hayden MK, Snydman DR, Zervos MJ, et al. Association between the presence of enterococcal virulence factors gelatinase, hemolysin, and enterococcal surface protein and mortality among patients with bacteremia due to *Enterococcus faecalis*. Clinical Infectious Diseases. 2002;**35**(5):570-575

[27] MacCallum WG, Hastings TW. A case of acute endocarditis caused by *Micrococcus zymogenes* (Nov. Spec.), with a description of the microorganism. The Journal of Experimental Medicine. 1899;**4**(5-6):521-534

[28] Bouza E, Menasalvas A, Munoz P, Vasallo FJ, Moreno MD, Fernandez MA. Infective endocarditis—A prospective study at the end of the twentieth century: New predisposing conditions, new etiologic agents, and still a high mortality. Medicine. 2001;**80**(5):298-307

[29] Murdoch DR, Corey GR, Hoen B, Miró JM, Fowler VG, Bayer AS, et al. Clinical presentation, etiology, and outcome of infective endocarditis in the 21st century: The international collaboration on endocarditis–prospective cohort study. Archives of Internal Medicine. 2009;**169**(5):463-473

[30] Dahl A, Rasmussen RV, Bundgaard H, Hassager C, Bruun LE, Lauridsen TK, et al. *Enterococcus faecalis* infective endocarditis: A pilot study of the relationship between duration of gentamicin treatment and outcome. Circulation. 2013;**127**(17):1810-1817

[31] Anderson DJ, Murdoch DR, Sexton DJ, Reller LB, Stout JE, Cabell CH, et al. Risk factors for infective endocarditis in patients with enterococcal bacteremia: A case-control study. Infection. 2004;**32**(2):72-77

[32] McDonald JR, Olaison L, Anderson DJ, Hoen B, Miro JM, Eykyn S, et al. Enterococcal endocarditis: 107 cases from the international collaboration on endocarditis merged database. The American Journal of Medicine. 2005;**118**(7):759-766

[33] Galloway-Peña J, Roh JH, Latorre M, Qin X, Murray BE. Genomic and SNP analyses demonstrate a distant separation of the hospital and community-associated clades of *Enterococcus faecium*. PLoS One. 2012;7(1):e30187

[34] Palmer KL, Godfrey P, Griggs A, Kos VN, Zucker J, Desjardins C, et al. Comparative genomics of enterococci: Variation in *Enterococcus faecalis*, clade structure in *E. faecium*, and defining characteristics of *E. gallinarum* and *E. casseliflavus*. MBio. 2012;**3**(1):e00318-e00311

[35] Baddour LM, Wilson WR, Bayer AS, Fowler VG Jr, Bolger AF, Levison ME, et al. Infective endocarditis: diagnosis, antimicrobial therapy, and management of complications: a statement for healthcare professionals from the Committee on Rheumatic Fever, Endocarditis, and Kawasaki Disease, Council on Cardiovascular Disease in the Young, and the Councils on Clinical Cardiology, Stroke, and Cardiovascular Surgery and Anesthesia, American Heart Association: endorsed by the Infectious Diseases Society of America. Circulation. 2005;**111**(23):e394-e434

[36] Matsumura S, Simor AE. Treatment of endocarditis due to vancomycin-resistant *Enterococcus faecium* with quinupristin/dalfopristin, doxycycline, and rifampin: A synergistic drug combination. Clinical Infectious Diseases. Dec 1, 1998;**27**(6):1554-1556

[37] Bethea JA, Walko CM, Targos PA. Treatment of vancomycin-resistant enterococcus with quinupristin/dalfopristin and high-dose ampicillin. The Annals of Pharmacotherapy. 2004;**38**(6):989-991

[38] Schlievert PM, Chuang-Smith ON, Peterson ML, Cook LC, Dunny GM. *Enterococcus faecalis* endocarditis severity in rabbits is reduced by IgG Fabs interfering with aggregation substance. PLoS One. 2010;**5**(10):e13194

[39] Singh KV, Nallapareddy SR, Sillanpää J, Murray BE. Importance of the collagen adhesin ace in pathogenesis and protection against *Enterococcus faecalis* experimental endocarditis. PLoS Pathogens. 2010;**6**(1):e1000716

[40] Nallapareddy SR, Singh KV, Murray BE. Contribution of the collagen

adhesin Acm to pathogenesis of *Enterococcus faecium* in experimental endocarditis. Infection and Immunity. 2008;**76**(9):4120-4128

[41] Heikens E, Bonten MJ, Willems RJ. Enterococcal surface protein Esp is important for biofilm formation of *Enterococcus faecium* E1162. Journal of Bacteriology. 2007;**189**(22):8233-8240

[42] Heikens E, Singh KV, Jacques-Palaz KD, van Luit-Asbroek M, Oostdijk EA, Bonten MJ, et al. Contribution of the enterococcal surface protein Esp to pathogenesis of *Enterococcus faecium* endocarditis. Microbes and Infection. 2011;**13**(14-15):1185-1190

[43] Kemp KD, Singh KV, Nallapareddy SR, Murray BE. Relative contributions of *Enterococcus faecalis* OG1RF sortase-encoding genes, srtA and bps (srtC), to biofilm formation and a murine model of urinary tract infection. Infection and Immunity. 2007;**75**(11): 5399-5404

[44] Nallapareddy SR, Singh KV, Sillanpää J, Zhao M, Murray BE. Relative contributions of Ebp Pili and the collagen adhesin ace to host extracellular matrix protein adherence and experimental urinary tract infection by *Enterococcus faecalis* OG1RF. Infection and Immunity. 2011;**79**(7):2901-2910

[45] Munita JM, Arias CA, Murray BE. Enterococcal endocarditis: Can we win the war? Current Infectious Disease Reports. 2012;**14**(4):339-349

[46] Chajęcka-Wierzchowska W, Zadernowska A, Łaniewska-Trokenheim Ł. Virulence factors of Enterococcus spp. presented in food. LWT. 2017;**75**:670-676

[47] Mohamed JA, Huang DB. Biofilm formation by enterococci. Journal of Medical Microbiology. 2007;**56**(12):1581-1588

[48] Singh KV, Qin X, Weinstock GM, Murray BE. Generation and testing of mutants of *Enterococcus faecalis* in a mouse peritonitis model. The Journal of Infectious Diseases. 1998;**178**(5):1416-1420

[49] Singh KV, Nallapareddy SR, Nannini EC, Murray BE. Fsr-independent production of protease (s) may explain the lack of attenuation of an *Enterococcus faecalis* fsr mutant versus a gelE-sprE mutant in induction of endocarditis. Infection and Immunity. 2005;**73**(8):4888-4894

[50] Engelbert M, Mylonakis E, Ausubel FM, Calderwood SB, Gilmore MS. Contribution of gelatinase, serine protease, and fsr to the pathogenesis of *Enterococcus faecalis* endophthalmitis. Infection and Immunity. 2004;**72**(6): 3628-3633

[51] Sifri CD, Mylonakis E, Singh KV, Qin X, Garsin DA, Murray BE, et al. Virulence effect of *Enterococcus faecalis* protease genes and the quorum-sensing locus fsr in *Caenorhabditis elegans* and mice. Infection and Immunity. 2002;**70**(10):5647-5650

[52] Zeng J, Teng F, Murray BE. Gelatinase is important for translocation of *Enterococcus faecalis* across polarized human enterocyte-like T84 cells. Infection and Immunity. 2005;**73**(3):1606-1612

[53] Qin X, Singh KV, Weinstock GM, Murray BE. Characterization of fsr, a regulator controlling expression of gelatinase and serine protease in *Enterococcus faecalis* OG1RF. Journal of Bacteriology. 2001;**183**(11):3372-3382

[54] Qin X, Singh KV, Weinstock GM, Murray BE. Effects of *Enterococcus faecalis* fsr genes on production of gelatinase and a serine protease and

virulence. Infection and Immunity. 2000;**68**(5):2579-2586

[55] Nakayama J, Cao Y, Horii T, Sakuda S, Akkermans AD, De Vos WM, et al. Gelatinase biosynthesis-activating pheromone: A peptide lactone that mediates a quorum sensing in *Enterococcus faecalis*. Molecular Microbiology. 2001;**41**(1):145-154

[56] Sava IG, Heikens E, Huebner J. Pathogenesis and immunity in enterococcal infections. Clinical Microbiology and Infection. 2010;**16**(6):533-540

[57] Coque TM, Patterson JE, Steckelberg JM, Murray BE. Incidence of hemolysin, gelatinase, and aggregation substance among enterococci isolated from patients with endocarditis and other infections and from feces of hospitalized and community-based persons. The Journal of Infectious Diseases. 1995;**171**(5):1223-1229

[58] Garsin DA, Sifri CD, Mylonakis E, Qin X, Singh KV, Murray BE, et al. A simple model host for identifying gram-positive virulence factors. Proceedings of the National Academy of Sciences USA. 2001;**98**(19):10892-10897

[59] Håkansson KO, Brugna M, Tasse L. The three-dimensional structure of catalase from *Enterococcus faecalis*. Acta Crystallographica, Section D: Biological Crystallography. 2004;**60**(8):1374-1380

[60] Zamocky M, Furtmüller PG, Obinger C. Evolution of catalases from bacteria to humans. Antioxidants & Redox Signaling. 2008;**10**(9):1527-1548

[61] Frankenberg L, Brugna M, Hederstedt L. *Enterococcus faecalis* heme-dependent catalase. Journal of Bacteriology. 2002;**184**(22):6351-6356

[62] Baureder M, Hederstedt L. Genes important for catalase activity in *Enterococcus faecalis*. PLoS One. 2012;**7**(5):e36725

[63] Hynes WL, Walton SL. Hyaluronidases of gram-positive bacteria. FEMS Microbiology Letters. 2000;**183**(2):201-207

[64] Arias CA, Panesso D, Singh KV, Rice LB, Murray BE. Cotransfer of antibiotic resistance genes and a hylEfm-containing virulence plasmid in *Enterococcus faecium*. Antimicrobial Agents and Chemotherapy. 2009;**53**(10):4240-4246

[65] Rice LB, Laktičova V, Carias LL, Rudin S, Hutton R, Marshall SH. Transferable capacity for gastrointestinal colonization in *Enterococcus faecium* in a mouse model. The Journal of Infectious Diseases. 2009;**199**(3):342-349

[66] Garsin DA, Frank KL, Silanpää J, Ausubel FM, Hartke A, Shankar N, et al. Pathogenesis and models of enterococcal infection. In: Enterococci: From Commensals to Leading Causes of Drug Resistant Infection [Internet]. Massachusetts Eye and Ear Infirmary; 2014

[67] Panesso D, Montealegre MC, Rincón S, Mojica MF, Rice LB, Singh KV, et al. The hyl Efm gene in pHyl Efm of *Enterococcus faecium* is not required in pathogenesis of murine peritonitis. BMC Microbiology. 2011;**11**(1):20

[68] Van Tyne D, Martin M, Gilmore M. Structure, function, and biology of the *Enterococcus faecalis* cytolysin. Toxins. 2013;**5**(5):895-911

[69] Shankar N, Baghdayan AS, Gilmore MS. Modulation of virulence within a pathogenicity island in vancomycin-resistant *Enterococcus faecalis*. Nature. 2002;**417**(6890):746

[70] Haas W, Shepard BD, Gilmore MS. Two-component regulator of

Enterococcus faecalis cytolysin responds to quorum-sensing autoinduction. Nature. 2002;**415**(6867):84

[71] Tendolkar PM, Baghdayan AS, Shankar N. Pathogenic enterococci: New developments in the 21st century. Cellular and Molecular Life Sciences: CMLS. 2003;**60**(12):2622-2636

[72] Clewell DB. Bacterial sex pheromone-induced plasmid transfer. Cell. 1993;**73**:9-12

[73] Ike YA, Clewell DB. Evidence that the hemolysin/bacteriocin phenotype of *Enterococcus faecalis* subsp. zymogenes can be determined by plasmids in different incompatibility groups as well as by the chromosome. Journal of Bacteriology. 1992;**174**(24):8172-8177

[74] Süßmuth SD, Muscholl-Silberhorn A, Wirth R, Susa M, Marre R, Rozdzinski E. Aggregation substance promotes adherence, phagocytosis, and intracellular survival of *Enterococcus faecalis* within human macrophages and suppresses respiratory burst. Infection and Immunity. 2000;**68**(9):4900-4906

[75] Todd EW. A comparative serological study of streptolysins derived from human and from animal infections, with notes on pneumococcal haemolysin, tetanolysin and staphylococcus toxin. The Journal of Pathology and Bacteriology. 1934;**39**:299-321

[76] Ike Y, Hashimoto H, Clewell DB. Hemolysin of *Streptococcus faecalis* subspecies zymogenes contributes to virulence in mice. Infection and Immunity. 1984;**45**(2):528-530

[77] Jett BD, Jensen HG, Nordquist RE, Gilmore MS. Contribution of the pAD1-encoded cytolysin to the severity of experimental *Enterococcus faecalis* endophthalmitis. Infection and Immunity. 1992;**60**(6):2445-2452

[78] Miyazaki S, Ohno A, Kobayashi I, Uji T, Yamaguchi K, Goto S. Cytotoxic effect of hemolytic culture supernatant from *Enterococcus faecalis* on mouse polymorphonuclear neutrophils and macrophages. Microbiology and Immunology. 1993;**37**(4):265-270

[79] Schlievert PM, Gahr PJ, Assimacopoulos AP, Dinges MM, Stoehr JA, Harmala JW, Hirt H, Dunny GM. Aggregation and binding substances enhance pathogenicity in rabbit models of Enterococcus faecalis endocarditis. Infection and Immunity. Jan 1, 1998;**66**(1):218-223

[80] Stevens SX, Jensen HG, Jett BD, Gilmore MS. A hemolysin-encoding plasmid contributes to bacterial virulence in experimental *Enterococcus faecalis* endophthalmitis. Investigative Ophthalmology & Visual Science. 1992;**33**(5):1650-1656

[81] Stein T, Heinzmann S, Solovieva I, Entian KD. Function of *Lactococcus lactis* nisin immunity genes nisI and nisFEG after coordinated expression in the surrogate host *Bacillus subtilis*. The Journal of Biological Chemistry. 2003;**278**(1):89-94

[82] Stein T, Heinzmann S, Düsterhus S, Borchert S, Entian KD. Expression and functional analysis of the subtilin immunity genes spaIFEG in the subtilin-sensitive host *Bacillus subtilis* MO1099. Journal of Bacteriology. 2005;**187**(3):822-828

[83] Coburn PS, Hancock LE, Booth MC, Gilmore MS. A novel means of self-protection, unrelated to toxin activation, confers immunity to the bactericidal effects of the *Enterococcus faecalis* cytolysin. Infection and Immunity. 1999;**67**(7):3339-3347

[84] Bassler BL, Losick R. Bacterially speaking. Cell. 2006;**125**(2):237-246

[85] Roux A, Payne SM, Gilmore MS. Microbial telesensing: Probing the

environment for friends, foes, and food. Cell Host & Microbe. 2009;**6**(2):115-124

[86] Wästfelt M, Stålhammar-Carlemalm M, Delisse AM, Cabezon T, Lindahl G. Identification of a family of streptococcal surface proteins with extremely repetitive structure. The Journal of Biological Chemistry. 1996;**271**(31):18892-18897

[87] Cucarella C, Solano C, Valle J, Amorena B, Lasa Í, Penadés JR. Bap, a Staphylococcus aureus surface protein involved in biofilm formation. Journal of Bacteriology. 2001;**183**(9):2888-2896

[88] Keane PF, Bonner MC, Johnston SR, Zafar A, Gorman SP. Characterization of biofilm and encrustation on ureteric stents in vivo. British Journal of Urology. 1994;**73**(6):687-691

[89] Sandoe JA, Witherden IR, Cove JH, Heritage J, Wilcox MH. Correlation between enterococcal biofilm formation in vitro and medical-device-related infection potential in vivo. Journal of Medical Microbiology. 2003;**52**(7):547-550

[90] Dautle MP, Wilkinson TR, Gauderer MW. Isolation and identification of biofilm microorganisms from silicone gastrostomy devices. Journal of Pediatric Surgery. 2003;**38**(2):216-220

[91] Dowidar N, Moesgaard F, Matzen P. Clogging and other complications of endoscopic biliary endoprostheses. Scandinavian Journal of Gastroenterology. 1991;**26**(11):1132-1136

[92] Eaton TJ, Gasson MJ. Molecular screening of *Enterococcus* virulence determinants and potential for genetic exchange between food and medical isolates. Applied and Environmental Microbiology. 2001;**67**(4):1628-1635

[93] Willems RJ, Homan W, Top J, van Santen-Verheuvel M, Tribe D, Manzioros X, et al. Variant esp gene as a marker of a distinct genetic lineage of vancomycin-resistant *Enterococcus faecium* spreading in hospitals. The Lancet. 2001;**357**(9259):853-855

[94] Leavis H, Top J, Shankar N, Borgen K, Bonten M, van Embden J, et al. A novel putative enterococcal pathogenicity island linked to the esp virulence gene of *Enterococcus faecium* and associated with epidemicity. Journal of Bacteriology. 2004;**186**(3): 672-682

[95] Leendertse M, Heikens E, Wijnands LM, van Luit-Asbroek M, Teske GJ, Roelofs JJ, et al. Enterococcal surface protein transiently aggravates *Enterococcus faecium*–induced urinary tract infection in mice. The Journal of Infectious Diseases. 2009;**200**(7):1162-1165

[96] Heikens E, Leendertse M, Wijnands LM, van Luit-Asbroek M, Bonten MJ, van der Poll T, et al. Enterococcal surface protein Esp is not essential for cell adhesion and intestinal colonization of *Enterococcus faecium* in mice. BMC Microbiology. 2009;**9**(1):19

[97] Tendolkar PM, Baghdayan AS, Gilmore MS, Shankar N. Enterococcal surface protein, Esp, enhances biofilm formation by *Enterococcus faecalis*. Infection and Immunity. 2004;**72**(10):6032-6039

[98] Dworniczek E, Wojciech L, Sobieszczanska B, Seniuk A. Virulence of enterococcus isolates collected in lower Silesia (Poland). Scandinavian Journal of Infectious Diseases. 2005;**37**(9):630-636

[99] Ramadhan AA, Hegedus E. Biofilm formation and esp gene carriage in enterococci. Journal of Clinical Pathology. 2005;**58**(7):685-686

[100] Kristich CJ, Li YH, Cvitkovitch DG, Dunny GM. Esp-independent biofilm formation by *Enterococcus faecalis*. Journal of Bacteriology. 2004;**186**(1):154-163

[101] Mohamed JA, Huang W, Nallapareddy SR, Teng F, Murray BE. Influence of origin of isolates, especially endocarditis isolates, and various genes on biofilm formation by *Enterococcus faecalis*. Infection and Immunity. 2004;**72**(6):3658-3663

[102] Kreft Á, Marre R, Schramm U, Wirth R. Aggregation substance of *Enterococcus faecalis* mediates adhesion to cultured renal tubular cells. Infection and Immunity. 1992;**60**(1):25-30

[103] Vanek NN, Simon SI, Jacques-Palaz K, Mariscalco MM, Dunny GM, Rakita RM. *Enterococcus faecalis* aggregation substance promotes opsonin independent binding to human neutrophils via a complement receptor type 3-mediated mechanism. FEMS Immunology and Medical Microbiology. 1999;**26**(1):49-60

[104] Waters CM, Hirt H, McCormick JK, Schlievert PM, Wells CL, Dunny GM. An amino-terminal domain of *Enterococcus faecalis* aggregation substance is required for aggregation, bacterial internalization by epithelial cells and binding to lipoteichoic acid. Molecular Microbiology. 2004;**52**(4):1159-1171

[105] Rakita RM, Vanek NN, Jacques-Palaz K, Mee M, Mariscalco MM, Dunny GM, et al. *Enterococcus faecalis* bearing aggregation substance is resistant to killing by human neutrophils despite phagocytosis and neutrophil activation. Infection and Immunity. 1999;**67**(11):6067-6075

[106] Schlievert PM, Gahr PJ, Assimacopoulos AP, Dinges MM, Stoehr JA, Harmala JW, et al. Aggregation and binding substances enhance pathogenicity in rabbit models of *Enterococcus faecalis* endocarditis. Infection and Immunity. 1998;**66**(1):218-223

[107] Ali SA, Hasan KA, Bin Asif H, Abbasi A. Environmental enterococci: I. prevalence of virulence, antibiotic resistance and species distribution in poultry and its related environment in Karachi, Pakistan. Letters in Applied Microbiology. 2014;**58**(5):423-432

[108] Ali SA, Bin-Asif H, Hasan KA, Rehman M, Abbasi A. Molecular assessment of virulence determinants, hospital associated marker (IS16gene) and prevalence of antibiotic resistance in soil borne *Enterococcus* species. Microbial Pathogenesis. 2017;**105**:298-306

[109] Zahid S, Bin-Asif H, Hasan KA, Rehman M, Ali SA. Prevalence and genetic profiling of tetracycline resistance (Tet-R) genes and transposable element (Tn916) in environmental *Enterococcus* species. Microbial Pathogenesis. 2017;**111**:252-261

[110] Hasan KA, Ali SA, Rehman M, Bin-Asif H, Zahid S. The unravelled *Enterococcus faecalis* zoonotic superbugs: Emerging multiple resistant and virulent lineages isolated from poultry environment. Zoonoses and Public Health. 2018;**65**(8):921-935

[111] Werner G, Fleige C, Geringer U, van Schaik W, Klare I, Witte W. IS element IS16 as a molecular screening tool to identify hospital-associated strains of *Enterococcus faecium*. BMC Infectious Diseases. 2011;**11**(1):80

[112] Lanthier M, Scott A, Lapen DR, Zhang Y, Topp E. Frequency of virulence genes and antibiotic resistances in *Enterococcus* spp.

isolates from wastewater and feces of domesticated mammals and birds, and wildlife. Canadian Journal of Microbiology. 2010;**56**(9):715-729

[113] Camargo IL, Gilmore MS, Darini AL. Multilocus sequence typing and analysis of putative virulence factors in vancomycin-resistant and vancomycin-sensitive *Enterococcus faecium* isolates from Brazil. Clinical Microbiology and Infection. 2006;**12**(11):1123-1130

[114] Biendo M, Adjide C, Castelain S, Belmekki M, Rousseau F, Slama M, et al. International Journal of Microbiology. 2010;**2010**:150464

[115] Duprè I, Zanetti S, Schito AM, Fadda G, Sechi LA. Incidence of virulence determinants in clinical *Enterococcus faecium* and *Enterococcus faecalis* isolates collected in Sardinia (Italy). Journal of Medical Microbiology. 2003;**52**(6):491-498

[116] Lopes MD, Simões AP, Tenreiro R, Marques JJ, Crespo MT. Activity and expression of a virulence factor, gelatinase, in dairy enterococci. International Journal of Food Microbiology. 2006;**112**(3):208-214

[117] Semedo T, Santos MA, Martins P, Lopes MF, Marques JJ, Tenreiro R, et al. Comparative study using type strains and clinical and food isolates to examine hemolytic activity and occurrence of the cyl operon in enterococci. Journal of Clinical Microbiology. 2003;**41**(6):2569-2576

[118] Almeida T, Brandão A, Muñoz-Atienza E, Goncalves A, Torres C, Igrejas G, et al. Identification of bacteriocin genes in enterococci isolated from game animals and saltwater fish. Journal of Food Protection. 2011;**74**(8):1252-1260

[119] Iweriebor BC, Gaqavu S, Obi LC, Nwodo UU, Okoh AI. Antibiotic susceptibilities of *Enterococcus* species isolated from hospital and domestic wastewater effluents in Alice, Eastern Cape Province of South Africa. International Journal of Environmental Research and Public Health. 2015;**12**(4):4231-4246

Chapter 7

Aspects of Photodynamic Inactivation of Bacteria

Faina Nakonechny and Marina Nisnevitch

Abstract

Increasing resistance of bacteria to antibiotics is a serious worldwide problem, and to combat resistant bacteria, new antibacterial approaches are to be developed. One alternative to traditional antibiotic therapy is photodynamic antimicrobial chemotherapy (PACT). PACT is based on excitation of photosensitizers (PS) capable of transferring the absorbed light energy to dissolved molecular oxygen causing generation of reactive oxygen species, which irreversibly damage bacterial cell components. The overall efficiency of PACT has been proven for Gram-positive and Gram-negative bacteria. The effectiveness of PACT can be increased by encapsulation of PS in liposomes providing more concentrated delivery of PS, enhanced cytotoxicity, improved pharmacokinetic properties, sustained release, and prolonged action of the PS. For continuous and reusable application, PS can be immobilized in polymers. Chemiluminescence, sonodynamic treatment, and radiofrequency irradiation allow to perform excitation of PS in the dark without external illumination, opening prospects for combating internal infections. Combination of PS with antibiotics can gain a synergistic effect, allowing in some cases to overcome the resistance of bacteria to antibiotics.

Keywords: photodynamic therapy (PDT), photodynamic antimicrobial chemotherapy (PACT), photosensitizer (PS), chemiluminescent antimicrobial chemotherapy (CPAT), sonodynamic antimicrobial chemotherapy (SACT), targeted drug delivery, liposomes, immobilization

1. Introduction

1.1 History of photodynamic therapy

The therapeutic properties of light were observed already in ancient Greece, Egypt, and India. However, they were not widely used for many centuries [1]. The history of modern photodynamic therapy (PDT) dates back to 1900, when Oscar Raab discovered the toxic properties of the dye acridine red on *Paramecium* spp. [2]. He and his supervisor, Hermann von Tappeiner, noticed a positive effect of illumination on the toxic activity of this dye. In his later work, von Tappeiner and his colleagues applied this approach to inactivation of bacteria [3] and to treatment of skin cancer [4]. In 1909, von Tappeiner introduced the term "Photodynamic Action" and showed that oxygen is essential for this procedure [5]. PDT has been studied and developed as an anticancer therapy for a long time and was approved by the Food and Drug Administration in the 1990s for various applications in this area of medicine [6–8]. The antimicrobial properties of this approach were unfairly forgotten for several decades. However, interest in antibacterial PDT has been rekindled and is continuously increasing because multidrug resistance of pathogenic microorganisms has become a serious threat to public health. Photodynamic antibacterial chemotherapy (PACT) has become a promising approach for combating bacterial infections, which are resistant to modern antibiotics.

1.2 Photosensitizers and their mechanism of action

PACT is based on the exposure of bacteria to photosensitive compounds—photosensitizers (PSs). When a PS located in the bacteria or on the bacterial surface is exposed to light (usually visible), it transfers from its low-energy ground state to an excited singlet state. Return of the PS to its ground state is accompanied by either emission of fluorescence or transition of the PS to a longer-living, higher-energy triplet state (PS*) via intersystem crossing. The PS* in turn reacts with surrounding molecules to form free radicals and hydrogen peroxide (Type I reaction) or transfers its energy to molecular oxygen to produce singlet oxygen and other highly reactive oxygen species (ROS; Type II reaction) [9, 10]. Type I and Type II reactions occur simultaneously, and the ratio at which they occur depends on both the PS type and the surrounding conditions. A detailed description of the photosensitization process can be found in the recent reviews of Castano et al. [11] and Cieplik [10]. ROSs formed in this process oxidize biomolecules, damage the cell membrane, and ultimately lead to cell death [12]. PACT usually proceeds predominantly through Type II processes. However, since Gram-negative bacteria are more susceptible to $OH^{\cdot}$ radicals than to singlet oxygen, the Type I reaction may be more efficient against such microorganisms [13, 14].

1.3 Photosensitizers for PACT

Hundreds of compounds are currently available for mediating PDT in various areas of medicine, where some have been shown to be suitable for antimicrobial applications. PSs employed for medical uses should be a single pure compound, stable at room temperature and inexpensive. The PS must have a strong absorption peak in the visible spectrum between 600 and 900 nm and should possess a high-triplet quantum yield that will provide high production of ROS upon illumination. It should not be toxic in the dark (especially to mammalian cells), mutagenic or carcinogenic [15–18]. In addition, when talking about PACT, it is very important that the PS will display preferential association with bacteria, accumulate within the cells, or bind to the bacterial cell envelope [14, 19].

PSs can generally be assigned to several chemical classes: tetrapyrroles (which include porphyrins, chlorins, bacteriochlorins, and phthalocyanines), synthetic dyes (phenothiazinium salts, Rose Bengal, squaraines, etc.), and naturally occurring compounds (such as riboflavin or curcumin). Cyclic tetrapyrroles present the most well-known class of clinically relevant PSs used mostly for anticancer applications [20]. This structure can be found naturally in such important biomolecules such as haem, chlorophyll, and bacteriochlorophyll. Unlike other types of PSs, most tetrapyrroles (except for bacteriochlorins) are more likely to react by a Type II reaction with the creation of singlet oxygen [16], whereas bacteriochlorins act *via* a Type I mechanism. Other well-known antimicrobial agents are phenothiazinium-based synthetic dyes, including methylene blue (MB) and toluidine blue O (TBO), which also act as anticancer agents in PDT. These structures can be synthesized more easily than tetrapyrroles but possess high-dark toxicity compared to other PSs [15, 21]. Another representative of synthetic dyes, Rose Bengal (RB), has already been used successfully in antimicrobial and anticancer applications for a long time [16]. Photodynamic active compounds isolated from plants arouse particular interest. These natural compounds include curcumin, extracted from the rhizomes of *Curcuma longa*, which was found effective in eradicating oral pathogens [22]. Another representative of this group is hypericin isolated from St. John's wort, which exhibits photodynamic activity against Gram-positive and Gram-negative bacteria. Detailed descriptions of all PS classes can be found in the reviews published by Hamblin and colleagues [15, 16].

2. Photosensitizer activation modes

2.1 Dark activity

The name photosensitizer implies the need for illumination in order to activate PS molecules and trigger their action. However, PSs possess some so-called "dark activity" even in the absence of illumination, leading to cell death in the dark [23–29]. This feature depends on the PS concentration and manifests itself in different ways for various PSs.

Shrestha demonstrated dark toxicity of RB against Gram-positive *Enterococcus faecalis*. Exposure of the cells to 10 μM RB in the absence of illumination for 15 min led to a 0.5 $\log_{10}$ reduction in cell concentration [26]. Furthermore, a marked dark toxicity of RB against clinical isolates of Gram-negative *Pseudomonas aeruginosa* was observed by Nakonieczna [27]. Brovko compared the activity of various PSs against several types of microorganisms and noted high dark toxicity of RB, as well as of phloxine B against Gram-positive *Bacillus sp*. and *Listeria monocytogenes* (more than 5 $\log_{10}$ reduction in the bacterial concentration after 30 min of treatment with the dye) [30]. The toxicity of malachite green in the dark against the same microorganisms was very low (<0.1 $\log_{10}$ reduction in concentration after 30 min of treatment with the dye). High concentrations (>500 μg/mL) of acriflavin neutral in the absence of light were significantly toxic to *E. coli* (more than 6 $\log_{10}$ reduction in concentration after 30 min of treatment with the dye, both under illumination and

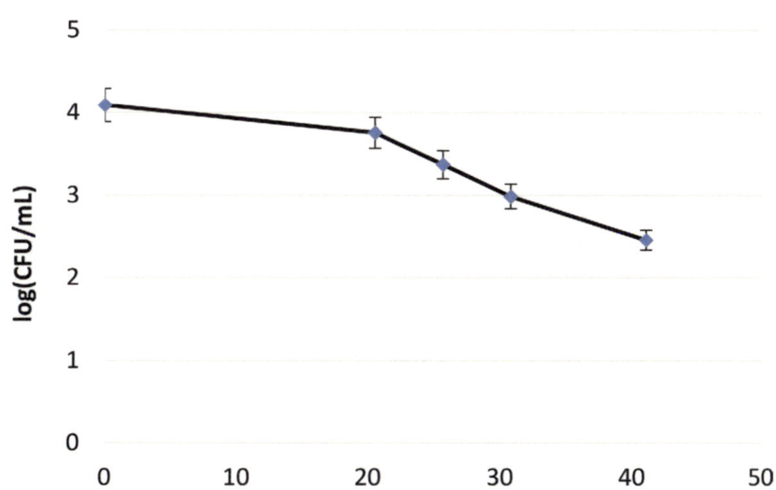

Figure 1.
Effect of RB concentration on its cytotoxic activity. S. aureus cells at the initial concentration of 104 CFU mL−1 were incubated for 3 min in dark conditions at various concentrations of RB. After the incubation, bacteria were tested by viable count. Error bars present standard deviations.

in the dark). However, illumination significantly enhanced its toxic effect against other tested microorganisms [30].

In our studies, we also noted the dark toxicity of various PSs against different types of bacteria (**Figures 1, 2, Table 1**). **Figure 1** shows the effect of various RB concentrations on *S. aureus* in the absence of light. The number of living cells decreases with increasing RB concentration in the dark. **Table 1** shows a comparison between dark and light toxicity of three PSs—malachite green oxalate (MGO), RB, and safranin O. The effect of MGO in the dark was the strongest, and a 0.87 μM concentration of MGO was sufficient for inhibiting the growth of *S. aureus*. The dark activity of RB and safranin O is noticeably weaker, and the minimal inhibitory

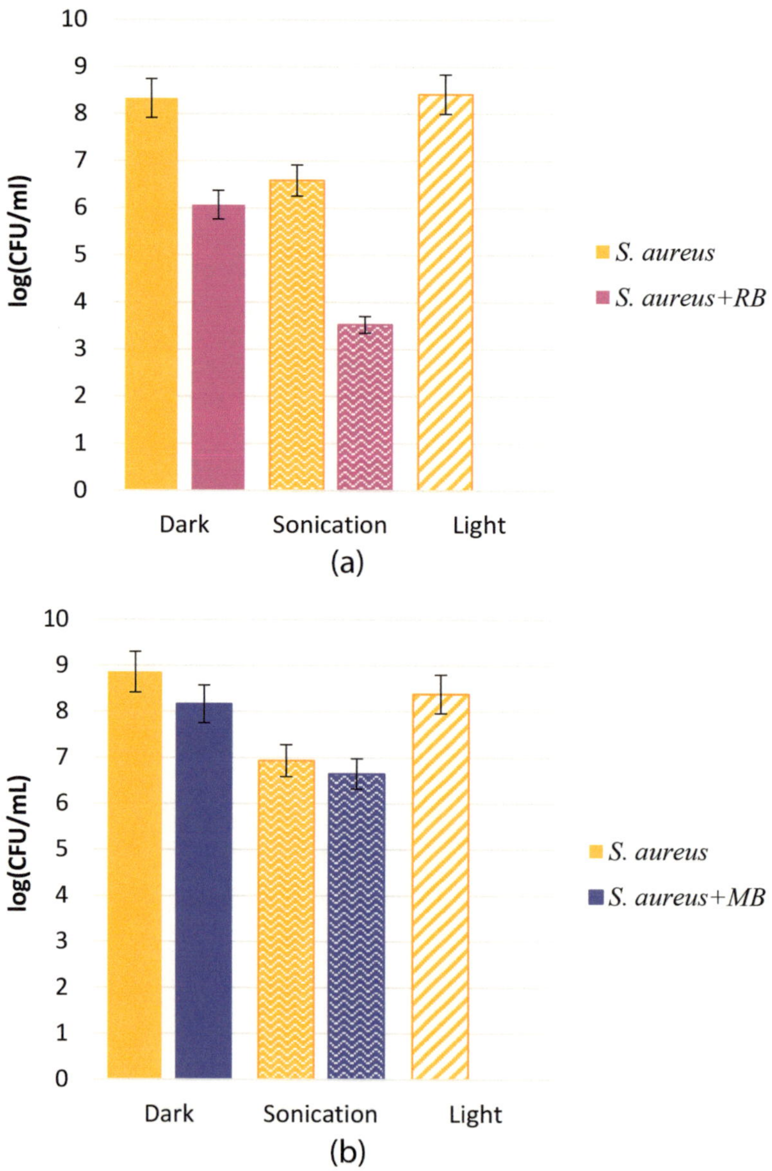

Figure 2.
SACT and PACT effect of MB on S. aureus. In SACT experiments, the cells at 108 CFU mL^{-1} concentration were incubated with (a) 5 μM RB or (b) 30 μM MB in the ultrasonic bath for 1 h in the dark. In PACT experiments, the cells were illuminated for 15 min by 1.6 mW cm^{-2} white light under the same conditions but without sonication. After the treatment, bacteria were tested by viable count. Error bars present standard deviations.

Photosensitizer	MIC, μM	
	Dark	Illumination
Malachite green oxalate	0.87	0.15
Rose Bengal	128	2
Safranin O	89	23

Table 1.
The MIC values of water-soluble PSs in the dark and under illumination. About 3 × 104 CFU mL^{-1} *of* S. aureus *were treated by malachite green oxalate, Rose Bengal, and Safranin O at doubled dilutions, illuminated at room temperature by white light of 1.6 mW cm*$^{-2}$ *intensity for 1 h, and incubated overnight in the dark by shaking at 37°C.*

concentrations (MIC) for these PSs against *S. aureus* are more than 100-fold higher. **Figure 2** shows that *S. aureus* cells were completely destroyed by RB at a concentration of 5 μM and MB at 30 μM under illumination. These PSs also showed a cytotoxic effect when applied at the same concentrations in the dark, where MB reduced the bacterial concentration by one and RB by two orders of magnitude.

2.2 Illumination

Although PSs are known to possess a certain dark activity, illumination noticeably increases their cytotoxic effect [6, 14]. An example of the difference in antibacterial activity of different PSs with and without illumination is shown in **Table 1**. In this experiment, the MIC of three PSs was determined for the bacterium *S. aureus* in the dark and after 1 h of illumination. As a result of illumination, the MIC of the examined PSs decreased approximately 6-fold for MGO, 64-fold for RB, and 4-fold for Safranin O.

The main light sources used today for activation of PSs are lasers, light-emitting diodes (LED), and gas discharge lamps (GDL) [10, 31, 32]. There is no absolute advantage of one of these light sources over the others. The choice of light source depends on the specific application. Laser is a high-intensity monochromatic source. It can be easily coupled to a single optical fiber and installed on different lighting devices. LED lamps are cheaper and provide a wide emission spectrum. GDLs are also cheaper than lasers—both in acquisition and in maintenance and have a wide emission spectrum. However, GDLs transmit more heat to the illuminated area than lasers and LEDs, which can lead to tissue damage. In general, the emission spectrum and light intensity are more important for the excitation of a specific PS than the particular light source type [10, 31, 32].

2.3 Sonodynamic excitation of photosensitizers

Illumination is undoubtedly the easiest and most effective way to activate PSs. However, its use is restricted, due to limited penetration of visible light into tissues. There is an ongoing search for alternative methods of PS excitation in the dark in order to overcome this problem. Ultrasonic activation seems to be attractive as an alternative to illumination. As with light activation, ultrasound can be selectively focused on a specific area, thus activating only PS molecules located in the affected area. Ultrasound can also easily penetrate into tissues, which opens prospects for its application in treatment of internal lesions and infections, without the need for invasive devices [33, 34]. Ultrasonic irradiation of PSs initiates the formation of highly active cytotoxic species—ROS and free radicals—which lead to the death of pathogenic cells. It was found that some well-known PSs also have sonosensitizing properties. Among them are porphyrins [35], RB [36, 37], chlorin e6 derivative, photodithazine [36], and curcumin [38]. Several studies found sonodynamic therapy (SDT) to be the promising treatment in various forms of cancerous tumors [39–43]. Sonodynamic therapy is also offered as treatment for atherosclerosis [44]. The applicability of sonodynamic antimicrobial chemotherapy (SACT) for the treatment of infectious diseases has been confirmed by various research groups [33, 34]. We have previously demonstrated the effectiveness of RB activated by ultrasonication for eradication of Gram-positive *S. aureus* and Gram-negative *E. coli* [29, 45, 46]. The effectiveness of SACT in inactivation of *S. aureus* by two other sensitizers—curcumin [38] and hematoporphyrin monomethyl ether [35]—was also reported. Alves et al. have recently reported on effective destruction of *Candida albicans* by photodithazine and RB in the dark under the ultrasonic excitation. A significant synergistic effect of the combination between PDT and SACT for combatting *C. albicans* biofilms was also found [36].

Figure 2 demonstrates the effect of ultrasonic activation that we showed on the antibacterial activity of two PSs—RB (**Figure 2a**) and MB (**Figure 2b**)—against *S. aureus* compared to photodynamic activation. **Figure 2a** shows that 15 min of sonication reduces the number of living cells by almost two orders of magnitude, from 2×10^8 to 4×10^6 CFU mL^{-1}. RB alone applied in the dark causes a two orders of magnitude decrease in the cell concentration. However, sonication in the presence of 5 μM RB exerts a much stronger effect, reducing the cell concentration by 5 orders of magnitude. It should be noted that RB at the same concentration under illumination by visible light of 1.6 mW cm^{-2} fluence causes complete eradication of *S. aureus* cells, whereas light alone does not cause any significant harm to these cells. However, MB applied under sonication at the concentration causing complete destruction of *S. aureus* cells in the light did not eradicate microbial cells more than sonication alone (**Figure 2**).

2.4 Activation of photosensitizers by radio waves

Another possible way for activating PSs in the dark is by using nonionizing radiofrequency electromagnetic waves. The ability of radiofrequency waves to heat human tissue has been known for a long time and has already been applied for local destruction of cancerous tumors [47, 48]. The effectiveness of this method can be significantly improved by using suitable sensitizers, which can be targeted to the affected area and activated by means of radiofrequency radiation for selective destruction of cells. Tamarov et al. proposed the use of crystalline silicon-based nanoparticles as sensitizers induced by 27 MHz radiofrequency waves for effective treatment of Lewis lung carcinoma *in vivo* [48]. Another approach involved using gold nanoparticles, which were heated by an electric field using 13.56 MHz radiofrequency, and effectively destroyed human pancreatic cancer cells *in vitro* [49]. The same frequency was used in other studies to activate fullerene [50] and transferrin [51] and to eradicate cancer tumors *in vitro* and *in vivo*. A possible mechanism of radiosensitization, according to Tamarov et al. [48] and Chung et al. [51], may be thermal activation of sensitizers by hyperthermia, caused by dissipation of electromagnetic energy, which leads to thermal damage of cancer cells.

In our studies, we tested the possibility of using radiofrequency radiation to sensitize PSs in order to destroy microorganisms [29]. For this purpose, we irradiated *S. aureus* cells in physiological saline alone and in the presence of RB with radio waves at different frequencies—from 1 to 20 GHz. *S. aureus* cells in physiological saline in the dark (without RB and without radiation), *S. aureus* cells treated with radio waves (in the absence of RB), and *S. aureus* cells in the presence of RB, but not exposed to radio waves, were used as controls. Radiofrequency radiation alone did not significantly affect the survival of *S. aureus*. RB in the dark applied at the same

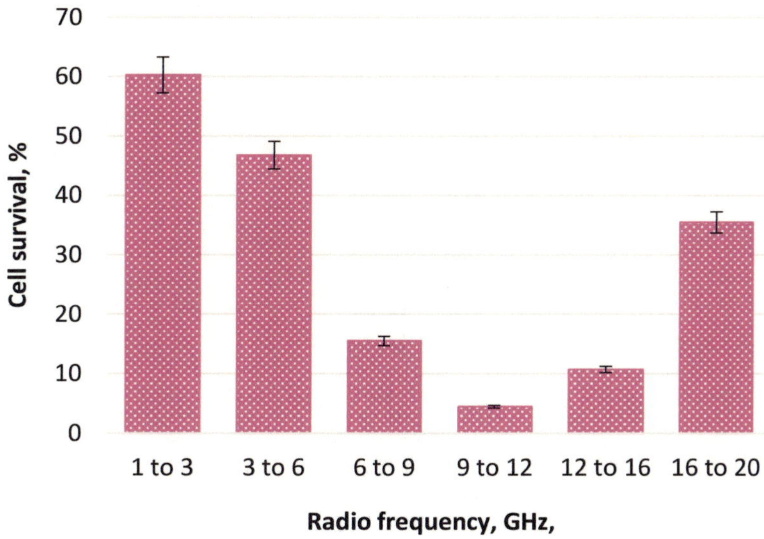

Figure 3.
Effect of RB at the 10 μM concentration under activation by radio waves at various frequency ranges on eradication of S. aureus at the initial cell concentration of 4.4 × 104 CFU mL^{-1} in the dark. Error bars present standard deviations.

concentration did not lead to any decrease in the bacterial concentration. However, exposure of *S. aureus* cells to radio waves in the presence of RB markedly reduced the number of live microorganisms. The rate of cell damage depended on the radio wave frequency. The most significant effect was observed in the frequency range of 9–12 GHz, where in the presence of RB, only 4.5% of the cells survived (**Figure 3**). For comparison, irradiation of cells treated by RB with radio waves in the frequency range of 1–3 GHz caused only a 40% reduction in the number of live cells.

To the best of our knowledge, our work was the first attempt to sensitize a PS by radio waves for destruction of bacteria. This topic naturally necessitates a broader and deeper study to understand the mechanisms of excitation and the possibilities of applying this method. The most likely mechanism of RB excitation by radio waves is conversion of electromagnetic energy into heat, which causes activation of RB, followed by energy transfer to dissolved oxygen and the formation of ROS, affecting the cells. We assume that when PSs are exposed to radiofrequency radiation, they actually behave like thermosensitizers excited by heat instead of light [29].

2.5 Chemiluminescent and bioluminescent excitation of photosensitizers

Another approach to overcoming the limitations of PACT in the treatment of deep infections is to replace the external light source by chemo- or bioluminescent light. Bioluminescence is a well-known phenomenon occurring in biological systems as a result of oxidation reactions of luciferins catalyzed by luciferases. This property is inherent in various microorganisms, worms, and insects, and the luciferins and luciferases of different organisms can be completely different. Bioluminescence is considered as a type of chemiluminescence, i.e., luminescence originating in the course of a chemical reaction. Bio- and chemiluminescence systems are used in various fields of medicine, pharmaceuticals, and bioanalytics [52, 53].

One of the well-studied and most effective chemical reactions involving light emission is oxidation of luminol [52, 54, 55]. Most applications of this reaction are associated with treatment of cancers [55–57]. Use of chemiluminescence as a light source for PACT has not been studied as extensively. Ferraz and colleagues

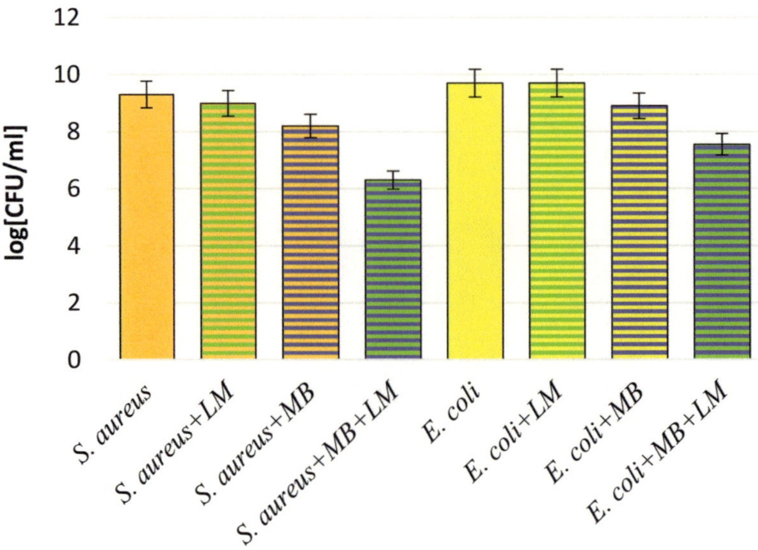

Figure 4.
Effect of chemiluminescent photodynamic antimicrobial treatment (CPAT) on the viability of S. aureus and E. coli. Cells were incubated with MB at 25 µM concentration in the presence of 0.7 mM luminol. After the treatment, bacteria were tested by viable count. Error bars present standard deviations.

evaluated the potential of chemiluminescent-excited photogem in killing *S. aureus* cells [58]. Our group demonstrated the effectiveness of chemiluminescent photodynamic antimicrobial therapy (CPAT) for destruction of *S. aureus* and *E. coli* by exposing these bacteria to the photosensitizer MB in the presence of luminol [46, 59, 60]. The results presented in **Figure 4** show that the rate of growth inhibition by MB increased in the presence of luminol compared to untreated cells or to cells exposed in the dark to MB only.

The dark effect of MB discussed in the above "Dark Activity" section can be seen in **Figure 4**, where the exposure of *S. aureus* and *E. coli* to 25 µM MB in the dark reduced the number of live cells by about 10-fold. Luminol alone had no toxic effect on the tested microorganisms. However, when combined with MB, it reduced the number of surviving bacteria by two additional orders of magnitude for *S. aureus* and 1.5 orders of magnitude in the case of *E. coli*. Thus, the use of chemiluminescence may expand the capabilities of PDT, allowing the use of PSs for the treatment of internal organs.

3. Encapsulation of photosensitizers in liposomes

Since PSs are usually inactive in the absence of excitation, focusing the beam of light, ultrasound or radio wave radiation on the affected area is the easiest way to achieve selective action of a PS. However, surrounding healthy tissues may also be affected by the PS, even under such focused processing. It is therefore very important to target the treatment directly to the infected site. Highly biocompatible and low immunogenic liposomes can serve as carriers for targeted delivery of PSs encapsulated into liposomes to the infected site [61–63].

Liposomes are spherical multi- or unilamellar vesicles consisting of phospholipids (e.g., phosphatidylcholines) with an internal hydrophilic cavity. They vary in composition, size, charge, and number of layers and can encapsulate and deliver both hydrophilic and hydrophobic compounds, which can be retained in the water core of liposomes or be encapsulated in the phospholipid bilayer, respectively.

A variety of methods have been developed for the production of liposomes with a controlled size and special properties. The most widely used method for producing liposomes is hydration of thin lipid films. In this case, lipids with or without active substances are dissolved in an organic solvent, which is evaporated on a rotary evaporator, producing a thin film on a flask wall. The lipid film is then rehydrated by an aqueous phase. Membrane extrusion and sonication methods are most commonly used for control of liposome size [64]. Advanced strategies for liposome preparation include charging the liposomes, attaching the ligands such as antibodies or lectins to their surface, or altering the physiological conditions such as increasing the temperature or changing the pH in the target tissues to produce heat-sensitive or pH-sensitive liposomes [65]. The works of Ghosh, Li, Bulbake, Abu Lila, and Alavi summarize the latest developments in the field of liposome design and optimization, including passive and active targeting, extended circulation, building multifunctional liposomes, and so on [62–66].

There exist several methods for PS encapsulation into liposomes (**Figure 5**). Hydrophilic PSs (e.g., MB, RB, or photofrin) are dissolved in aqueous buffer and are included into the internal cavity of liposomes. Hydrophobic compounds (such as temoporfin and bacteriochlorin a) are integrated in the phospholipid bilayer [62, 67]. Several groups have shown that encapsulation of PSs in liposomes improves their effectiveness against cancer *in vivo*. Back in 1983, Jori and colleagues reported that hematoporphyrin and its derivatives incorporated into liposomes on the basis of dipalmitoyl-phosphatidyl-choline are effective for systemic delivery of PSs to tumors in rats [68]. Enhancement of the photodynamic effects of photofrin encapsulated in a liposome carrier was later demonstrated on a human glioma implanted in rat brain [69]. A variety of PSs (temoporfin, zinc phthalocyanine, benzoporphyrin derivative monoacid, etc.) in various liposomal formulations, such

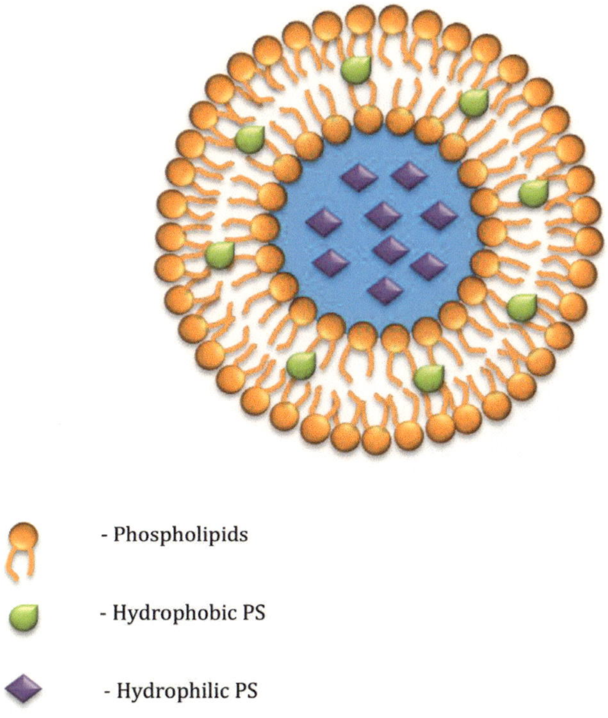

Figure 5.
Schematic representation of a liposome with PS entrapped in the internal aqueous phase and within the external phospholipid bilayer.

as dipalmitoylphosphatidylcholine, dipalmitoylphosphatidylglycerol, dioleoylphosphatidylcholine, and others, were found to be effective on HT29 and Meth A tumor models *in vivo* [62]. However, the only clinically approved liposomal PS drug to date is Visudyne, developed by QLT in Vancouver, and produced by Novartis AG, Switzerland. This formulation is produced from a derivative of benzoporphyrin monoacid encapsulated in unilamellar dimyristoylphosphatidylcholine/egg phosphatidylglycerol liposomes. The liposomes in this drug not only dissolve the

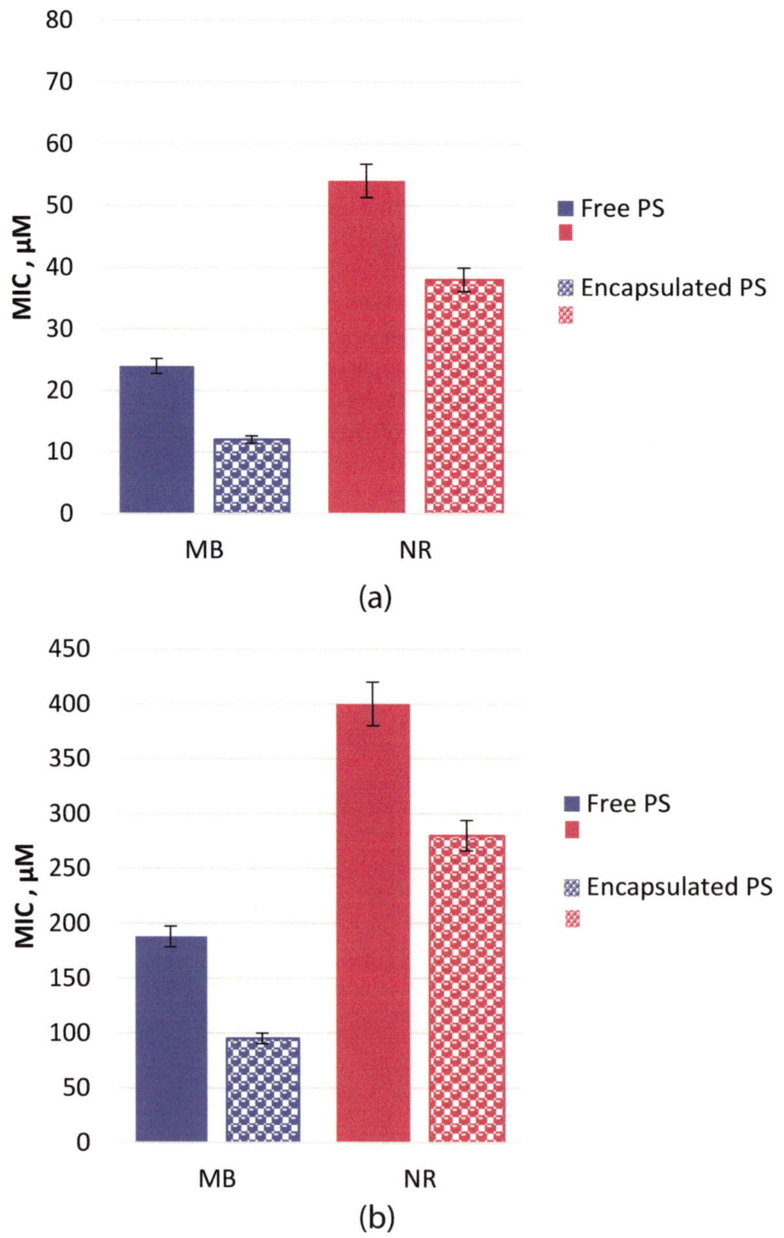

Figure 6.
MIC values of free and liposome encapsulated MB and NR determined against (a) S. aureus and (b) E. coli. Liposomes were prepared from dipalmitoyl phosphatidylcholine and dimyristoyl phosphatidylglycerol at 15 mg/mL total lipid concentration by sonication for 10 sec. Bacteria at 3×104 *CFU* mL^{-1} *concentration were treated by MB and NR at doubled dilutions, illuminated at room temperature by white light of 1.6 mW* cm^{-2} *intensity for 1 h, and incubated overnight in the dark by shaking at 37°C. Error bars present standard deviations.*

lipophilic PS for intravenous administration but also contribute to its enhanced absorption in tumor tissues [62, 64].

Liposomal PS preparations are suitable for antibacterial applications. This approach ensures the delivery of the compound at a higher concentration, thus increasing the cytotoxicity of the drug. In addition, the local use of liposomal preparations provides a slow release of active components, which helps prolong their effect in infected tissues. In Gram-negative bacteria, fusion between liposomes and the outer cell membranes leads to the delivery of concentrated liposome contents directly into the cytoplasm [70–72]. In Gram-positive bacteria, the PS is probably released when liposomes interact with the external peptidoglycan and diffuse through the cell wall [72–74]. Various researchers have demonstrated the effectiveness of liposomal formulations of various PSs against Gram-positive and Gram-negative microorganisms and also against fungal infections *in vitro* and *in vivo*. Ferro et al. showed high efficacy of porphyrin incorporated into cationic liposomes against *S. aureus*, compared to the free drug [75, 76]. Tsai also showed an increase in the bactericidal efficacy of hematoporphyrin against a number of Gram-positive bacteria, including *S. aureus*, as a result of its incorporation into liposomes [77]. Yang proved the efficacy of chlorine e6 encapsulated in cationic liposomes against susceptible and drug-resistant clinical isolates of *C. albicans* both *in vitro* and for infected burn wounds *in vivo* [78].

In our studies, we tested the effect of different PSs in different liposome formulations on Gram-positive and Gram-negative bacteria. **Figure 6** presents a comparison between the MICs of free and dipalmitoyl phosphatidylcholine and dimyristoyl phosphatidylglycerol liposome-encapsulated MB and NR against *S. aureus* (**Figure 6a**) and *E. coli* (**Figure 6b**).

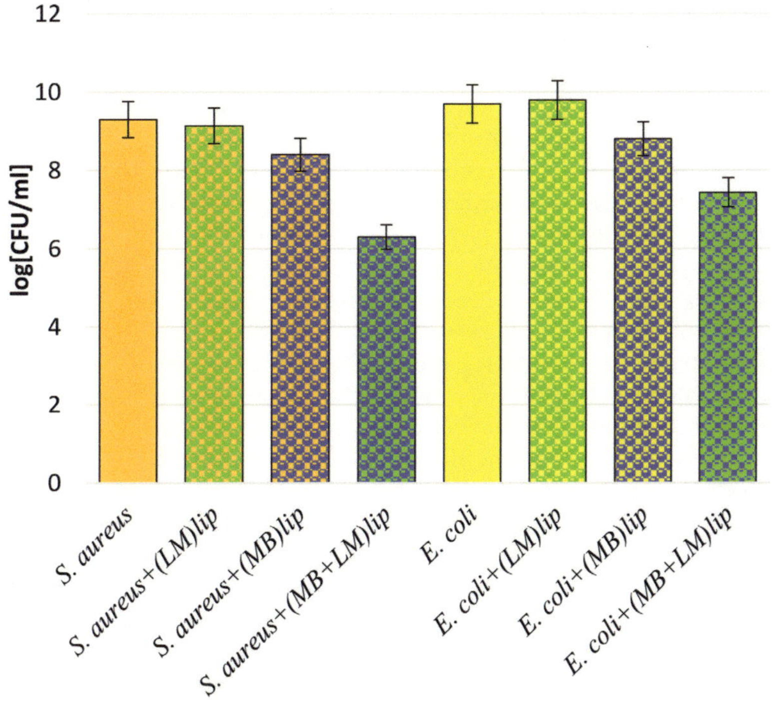

Figure 7.
Chemiluminescent photodynamic antimicrobial treatment effect on the viability of S. aureus and E. coli. Cells were incubated with 25 μM MB liposome (lip) encapsulated together with 0.7 mM luminol (LM). After the treatment, the bacteria were tested by viable count. Error bars present standard deviations.

As can be seen from the results, incorporation into liposomes significantly increased the antibacterial activity of MB and NR. Following encapsulation, the MIC of MB decreased by approximately 2-fold and that of NR by about 1.4-fold for both tested microorganisms (**Figure 6**). We tested the effect of liposome composition on the delivery of these PSs to cells and determined the conditions for efficient use of encapsulated PSs [74].

In addition, we tried to apply liposomal forms of PSs to CPAT by encapsulating not only PSs in liposomes but also luminol and introduced to activate PSs in sites inaccessible to external lighting [59]. We monitored the survival of the cells following their exposure to either liposomal MB or luminol, as well as to liposomes containing both compounds together (**Figure** 7) when the experiments were carried out in the dark.

It can be seen (**Figure** 7) that luminol itself did not lead to cell damage. MB in the liposomal form exhibited certain dark activity, similar to that in a free form discussed in the "Dark Activity" section. The addition of luminol to MB liposomes markedly increased its antibacterial activity toward *S. aureus* and *E. coli*. Liposomes were not targeted in this study. Targeting of liposomes can lead to an additional increase in the efficiency and specificity of this technique.

4. Immobilization

New prospects of using PSs are opened by the immobilization of PSs onto a solid phase. This approach may allow repeated or continuous use of PSs. PSs can be immobilized by adsorption and covalent bonding onto solid supports and by ionic bonding to ion-exchange resins or incorporation into polymer films. The photodynamic properties of immobilized PSs are reported to be retained for a long time [79–83]. PSs studied in the immobilized form include RB, MB, and TBO; the porphyrin derivatives 5,10,15,20-tetrakis (p-hydroxy phenyl) porphyrin, 5,10,15,20-tetrakis (p-aminophenyl) porphyrin, and zinc (II) phthalocyanine tetrasulfonic acid; and the ruthenium salts tris (4,4′-diphenyl-2,2′-bipyridine) ruthenium (II), tris (4,7-diphenyl-1,10-phenanthroline) ruthenium (II), tris (1,10-phenanthrolinyl-4,7-bis (benzenesulfonate) ruthenate (II), and tris (4,40-dinonyl-1,10-phenan throline) ruthenium (II). Solid supports applied for immobilization of PSs include polyethylene, polypropylene, polystyrene, polycarbonate, polymethyl methacrylate, polyester isophthalic resin, silicone, cationic nylon, porous silicones, poly (vinylidene difluoride), cellulose membranes, and chitosan [82–88]. Immobilized PSs demonstrated antibacterial properties against Gram-negative and Gram-positive bacteria in batch and continuous regimes and under reuse. Immobilized PSs were found more stable and resistant to photobleaching than in a free form [82, 86, 88].

Our group immobilized PSs in polymers using several techniques. The first method included mixing solutions of PSs in chloroform with solutions of polymers in the same solvent, followed by evaporation of the solvent, which yielded thin polymeric films with homogeneously incorporated PSs. This technique was applied to RB and MB immobilized onto polystyrene, polycarbonate, and polymethyl methacrylate [88–90]. In all cases, the obtained polymer films showed high antibacterial activity against Gram-positive and Gram-negative bacteria when exposed to an external source of white light. However, since this method involves using an organic solvent, it cannot be considered environmentally friendly. The second method is based on dissolution of PSs in a melted polymer under extrusion and does not require any additional chemical reagents [91]. The photosensitizers RB, Rose Bengal lactone, MB, and hematoporphyrin were immobilized in polyethylene and polypropylene using this method. The antibacterial efficiency of immobilized

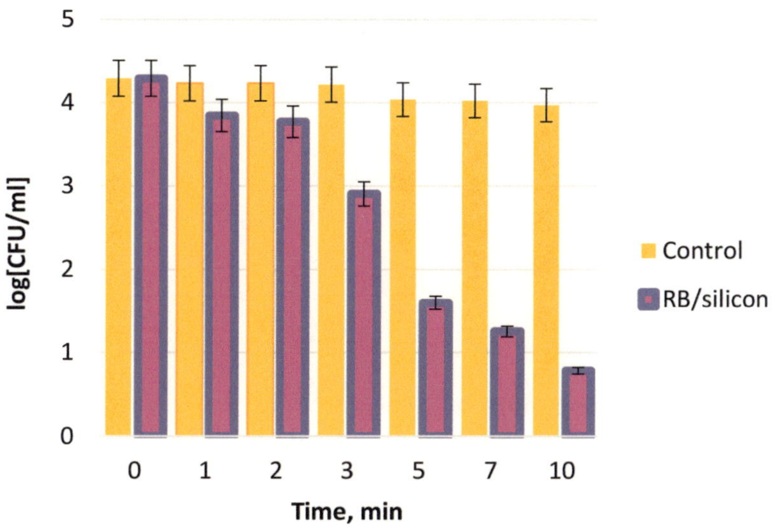

Figure 8.
Antibacterial activity of silicon-immobilized RB (5% w/w) under ultrasonic treatment in the dark. Control—S. aureus cells treated by ultrasound only. After the treatment, bacteria were tested by viable count. Error bars present standard deviations.

PSs obtained as polymeric strips and beads was tested against *S. aureus* and *E. coli* in batch and continuous regimes under white fluorescent light. All immobilized PSs significantly reduced the concentration of the tested microorganisms, up to their complete eradication [91].

Another immobilization technique was based on polymerization of silicon in the presence of RB as the photosensitizer. Silicon tablets produced by this method contained evenly distributed RB that was not bound to the support by covalent bonds [29]. The antibacterial activity of the immobilized RB was tested under illumination and using ultrasonic activation in the dark (**Figure 8**). **Figure 8** demonstrates the effect of immobilized RB on *S. aureus* cells when subjected to ultrasound in the dark. Silicone alone did not affect the microorganisms with and without sonication. However, the number of alive cells in samples subjected to immobilized RB under sonication decreased with sonication time and decreased by more than three orders of magnitude after 10 min of treatment.

Further development of immobilization methods and different PSs and polymers may expand the possibilities of this approach and yield the applications in various fields, such as the production of antibacterial surfaces and water disinfection.

5. Conclusions

Numerous studies show that photodynamic antibacterial chemotherapy is a powerful tool for killing microorganisms. Since this method requires external illumination, it can be successfully applied only to the treatment of local superficial skin and oral cavity infections. Development of new modes of PS excitation by ultrasound, radio waves, chemiluminescent, and bioluminescent light opens new prospects for their use in treating internal infections. Encapsulation of PSs in liposomes may solve the problem of using hydrophobic PSs with poor solubility in the aqueous phase. It can also provide delivery of a concentrated PS directly to the target site, thus increasing efficiency and reducing side effects of the treatment. Immobilization of PSs in a solid phase enables using them repeatedly or in a continuous mode. It can be assumed that PSs have a good potential for various clinical and nonclinical applications.

Acknowledgements

This work was supported by the Research Authority of the Ariel University, Ariel, Israel.

Conflict of interest

The authors declare no conflict of interest.

Author details

Faina Nakonechny and Marina Nisnevitch*
Department of Chemical Engineering, Biotechnology and Materials, Ariel University, Ariel, Israel

*Address all correspondence to: marinan@ariel.ac.il

References

[1] Daniell MD, Hill JS. A history of photodynamic therapy. The Australian and New Zealand Journal of Surgery. 1991;**61**:340-348. DOI: 10.1111/j.1445-2197.1991.tb00230.x

[2] Raab O. Uber die Wirkung fluorescierender Stoffe auf Infusoria. Zeitschrift für Biologie. 1900;**39**:524

[3] Jodlbauer A, Tappeiner von H. Uber die Wirkung photodynamischer (fluoreszierender) Stoffe auf Bakterien. Münchener Medizinische Wochenschrift (1950). 1904;**51**:1096-1097

[4] Jesionek A, Tappeiner von H. Zur Behandlung der Hautcarcinome mit fluoreszierenden Stoffen. Deutsches Archiv für Klinische Medizin. 1905;**85**:223-239

[5] Tappeiner von H. Die photodynamische Erscheinung (Sensibilisierung durch fluoreszierende Stoffe). Ergebnisse der Physiologie. 1909;**8**:698-741. Available from: https://link.springer.com/article/10.1007/BF02321096

[6] Ormond AB, Freeman HS. Dye sensitizers for photodynamic therapy. Materials (Basel). 2013;**6**:817-840. DOI: 10.3390/ma6030817

[7] Debele TA, Peng S, Tsai HC. Drug carrier for photodynamic cancer therapy. International Journal of Molecular Sciences. 2015;**16**:22094-22136. DOI: 10.3390/ijms160922094

[8] Zhang J, Jiang C, Figueiró Longo JP, Azevedo RB, Zhang H, Muehlmann LA. An updated overview on the development of new photosensitizers for anticancer photodynamic therapy. Acta Pharmaceutica Sinica B. 2018;**8**:137-146. DOI: 10.1016/j.apsb.2017.09.003

[9] Macdonald IJ, Dougherty TJ. Basic principles of photodynamic therapy. Journal of Porphyrins and Phthalocyanines. 2001;**5**:105-129. DOI: 10.1002/jpp.328

[10] Cieplik F, Deng D, Crielaard W, Buchalla W, Hellwig E, Al-Ahmad A, et al. Antimicrobial photodynamic therapy - what we know and what we don't. Critical Reviews in Microbiology. 2018;**44**:571-589. DOI: 10.1080/1040841X.2018.1467876

[11] Castano AP, Demidova TN, Hamblin MR. Mechanisms in photodynamic therapy: Part one-photosensitizers, photochemistry and cellular localization. Photodiagnosis and Photodynamic Therapy. 2004;**1**:279-293. DOI: 10.1016/S1572-1000(05)00007-4

[12] Maisch T. Anti-microbial photodynamic therapy: Useful in the future? Lasers in Medical Science. 2007;**22**:83-91. DOI: 10.1007/s10103-006-0409-7

[13] Huang L, Xuan Y, Koide Y, Zhiyentayev T, Tanaka M, Hamblin MR. Type I and type II mechanisms of antimicrobial photodynamic therapy: An in vitro study on gram-negative and gram-positive bacteria. Lasers in Surgery and Medicine. 2012;**44**:490-499. DOI: 10.1002/lsm.22045

[14] Liu Y, Qin R, Zaat SAJ, Breukink E, Heger M. Antibacterial photodynamic therapy: Overview of a promising approach to fight antibiotic-resistant bacterial infections. Journal of Clinical and Translational Research. 2015;**1**:140-167. DOI: 10.18053/jctres.201503.002

[15] Yin R, Hamblin MR. Antimicrobial photosensitizers: Drug discovery under the spotlight. Current Medicinal Chemistry. 2015;**22**:2159-2185. DOI: 10.2174/0929867322666150319120134

[16] Abrahamse H, Hamblin MR. New photosensitizers for photodynamic

therapy. The Biochemical Journal. 2016;**473**:347-364. DOI: 10.1042/BJ20150942

[17] Carrera ET, Dias HB, Corbi SCT, Marcantonio RAC, Bernardi ACA, Bagnato VS, et al. The application of antimicrobial photodynamic therapy (aPDT) in dentistry: A critical review. Laser Physics. 2016;**26**:123001. DOI: 10.1088/1054-660X/26/12/123001

[18] Hamblin MR. Antimicrobial photodynamic inactivation: A bright new technique to kill resistant microbes. Current Opinion in Microbiology. 2016;**33**:67-73. DOI: 10.1016/j.mib.2016.06.008

[19] Tim M. Strategies to optimize photosensitizers for photodynamic inactivation of bacteria. Journal of Photochemistry and Photobiology. B. 2015;**150**:2-10. DOI: 10.1016/j.jphotobiol.2015.05.010

[20] Nyman ES, Hynninen PH. Research advances in the use of tetrapyrrolic photosensitizers for photodynamic therapy. Journal of Photochemistry and Photobiology. B. 2004;**73**:1-28. DOI: 10.1016/j.jphotobiol.2003.10.002

[21] Harris F, Chatfield LK, Phoenix DA. Phenothiazinium based photosensitisers - photodynamic agents with a multiplicity of cellular targets and clinical applications. Current Drug Targets. 2005;**6**:615-627. DOI: 10.2174/1389450054545962

[22] Leite DP, Paolillo FR, Parmesano TN, Fontana CR, Bagnato VS. Effects of photodynamic therapy with blue light and curcumin as mouth rinse for oral disinfection: A randomized controlled trial. Photomedicine and Laser Surgery. 2014;**32**:627-632. DOI: 10.1089/pho.2014.3805

[23] Bond JS, Francis SH, Park JH. An essential histidine in the catalytic activities of 3-phosphoglyceraldehyde dehydrogenase. The Journal of Biological Chemistry. 1970;**245**:1041-1053. Available from: http://www.jbc.org/content/245/5/1041.long

[24] Coulson AF, Yonetani T. Interaction of rose bengal with apo-hemoproteins. An essential histidine residue in cytochrome c peroxidase. European Journal of Biochemistry. 1972;**26**:125-131. DOI: 10.1111/j.1432-1033.1972.tb01748.x

[25] Hsieh YH, Huang YJ, Jin JS, Yu L, Yang H, Jiang C, et al. Mechanisms of Rose Bengal inhibition on SecA ATPase and ion channel activities. Biochemical and Biophysical Research Communications. 2014;**454**:308-312. DOI: 10.1016/j.bbrc.2014.10.070

[26] Shrestha A, Hamblin MR, Kishen A. Photoactivated rose bengal functionalized chitosan nanoparticles produce antibacterial/biofilm activity and stabilize dentin-collagen. Nanomedicine. 2014;**10**:491-501. DOI: 10.1016/j.nano.2013.10.010

[27] Nakonieczna J, Wolnikowska K, Ogonowska P, Neubauer D, Bernat A, Kamysz W. Rose bengal-mediated photoinactivation of multidrug resistant Pseudomonas aeruginosa is enhanced in the presence of antimicrobial peptides. Frontiers in Microbiology. 2018;**9**:1949. DOI: 10.3389/fmicb.2018.01949

[28] Goulart Rde C, Bolean M, Paulino Tde P, Thedei G Jr, Souza SL, Tedesco AC, et al. Photodynamic therapy in planktonic and biofilm cultures of Aggregatibacter actinomycetemcomitans. Photomedicine and Laser Surgery. 2010;**28**(Suppl. 1):S53-S60. DOI: 10.1089/pho.2009.2591

[29] Nakonechny F, Barel M, David A, Koretz S, Litvak B, Ragozin E, et al. Dark antibacterial activity of Rose Bengal. International Journal of

Molecular Sciences. 2019;**20**:E3196. DOI: 10.3390/ijms20133196

[30] Brovko LY, Meyer A, Tiwana AS, Chen W, Liu H, Filipe CD, et al. Photodynamic treatment: A novel method for sanitation of food handling and food processing surfaces. Journal of Food Protection. 2009;**72**:1020-1024. DOI: 10.4315/0362-028X-72.5.1020

[31] Wilson BC, Patterson MS. The physics, biophysics and technology of photodynamic therapy. Physics in Medicine and Biology. 2008;**53**:R61-R109. DOI: 10.1088/0031-9155/53/9/R01

[32] Brancaleon L, Moseley H. Laser and non-laser light sources for photodynamic therapy. Lasers in Medical Science. 2002;**17**:173-186. DOI: 10.1007/s101030200027

[33] Harris F, Dennison SR, Phoenix DA. Using sound for microbial eradication - light at the end of the tunnel? FEMS Microbiology Letters. 2014;**356**:20-22. DOI: 10.1111/1574-6968.12484

[34] Serpe L, Giuntini F. Sonodynamic antimicrobial chemotherapy: First steps towards a sound approach for microbe inactivation. Journal of Photochemistry and Photobiology. B. 2015;**150**:44-49. DOI: 10.1016/j.jphotobiol.2015.05.012

[35] Zhuang D, Hou C, Bi L, Han J, Hao Y, Cao W, et al. Sonodynamic effects of hematoporphyrin monomethyl ether on Staphylococcus aureus in vitro. FEMS Microbiology Letters. 2014;**361**:174-180. DOI: 10.1111/fml.2014.361

[36] Alves F, Pavarina AC, Mima EGO, McHale AP, Callan JF. Antimicrobial sonodynamic and photodynamic therapies against *Candida albicans*. Biofouling. 2018;**34**:357-367. DOI: 10.1080/08927014.2018.1439935

[37] Costley D, Nesbitt H, Ternan N, Dooley J, Huang YY, Hamblin MR, et al. Sonodynamic inactivation of gram-positive and gram-negative bacteria using a Rose Bengal-antimicrobial peptide conjugate. International Journal of Antimicrobial Agents. 2017;**49**:31-36. DOI: 10.1016/j.ijantimicag.2016.09.034

[38] Wang X, Ip M, Leung AW, Xu C. Sonodynamic inactivation of methicillin-resistant *Staphylococcus aureus* in planktonic condition by curcumin under ultrasound sonication. Ultrasonics. 2014;**54**:2109-2114. DOI: 10.1016/j. ultras.2014.06.017

[39] Sadanala KC, Chaturvedi PK, Seo YM, Kim JM, Jo YS, Lee YK, et al. Sono-photodynamic combination therapy: A review on sensitizers. Anticancer Research. 2014;**34**:4657-4664. Available from: http://ar.iiarjournals.org/content/34/9/4657.full.pdf

[40] Costley D, Mc Ewan C, Fowley C, McHale AP, Atchison J, Nomikou N, et al. Treating cancer with sonodynamic therapy: A review. International Journal of Hyperthermia. 2015;**31**:107-117. DOI: 10.3109/02656736.2014.992484

[41] McHale AP, Callan JF, Nomikou N, Fowley C, Callan B. Sonodynamic therapy: Concept, mechanism and application to cancer treatment. Advances in Experimental Medicine and Biology. 2016;**880**:429-450. DOI: 10.1007/978-3-319-22536-4_22

[42] Pan X, Wang H, Wang S, Sun X, Wang L, Wang W, et al. Sonodynamic therapy (SDT): A novel strategy for cancer nanotheranostics. Science China. Life Sciences. 2018;**61**:415-426. DOI: 10.1007/s11427-017-9262-x

[43] Rengeng L, Qianyu Z, Yuehong L, Zhongzhong P, Libo L. Sonodynamic therapy, a treatment developing from photodynamic therapy. Photodiagnosis and Photodynamic Therapy. 2017;**19**:159-166. DOI: 10.1016/j.pdpdt.2017.06.003

[44] Geng C, Zhang Y, Hidru TH, Zhi L, Tao M, Zou L, et al. Sonodynamic therapy: A potential treatment for atherosclerosis. Life Sciences. 2018;**207**:304-313. DOI: 10.1016/j.lfs.2018.06.018

[45] Nakonechny F, Nisnevitch M, Nitzan Y, Nisnevitch M. Sonodynamic excitation of Rose Bengal for eradication of gram-positive and gram-negative bacteria. BioMed Research International. 2013;**2013**:684930. DOI: 10.1155/2013/684930

[46] Nisnevitch M, Nakonechny F. Sensitivity of bacteria to photodynamic chemotherapy. In: Nisnevitch M, editor. Prokaryotes: Physiology, Biochemistry and Cell Behavior. New York, NY, USA: Nova Science Publishers Inc.; 2015. pp. 197-220

[47] Mirza AN, Fornage BD, Sneige N, Kuerer HM, Newman LA, Ames FC, et al. Radiofrequency ablation of solid tumors. Cancer Journal. 2001;**7**:95-102

[48] Tamarov KP, Osminkina LA, Zinovyev SV, Maximova KA, Kargina JV, Gongalsky MB, et al. Radio frequency radiation-induced hyperthermia using Si nanoparticle-based sensitizers for mild cancer therapy. Scientific Reports. 2014;**4**:7034. DOI: 10.1038/srep07034

[49] Amini SM, Kharrazi S, Rezayat SM, Gilani K. Radiofrequency electric field hyperthermia with gold nanostructures: Role of particle shape and surface chemistry. Artificial Cells, Nanomedicine, and Biotechnology. 2018;**46**:1452-1462. DOI: 10.1080/21691401.2017.1373656

[50] Shi J, Wang L, Gao J, Liu Y, Zhang J, Ma R, et al. A fullerene-based multi-functional nanoplatform for cancer theranostic applications. Biomaterials. 2014;**35**:5771-5784. DOI: 10.1016/j.biomaterials.2014.03.071

[51] Chung HJ, Lee HK, Kwon KB, Kim HJ, Hong ST. Transferrin as a thermosensitizer in radiofrequency hyperthermia for cancer treatment. Scientific Reports. 2018;**8**:13505. DOI: 10.1038/s41598-018-31232-9

[52] Merényi G, Lind J, Eriksen TE. Luminol chemiluminescence: Chemistry, excitation, emitter. Journal of Bioluminescence and Chemiluminescence. 1990;**5**:53-56. DOI: 10.1002/bio.1170050111

[53] Gu X, Tang BZ. No UV irradiation needed! Chemiexcited AIE dots for cancer theranostics. Chem. 2017;**3**:917-927. DOI: 10.1016/j.chempr.2017.11.013

[54] Rose AL, Waite TD. Chemiluminescence of luminol in the presence of iron(II) and oxygen: Oxidation mechanism and implications for its analytical use. Analytical Chemistry. 2001;**73**:5909-5920. DOI: 10.1021/ac015547q

[55] Magalhães CM, da Silva JCGE, daSilva LP. Chemiluminescence and bioluminescence as an excitation source in the photodynamic therapy of cancer: A critical review. ChemPhysChem. 2016;**17**:2286-2294. DOI: 10.1002/cphc.201600270

[56] Laptev R, Nisnevitch M, Siboni G, Malik Z, Firer MA. Intracellular chemiluminescence activates targeted photodynamic destruction of leukemic cells. British Journal of Cancer. 2006;**95**:189-196. DOI: 10.1038/sj.bjc.6603241

[57] Davis RW 4th, Snyder E, Miller J, Carter S, Houser C, Klampatsa A, et al. Luminol chemiluminescence reports photodynamic therapy-generated neutrophil activity in vivo and serves as a biomarker of therapeutic efficacy. Photochemistry and Photobiology. 2019;**95**:430-438. DOI: 10.1111/php.13040

[58] Ferraz RC, Fontana CR, Ribeiro AP, Trindade FZ, Bartoloni FH, Baader JW, et al. Chemiluminescence as a PDT light source for microbial control. Journal of Photochemistry and Photobiology. B. 2011;**103**:87-92. DOI: 10.1016/j.jphotobiol.2011.01.018

[59] Nakonechny F, Firer MA, Nitzan Y, Nisnevitch M. Intracellular antimicrobial photodynamic therapy: A novel technique for efficient eradication of pathogenic bacteria. Photochemistry and Photobiology. 2010;**86**:1350-1355. DOI: 10.1111/j.1751-1097.2010.00804.x

[60] Nakonechny F, Nisnevitch M, Nitzan Y, Firer MA. New techniques in antimicrobial photodynamic therapy: Scope of application and overcoming drug resistance in nosocomial infections. In: Science against Microbial Pathogens: Communicating Current Research and Technological Advances; Méndez-Vilas, A., Ed.; Formatex Research Center Publisher: Badajoz, Spain; 2011. Volume 1, p. 684-691. ISNB: 13: 978-84-939843-1-1

[61] Derycke AS, de Witte PA. Liposomes for photodynamic therapy. Advanced Drug Delivery Reviews. 2004;**56**:17-30. DOI: 10.1016/j.addr.2003.07.014

[62] Ghosh S, Carter KA, Lovell JF. Liposomal formulations of photosensitizers. Biomaterials. 2019;**218**:119341. DOI: 10.1016/j.biomaterials.2019.119341

[63] Li M, Du C, Guo N, Teng Y, Meng X, Sun H, et al. Composition design and medical application of liposomes. European Journal of Medicinal Chemistry. 2019;**164**:640-653. DOI: 10.1016/j.ejmech.2019.01.007

[64] Bulbake U, Doppalapudi S, Kommineni N, Khan W. Liposomal formulations in clinical use: An updated review. Pharmaceutics. 2017;**9**, pii: E12. DOI: 10.3390/pharmaceutics9020012

[65] Abu Lila AS, Ishida T. Liposomal delivery systems: Design optimization and current applications. Biological & Pharmaceutical Bulletin. 2017;**40**:1-10. DOI: 10.1248/bpb.b16-00624

[66] Alavi M, Karimi N, Safaei M. Application of various types of liposomes in drug delivery systems. Advanced Pharmaceutical Bulletin. 2017;7:3-9. DOI: 10.15171/apb.2017.002

[67] Damoiseau X, Schuitmaker HJ, Lagerberg JW, Hoebeke M. Increase of the photosensitizing efficiency of the Bacteriochlorin a by liposome-incorporation. Journal of Photochemistry and Photobiology. B. 2001;**60**:50-60. DOI: 10.1016/S1011-1344(01)00118-X

[68] Jori G, Tomio L, Reddi E, Rossi E, Corti L, Zorat PL, et al. Preferential delivery of liposome-incorporated porphyrins to neoplastic cells in tumour-bearing rats. British Journal of Cancer. 1983;**48**:307-309. DOI: 10.1038/bjc.1983.186

[69] Jiang F, Lilge L, Grenier J, Li Y, Wilson MD, Chopp M. Photodynamic therapy of U87 human glioma in nude rat using liposome-delivered photofrin. Lasers in Surgery and Medicine. 1998;**22**:74-80. DOI: 10.1002/(SICI)1096-9101(1998)22:2<74::AID-LSM2>3.0.CO;2-T

[70] Mugabe C, Halwani M, Azghani AO, Lafrenie RM, Omri A. Mechanism of enhanced activity of liposome-entrapped aminoglycosides against resistant strains of Pseudomonas aeruginosa. Antimicrobial Agents and Chemotherapy. 2006;**50**:2016-2022. DOI: 10.1128/AAC.01547-05

[71] Sachetelli S, Khalil H, Chen T, Beaulac C, Sénéchal S, Lagacé J. Demonstration of a fusion mechanism between a fluid bactericidal liposomal formulation and bacterial cells.

Biochimica et Biophysica Acta. 2000;**1463**:254-266. DOI: 10.1016/S0005-2736(99)00217-5

[72] Nakonechny F, Nitzan Y, Nisnevitch M. Olive oil-based delivery of photosensitizers for bacterial eradication. In: Olive Oil-Constituents, Quality, Health Properties and Bioconversions; Boskou, D., Ed.; InTech Open Access Publisher: Rijeka, Croatia; 2012. p. 471-492. DOI: 10.5772/30623

[73] Storm G, Crommelin DJA. Liposomes: quo vadis? Pharmaceutical Science & Technology Today. 1998;**1**:19-31. DOI: 10.1016/S1461-5347(98)00007-8

[74] Nisnevitch M, Nakonechny F, Nitzan Y. Photodynamic antimicrobial chemotherapy by liposome-encapsulated water-soluble photosensitizers. Russian Journal of Bioorganic Chemistry. 2010;**36**:363-369. DOI: 10.1134/S106816201003012X

[75] Ferro S, Ricchelli F, Mancini G, Tognon G, Jori G. Inactivation of methicillin-resistant Staphylococcus aureus (MRSA) by liposome-delivered photosensitising agents. Journal of Photochemistry and Photobiology. B. 2006;**83**:98-104. DOI: 10.1016/j.jphotobiol.2005.12.008

[76] Ferro S, Ricchelli F, Monti D, Mancini G, Jori G. Efficient photoinactivation of methicillin-resistant Staphylococcus aureus by a novel porphyrin incorporated into a poly-cationic liposome. The International Journal of Biochemistry & Cell Biology. 2007;**39**:1026-1034. DOI: 10.1016/j.biocel.2007.02.001

[77] Tsai T, Yang YT, Wang TH, Chien HF, Chen CT. Improved photodynamic inactivation of gram-positive bacteria using hematoporphyrin encapsulated in liposomes and micelles. Lasers in Surgery and Medicine. 2009;**41**:316-322. DOI: 10.1002/lsm.20754

[78] Yang YT, Chien HF, Chang PH, Chen YC, Jay M, Tsai T, et al. Photodynamic inactivation of chlorin e6-loaded CTAB-liposomes against Candida albicans. Lasers in Surgery and Medicine. 2013;**45**:175-185. DOI: 10.1002/lsm.22124

[79] Paczkowski J, Neckers DC. Photochemical properties of rose bengal. 11. Fundamental studies in heterogeneous energy transfer. Macromolecules. 1985;**18**:2412-2418. DOI: 10.1021/ma00154a013

[80] Faust D, Funken K, Horneck G, Milow B, Ortner J, Sattlegger M, et al. Immobilized photosensitizers for solar photochemical applications. Solar Energy. 1999;**65**:71-74. DOI: 10.1016/S0038-092X(98)00099-1

[81] Nowakowska M, Kępczyński M, Dąbrowska M. Polymeric photosensitizers. 5. Synthesis and photochemical properties of poly[(N-isopropylacrylamide)-co-(vinylbenzyl chloride)] containing covalently bound rose bengal chromophores. Macromolecular Chemical Physics. 2001;**202**:1679-1688. DOI: 10.1002/1521-3935(20010601)202:9<1679::AID-MACP1679>3.0.CO;2-R

[82] Bonnett R, Krysteva MA, Lalov IG, Artarsky SV. Water disinfection using photosensitizers immobilized on chitosan. Water Research. 2006;**40**:1269-1275. DOI: 10.1016/j.watres.2006.01.014

[83] Jiménez-Hernández ME, Manjón F, García-Fresnadillo D, Orellana G. Solar water disinfection by singlet oxygen photogenerated with polymer-supported Ru(II) sensitizers. Solar Energy. 2006;**80**:1382-1387. DOI: 10.1016/j.solener.2005.04.027

[84] Villén L, Manjón F, García-Fresnadillo D, Orellana G. Solar water disinfection by photocatalytic singlet oxygen production in heterogeneous medium. Applied Catalysis. 2006;**69**:1-9. DOI: 10.1016/j.apcatb.2006.05.015

[85] Manjón F, García-Fresnadillo D, Orellana G. Water disinfection with Ru(II) photosensitisers supported on ionic porous silicones. Photochemical & Photobiological Sciences. 2009;**8**:926-932. DOI: 10.1039/b902014d

[86] Manjón F, Santana-Magaña M, García-Fresnadillo D, Orellana G. Singlet oxygen sensitizing materials based on porous silicone: Photochemical characterization, effect of dye reloading and application to water disinfection with solar reactors. Photochemical & Photobiological Sciences. 2010;**9**:838-845. DOI: 10.1039/c0pp00026d 15

[87] Cahan R, Schwartz R, Langzam Y, Nitzan Y. Light-activated antibacterial surfaces comprise photosensitizers. Photochemistry and Photobiology. 2011;**87**:1379-1386. DOI: 10.1111/j.1751-1097.2011.00989.x

[88] Valkov A, Nakonechny F, Nisnevitch M. Polymer-immobilized photosensitizers for continuous eradication of bacteria. International Journal of Molecular Sciences. 2014 25;**15**:14984-14996. DOI: 10.3390/ijms150914984

[89] Valkov A, Nakonechny F, Nisnevitch M. Antibacterial properties of Rose Bengal immobilized in polymer supports. Applied Mechanics and Materials. 2015;**719**:21-24. DOI: 10.4028/www.scientific.net/AMM.719-720.21

[90] Nakonechny F, Pinkus A, Hai S, Yehosha O, Nitzan Y, Nisnevitch M. Eradication of gram-positive and gram-negative bacteria by photosensitizers immobilized in polystyrene. Photochemistry and Photobiology. 2013;**89**:671-678. DOI: 10.1111/php.12022

[91] Valkov A, Raik KA, Mualem-Sinai Y, Nakonechny F, Nisnevitch M. Water disinfection by immobilized photosensitizers. Water. 2019;**11**:26. DOI: 10.3390/w11010026

Chapter 8

Probiotics and Bioremediation

Qomarudin Helmy, Edwan Kardena and Sri Gustiani

Abstract

Increased environmental contamination leads to a progressive decline in environmental quality. Probiotics play a role as remediation agents which are expected to be able to help the host in responding to environmental changes. Probiotics are live microbial feed supplements that favorably affect host (humans, animals, plants) by increasing the balance of intestinal microbes. Probiotics are used in livestock/aquaculture to improve growth performance through improved utilization of nutrients, reducing disease, also developing the immune system. In addition to feed supplements, certain types of probiotics act as bioremediation or decomposing agents of hazardous substances. The bioremediation system is the use of microorganisms (bacteria, fungi, yeast, and algae) or microbial products to degrade, reduce, or remove pollutants in the environment although in some cases plants are also utilized for this purpose called phytoremediation. When bioremediation occurs, enzymes produced by microorganisms modify toxic pollutants by changing the chemical structure of pollutants. This event is called biotransformation. In many cases, biotransformation leads to biodegradation, when toxic pollutants are degraded, the structure becomes simpler, and ultimately becomes harmless and non-toxic metabolites, that called mineralization.

Keywords: probiotics, bioremediation, aquaculture, livestock, agriculture

1. Probiotics: small creatures that do big things

Bacteria are often associated with diseases or with something that is frightening and disgusting. But along with advances in technology, bacteria today are not considered only as an enemy but can also be friends. Basically bacteria are microorganisms that is part of nature that are found almost everywhere on the earth's surface, atmosphere, also the upper atmosphere [1–3]. The shape is very small, cannot be seen by the eye but its existence is very important to maintain environmental balance. Microorganisms are usually considered to include all prokaryotes, protists and microalgae. Fungi, especially those that are small and do not form hyphae, can also be considered as a part of it, although many do not agree. Most people assume that what can be considered microorganisms are all very small organisms that can be bred in petri dishes or incubators in the laboratory and capable of reproducing themselves by mitosis [4].

Collection of beneficial microbes known as probiotics, a word derived from the Greek word that means for life. Among the researchers there were slight differences of opinion regarding what could be called probiotics, including:

- Probiotics are compounds produced by a microorganism that can stimulate the growth of other microorganisms, so it is the opposite of antibiotics [5].

- Probiotics are extracts from tissues that can stimulate the growth of microorganisms [6].

- Probiotics are organisms and substrates that have an influence on the balance of microbiota in the digestive system [7].

- Probiotics are living microorganisms that when consumed by the host will have a beneficial effect on them by improving the microbiota environment that exists in the digestive system [8].

- The current definition of probiotics was formulated by FAO/WHO in 2001 as living microorganisms which, if given in sufficient quantities, provide health benefits to the host [9].

Probiotics can be found in several products depending on the intended use. They can be marketed as food, medical food, food supplements or medicine. In general, probiotics are marketed as food supplements e.g. pill/tablet products or as food items, e.g., Yogurt, Kefir. Some probiotic organisms including *Lactobacillus* sp., *Streptococcus* sp., and *Bifidobacterium* sp. have been "generally recognized as safe" status, which means that they are additives that are permitted in food substances. The term probiotic was originally intended as a supplement used to improve or prolong human life or health by providing a sufficient amount of good microorganisms into the body through food. Having recognized the positive benefits of probiotics on human health, many practitioners and researchers have begun to try and apply probiotics to livestock, agriculture, and fisheries.

2. Probiotics in Aquaculture, Animal Husbandry, and Agriculture

Probiotics are actually used not only for human, but also fisheries, agriculture, and animal husbandry with the aim of increasing yield productivity and decomposing waste. In livestock farming, increased production can be achieved by optimizing environmental conditions, getting the right stocking density, improving seed quality and providing good quality feed. In addition, it can also be done through efforts to reduce mortality rates and increase individual growth rates. In intensive livestock farming, feed is the highest component of production costs, about 60–70% of operational costs [10]. The digestive system plays a vital role in the extraction of nutrients from feed and their absorption to be used by body cells. The main key that occurs in the digestive system is its ability to digest food which allows nutrients to be absorbed by the body. The digestibility value of a food ingredient illustrates the ability of livestock to digest a food and the digestibility itself is determined by the quality of the food provided. The digestive system plays a vital role in the extraction of nutrients from feed and their absorption to be used by body cells. The main key that occurs in the digestive system is its ability to digest food which allows nutrients to be absorbed by the body. Digestion is a chemical process and fermentation by microbiota in the intestine. The interaction between microbiota and nutrition is very complex. Microbiota and its metabolic products can affect the digestion and absorption of nutrients by the host. In other words, microbial balance in the digestive system plays an important role for health, feed digestibility, and production efficiency.

There are three suggested probiotic working mechanisms, namely:

1. Suppress harmful microbial populations through competition by producing antimicrobial compounds or through nutrition competition and attachment sites on the intestine wall,

2. Stimulating immunity through increased levels of antibodies or macrophage activity, and

3. Changing microbial metabolism by increasing or decreasing the activity of certain enzymes.

The growing livestock farming industry has a relationship with the presence of bacteria in the environment. An intensive cultivation system without good management will have a negative effect. For example, in intensive aquaculture farming with excessive feeding, maintenance of fish with high stocking densities without good management will cause disease for the fish. The resulting waste in the form of high organic matter will cause bacteria to flourish and ultimately reduce fish production and also damage the environment. Other problems with increasing intensive aquaculture activities are eutrophication, increased sedimentation and excessive plankton growth. This can result in high levels of ammonia in water and lack of oxygen.

Biological processes that occur in nature including the cultivation environment must be maintained in balance so that the quality of the environment is maintained well. One appropriate way is to use probiotics. Probiotics itself is a collection of beneficial microorganisms that help the process of absorption of food, can increase endurance and help improve environmental quality. Microorganisms that exist in probiotics are usually bacteria such as *Lactobacillus* sp., *Bacillus* sp., *Micrococcus* sp., Nitrobacter and Nitrosomonas. Probiotics can be given in various ways, through aquaculture feed or directly stocked into the aquatic environment. In probiotics itself there are several types of bacteria that can help the digestive process better such as *Lactobacillus* sp. These bacteria are able to maximize the absorption of food in the intestines of fish so that not much food is wasted and fish growth becomes faster. Under these conditions the amount of feed given is more efficient but provides maximum results. This can occur because the bacteria with its mechanism produces digestive enzymes such as amylase, protease, lipase to break down carbohydrates, proteins, fats in the feed so that it is more quickly absorbed by the body. Other probiotics with the type Nitrobacter and Nitrosomonas are able to decipher the rest of the remaining feed or feces that settles at the bottom so that the quality of water in ponds or ponds for the better. Stools and food waste that accumulate and not decompose will result in high levels of ammonia and sulfides which can be toxic to aquatic organisms.

The presence of beneficial microorganisms in aquaculture activities has been proven through several studies [11–15]. Fish feed with additional probiotics turned out to be able to increase the growth of catfish better than ordinary feed. Application of microorganisms to the environment is also shown by giving probiotics in the water from shrimp ponds that show a decrease in ammonia levels and maintain pH in neutral conditions compared to without giving probiotics. The antimicrobial role released by probiotic bacteria also has a role to prevent the emergence of disease, thereby increasing the resistance of fish or livestock to disease [16–20]. The presence of beneficial microbes that are naturally present in the intestine provides an opportunity and possibility to isolate and reproduce them, which are then reintroduced into the digestive system and used as probiotics (**Table 1**).

The application of probiotics in agriculture has been widely used, especially as a starter in making organic fertilizer. With this probiotic can accelerate the manufacture of organic fertilizer, compost for example naturally will take 3-6 months, but with probiotic technology only takes 3-4 weeks. Inoculation of probiotic microorganisms can be used as an alternative in overcoming the scarcity of inorganic fertilizers due to the high price and availability of fertilizers in certain areas. Inoculation

Subject	Probiotics organism	Results	Application	Reference
Aquaculture				
Shrimp (*Litopenaeus vannamei*)	Bacteria: *Bacillus thuringiensis*, *Bacillus megaterium*, *Bacillus polymyxa*, *Bacillus licheniformis* and *Bacillus subtilis* Yeast: *Debaryomyces hansenii*, *Rhodotorula* sp. Algae: *Chaetoceros* sp.	Post-larval stage of *L. vannamei* treated with either bacteria and yeasts or bacteria, yeasts and Chaetoceros exhibited increases ($P < 0.05$) in growth and survival as compared to controls.	Food pellets in the form of microencapsulated beads using sodium alginate	[11]
Striped catfish (*Pangasianodon hypophthalmus*)	*Bacillus amyloliquefaciens* 54A and *B. pumilus* 47B	The average weight gain of fish fed probiotics at 5×10^8 CFU/g significant higher than control after 90 days of feeding, but there was not significant effect on feed conversion ratio and specific growth rate.	Mixed with food pellets	[12]
Salmon (*Salmo salar* L.) and Trout (*Oncorhynchus mykiss*)	*Carnobacterium inhibens* K1	Increase appetite and feeding efficiency and increase resistance to *A. salmonicida*, *V. ordalli* and *Y. ruckeri*	Mixed with food pellets	[13]
Fresh water prawn (*Macrobrachium rosenbergii*)	Commercial probiotic: Zymetin (*Streptococcus faecalis*, *Clostridium butyricum*, *Bacillus mesentericus*, Beer yeast). Super PS (*Rhodobacter* sp. and *Rhodococcus* sp.)	The production of probiotics treated pond was always higher than without probiotics treated ponds, highest growth and production were found when Zymetin and Super PS were used together.	Mixed with food pellets	[14]
Sea bream larvae (*Sparus aurata*)	*Lactobacillus fructivorans* and *Lactobacillus plantarum*	The combination of probiotics (80:20 w/w) through dry or live feed promoted the sea bream larvae (*Sparus aurata*) intestinal microflora changes that contribute to reducing mortality in both sea bream larvae and fries.	Live feed and mixed with food pellets	[15]

Subject	Probiotics organism	Results	Application	Reference
Animal husbandry				
Pig	*Enterococcus faecium* SF 68	Positive effect of probiotics consumption on the digestive tract thus decreasing the diarrheic diseases that are frequent in the pig husbandry.	Mixed with diet formulation	[16]
Weaned piglets	*Lactobacillus reuteri*, *Bacillus subtilis* and *Bacillus licheniformis*	Digestibility of dry matter, crude protein, and crude fat increased upon treatment with probiotic and significant reduction of fecal Salmonella and *E. coli* counts with an increase of probiotics counts.	Mixed with diet formulation	[17]
Broiler chicks	*Lactobacillus acidophilus* (LASW), *L. fermentum* (LF33), *L. plantarum* (LPL05), and *Enterococcus faecium* (TM39)	LAB administration showed effectiveness in antagonistic effect against *Salmonella* colonization, invasion, and the induced inflammation.	Mixed with diet formulation	[18]
Dairy cows	Propionibacterium, *S. cerevisiae*, *L. acidophilus*, *L. ecidophilus*, *L. casei* and *Enterococcus faecium*	Significantly improves milk yield as well as the milk composition.	Direct-feds microbials	[19]
Crossbred cows	*L. acidophilus*, *S. cerevisiae*, *S. boulardii* and *Propionibacterium frendenreichii*	Probiotics proved to be effective in increasing milk production of lactating cows. Milk fat, milk protein and solid-not-fat content tended to be higher in cows supplemented with probiotics.	Direct-feds microbials	[20]
Agriculture				
Wheat	*Enterobacter*, *Serratia*, *Microbacterium*, *Pseudomonas* and *Achromobacter*	Halotolerant bacterial consortia significantly (P ≥ 0.05) increased the emergence, growth, biomass and Super Oxide Dismutase activity of wheat seedlings exposed to salt stress.	Biofertilizers	[21]

Subject	Probiotics organism	Results	Application	Reference
Cotton	Rhizobacteria	The bacterial inoculum (50 g/kg of seed) significantly increased seed cotton yield (21%), plant height (5%) and microbial population in soil (41%) over their respective controls	Biofertilizers	[22]
Arabidopsis and cucumber (*Cucumis sativus* L.)	*Trichoderma asperelloides* T203	*Trichoderma* spp. stimulate plant growth prior to salt stress imposition and significantly improved seed germination.	Biofertilizers	[23]
Iceberg lettuce (*Lactuca sativa* L.) and rocket (*Eruca sativa* Mill.)	Trichoderma strains: *T. virens* (GV41) and *T. harzianum* (T22)	Trichoderma able to manage the nutrient content of leafy horticulture crops cultivated in low fertility soils, and assist vegetable growers in reducing the use of synthetic fertilizers, and optimize N use efficiency.	Biofertilizers	[24]
Saffron (*Crocus sativus* L.)	*R. intraradices* and *Funneliformis mosseae*	Soilless cultivation systems enhanced by the symbiosis with arbuscular mycorrhizal fungi able to produce high quality saffron.	Biofertilizers	[25]

Table 1.
Reports of the use of probiotics in aquaculture, animal husbandry, and agriculture.

of probiotic microorganisms that are able to dissolve the P element in acid soils can increase the production of agricultural commodities as reported by many researchers [21–25].

3. Bioremediation: healthy environment healthy life

Application of microorganisms, in this case, bacteria to improve the environment is actually not only probiotics but also the bioremediation process. In the bioremediation, microorganisms are also used to neutralize, detoxify, remove, clean up, break down, and/or decompose waste and other hazardous pollutants into less toxic or non-toxic substances. Microorganisms in the aquaculture environment are in direct contact with animals, with gills and food supplied, having easy access to the digestive tract of animals. Among the microorganisms that exist in the aquatic environment are microorganisms that are potentially pathogenic, opportunistic,

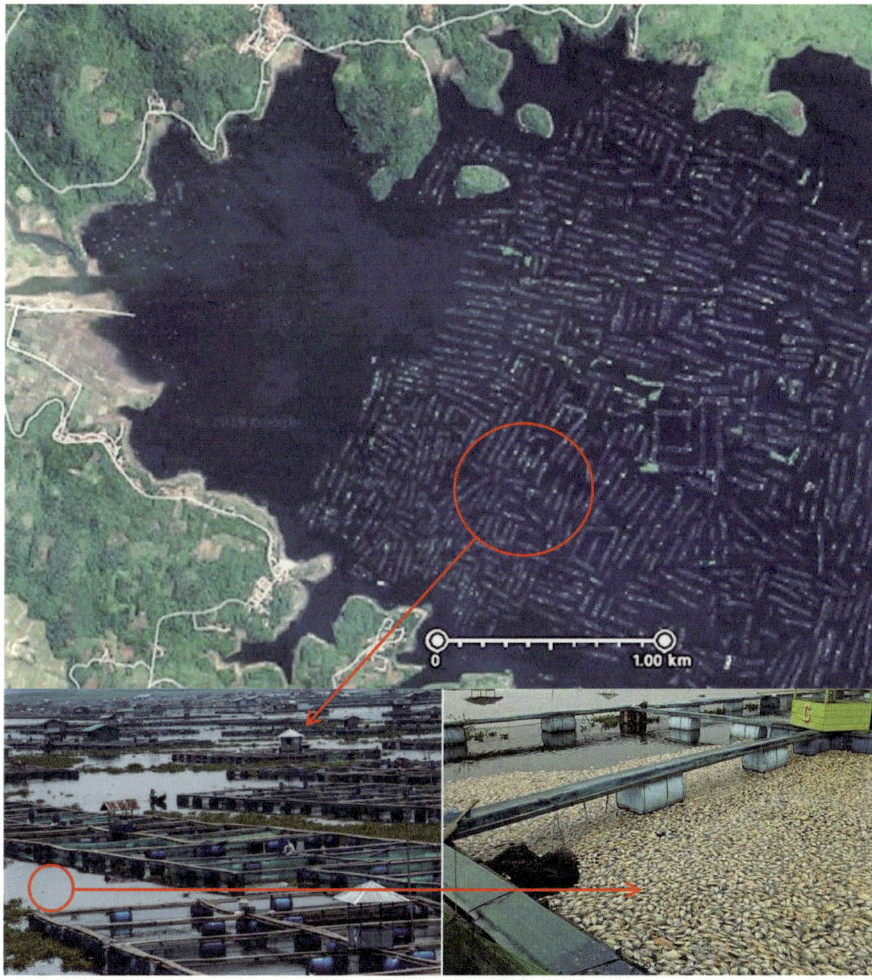

Figure 1.
Aerial image of Cirata dam located in West Java, Indonesia (upper) showing uncontrolled density of floating cage aquaculture causing rapid degradation in its water quality (lower-left); deaths of thousands of tons of fish in floating cages aquaculture due to lack of oxygen and rising toxic gas (lower-right) causing economic loss of approx. USD 28.5 million [29]. Image courtesy of Google maps [31–33].

those who take advantage of situations of animal stress in high stocking density, poor nutrition to cause infections, low growth and feed efficiency rates, and even death (**Figure 1**).

Mass fish deaths often occur in Indonesian waters with the greatest frequency and quantity occurring in lake/reservoir waters, followed by rivers and finally in sea/coast waters. The most common cause of mass death of fish is hypoxia, mainly due to the up-welling phenomenon, namely the increase in the mass of the bottom water to the surface, so that the waste generated from aquaculture activities in the form of organic material rises to the surface [26]. Up-welling events usually take place in the rainy season which results in differences in water density between surface water and bottom waters. Rising water will mix and carry organic materials and toxic gases (such as H_2S and ammonia) that can cause poisoning to fish or cause a decrease in dissolved oxygen levels due to plankton blooming that is triggered by excess of nitrate and phosphate nutrients. In marine waters, mass fish deaths have also occurred, such as in November 2015 in Jakarta Bay or earlier in 2004 and 2005, and in Lampung Bay in 2013. The mass deaths of sea fish are usually caused by the phenomenon of algal blooming (red tide), namely the plankton population explosion that was followed by mass death and caused a decrease in oxygen content due to algal respiration processes [27, 28]. As a result, fish become oxygen deficient or there is a blockage in the fish's respiratory organs (gills) by plankton. Besides causing economic losses because most of the fish that die suddenly cannot be utilized anymore, the remaining fish carcasses also pollute the environment due to their suboptimal handling. Potential economic loss of aquaculture sudden death in Cirata dam incident in 2013 is IDR 427,6 Billion or USD 28.5 Million [29]. For this reason, the use of probiotics for aquaculture aims not only at direct benefits for animals but also their effects on the fishing environment. There are two types of probiotic applications, first with direct feed (mixed into diet formulation) and secondly through the environment (mixed with water and/or sediment). Application of the first method can improve feed quality by adding additives in the form of probiotics containing beneficial microbes and decomposers into the feed that can function to improve feed quality by converting them into compounds that are more easily absorbed by the intestine thereby increasing feed digestibility. While the second application will improve the quality of the surrounding environment, e.g., decompose toxic substance such as ammonia, sulphide, fish excreta, also excess feed that potentially attract pathogens and other unwanted micro- and macroflora. Benefits observed in probiotic supplementation in aquaculture include:

1. Increasing the nutritional value of food and its absorption by increasing the extent of the absorption area;

2. Increased activity of digestive enzymes;

3. Factors driving growth;

4. Pathogens inhibition;

5. Increased immune response; and

6. Improving the quality of fisheries water.

Meanwhile, one way to improve soil environmental quality is by introducing microbes to the soil through compost. Making compost from organic materials, in the form of by-products of agricultural activities as well as household, market and municipal waste needs to be encouraged. More effective compost making technology can be done by using a starter or microbes that accelerates the decomposition of organic material which is beneficial for plants. Giving compost on agricultural land provides a double benefit, namely improving soil microbial composition and increasing soil organic matter content. In order to make effective use of microbes in improving agriculture, animal husbandry and fisheries, it is necessary to develop microbial cocktails for each species of plant, livestock or fish [30]. The cocktail must have key features such as:

1. Able to be prepared on a large enough scale,

2. Remain stable and viable for a long time,

3. Good viability and growth, and

4. Give a positive impact on the host.

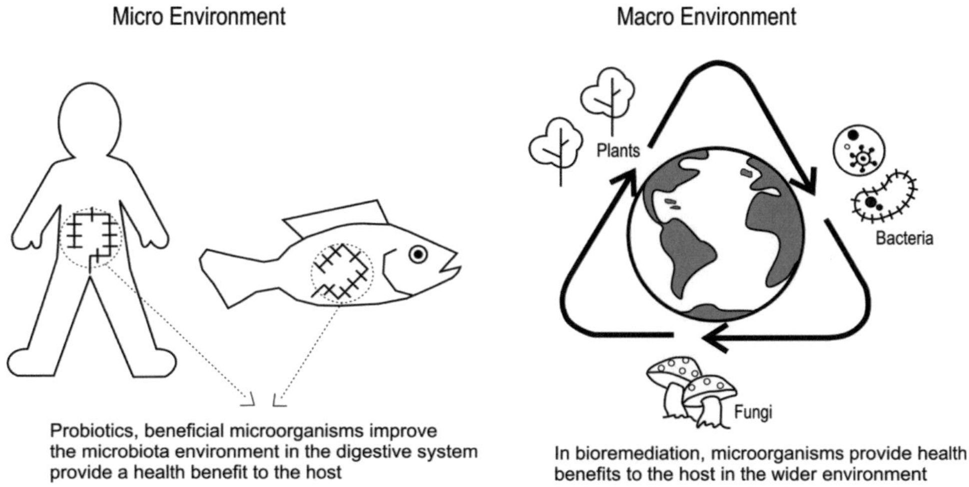

Figure 2.
The relationship between probiotics and bioremediation on a broader scale.

The working mechanism of bioremediation involves several technical aspects such as biotransformation, biodegradation, mineralization, phytohydraulics, bioaccumulation, and biovolatilization where the degrading microbes or plants remove, transform, modify, and/or convert a complex compound of pollutants into simpler and less toxic compounds. This bioremediation system has been successfully applied in cleaning contaminated sites, agricultural land, sediments, ground water, surface water, and sea water. Bioremediation through microorganisms generally involves the application of aerobic and anaerobic bacteria and fungi to restore the environment. Rhizoremediation is a remediation technique of soil contaminated pollutants by the action of plants (phytoremediation) and their symbiotic rhizosphere microbes. Plant growth-promoting microbes have been used for the restoration of infertile marginal land by increasing crop productivity. The application of probiotics on a broader scale is bioremediation. **Figure 2** shows the relationship between probiotics and bioremediation, where probiotics are defined as living microorganisms which, if given in sufficient quantities, provide health benefits to the host (humans, and/or animals), while bioremediation is augmentation or stimulation of microorganisms in sufficient quantities and manners, providing health benefits to the host (remediate or restoring polluted environment).

Author details

Qomarudin Helmy[1,2*], Edwan Kardena[1,2] and Sri Gustiani[3]

1 Water and Wastewater Engineering Research Group, Institut Teknologi Bandung, Indonesia

2 Bioscience and Biotechnology Research Center, Institut Teknologi Bandung, Indonesia

3 Center for Textile, Indonesian Ministry of Industry, Bandung, Indonesia

*Address all correspondence to: kihelmy@gmail.com

References

[1] Wainwright M, Wickramasinghe NC, Narlikar JV, Rajaratnam P. Microorganisms cultured from stratospheric air samples obtained at 41 km. FEMS Microbiology Letters. 2003;**218**:161-165

[2] Shivaji S, Chaturvedi P, Suresh K, Reddy GSN, Rajaratnam P, Wainwright M, et al. *Bacillus aerius* sp. nov., *Bacillus aerophilus* sp. nov., *Bacillus stratosphericus* sp. nov. and *Bacillus altitudinis* sp. nov., isolated from cryogenic tubes used for collecting air samples from high altitudes. International Journal of Systematic and Evolutionary Microbiology. 2006;**56**:1465-1473

[3] Shivaji S, Chaturvedi P, Begum Z, Pindi PK, Manorama R, Padmanaban DA, et al. *Janibacter hoylei* sp. nov., *Bacillus isronensis* sp. nov. and *Bacillus aryabhattai* sp. nov., isolated from cryotubes used for collecting air from the upper atmosphere. International Journal of Systematic and Evolutionary Microbiology. 2009;**59**:2977-2986

[4] Madigan MT, Martinko JM, Stahl DA, Clark DP. Brock Biology of Microorganisms. 13th ed. Benjamin Cummings, San Francisco; 2010:1152

[5] Lilly DM, Stillwell RH. Probiotics: Growth promoting factors produce by microorganisms. Science. 1965;**147**:747-748

[6] Sperti GS. Probiotics. West Point, Connecticut: AVI Publishing Co.; 1971

[7] Parker RB. Probiotics, the other half of the antibiotic story. Animal Nutrition and Health. 1974;**29**:109-121

[8] Fuller R. Probiotics in man and animals. The Journal of Applied Bacteriology. 1989;**66**:365-378

[9] Ebner S, Smug LN, Kneifel W, Salminen SJ, Sanders ME. Probiotics in dietary guidelines and clinical recommendations outside the European Union. World Journal of Gastroenterology. 2014;**20**(43):16095-16100

[10] Mansyur A, Tangko AM. Probiotik: Pemanfaatannya untuk pakan ikan berkualitas rendah. Media Akuakultur. 2008;**3**(2):145-149

[11] Nimrat S, Boonthai T, Vuthiphandchai V. Effects of probiotic forms, compositions of and mode of probiotic administration on rearing of Pacific white shrimp (*Litopenaeus vannamei*) larvae and postlarvae. Animal Feed Science and Technology. 2011;**169**:244-258

[12] Thy HTT, Tri NN, Quy OM, Fotedar R, Kannika K, Unajak S, et al. Effects of the dietary supplementation of mixed probiotic spores of *Bacillus amyloliquefaciens* 54A, and *Bacillus pumilus* 47B on growth, innate immunity and stress responses of striped catfish (*Pangasianodon hypophthalmus*). Fish & Shellfish Immunology. 2017;**60**:391-399

[13] Robertson PAW, o'Dowd C, Burrells C, Williams P, Austin B. Use of *Carnobacterium* sp. as a probiotic for Atlantic salmon (*Salmo salar* L.) and rainbow trout (*Oncorhynchus mykiss*, Walbaum). Aquaculture. 2000;**185**:235-243

[14] Ghosh AK, Bir J, Azad AK, Hasanuzzaman AF, Islam MS, Huq KA. Impact of commercial probiotics application on growth and production of giant fresh water prawn (*Macrobrachium rosenbergii* De Man, 1879). Aquaculture Reports. 2016;**4**:112-117

[15] Carnevali O, Zamponi MC, Sulpizio R, Rollo A, Nardi M, Orpianesi C, et al. Administration of probiotic strain

to improve sea bream wellness during development. Aquaculture International. 2004;**12**:377-386

[16] Pospiskova P, Zornikova G, Kolarova M, Sladek Z, Komprda T, Gersiova J. Effect of the probiotic in the pig nutrition on the pathogenic bacteria count in the gut. Acta Universitatis Agriculturae et Silviculturae Mendelianae Brunensis. 2013;**61**:1839-1843

[17] Ahmed ST, Hoon J, Mun H, Yang C. Evaluation of Lactobacillus and Bacillus-based probiotics as alternatives to antibiotics in enteric microbial challenged weaned piglets. African Journal of Microbiology Research. 2014;**8**:96-104

[18] Chen C, Tsen H, Lin C, Yu B, Chen C. Oral administration of a combination of select lactic acid bacteria strains to reduce the Salmonella invasion and inflammation of broiler chicks. Poultry Science. 2012;**91**:2139-2147

[19] Tesfaye A, Hailu Y. The effects of probiotics supplementation on milk yield and composition of lactating dairy cows. The Journal of Phytopharmacology. 2019;**8**(1):12-17

[20] Vibhute VM, Shelke RR, Chavan SD, Nage SP. Effect of probiotics supplementation on the performance of lactating crossbred cows. Veterinary World. 2011;**4**(12):557-561

[21] Barra PJ, Inostroza NG, Acuna JJ, Mora ML, Crowley DE, Jorquera MA. Formulation of bacterial consortia from avocado (*Persea americana* Mill.) and their effect on growth, biomass and superoxide dismutase activity of wheat seedlings under salt stress. Applied Soil Ecology. 2016;**102**:80-91

[22] Anjum MA, Sajjad MR, Akhtar N, Qureshi MA, Iqbal A, Jami AR, et al. Response of cotton to plant growth promoting rhizobacteria (PGPR) inoculation under different levels of nitrogen. Journal of Agricultural Research. 2007;**45**(2):135-143

[23] Brotman Y, Landau U, Cuadros-Inostroza A, Takayuki T, Fernie AR, Chet I, et al. Trichoderma-plant root colonization: Escaping early plant defense responses and activation of the antioxidant machinery for saline stress tolerance. PLoS Pathogens. 2013;**9**(4):1-15

[24] Fiorentino N, Ventorino V, Woo SL, Pepe O, De Rosa A, Gioia L, et al. Trichoderma-based biostimulants modulate rhizosphere microbial populations and improve N uptake efficiency, yield, and nutritional quality of leafy vegetables. Frontiers in Plant Science. 2018;**9**:1-15

[25] Caser M, Demasi S, Marisa I, Victorino M, Donno D, Faccio A, et al. Arbuscular mycorrhizal fungi modulate the crop performance and metabolic profile of saffron in soilless cultivation. Agronomy. 2019;**9, 9**:1-1

[26] Dwiyitno, Irianto DE, Yennie Y, Ariyani F, Indriati N, Yusuf G, et al. Utilization of Fish After Mass Death, Policy Brief No. PB03-5-2017. Research Center for Product Processing and Marine and Fisheries Biotechnology, Indonesian Ministry of Marine and Fisheries, Jakarta; 2017

[27] La VT, Cooke SJ. Advancing the science and practice of fish kill investigations. Reviews in Fisheries Science. 2011;**18**(1):21-33

[28] Irawan A, Hasani Q, Yuliyanto H. Fenomena Harmful Algal Blooms (HABs) di Pantai Ringgung Teluk Lampung, pengaruhnya dengan tingkat kematian ikan yang dibudidayakan pada Karamba Jaring Apung. Jurnal Penelitian Pertanian Terapan. 2015;**15**(1):48-53

[29] Anhar FP. Economic losses estimation due to water pollution in Cirata Reservoir, Cianjur Regency, West Java [Thesis]. Indonesia: Bogor Agricultural Institute; 2013

[30] Kompiang IP. Pemanfaatan mikroorganisme sebagai probiotik untuk meningkatkan produksi ternak unggas di indonesia. Pengembangan Inovasi Pertanian. 2009;**2**(3):177-191

[31] Google Maps Image. 2019. Coordinate: 6°43′21.0″S 107°16′04.2″E [Accessed: 10 October 2019]

[32] Sudarno A. Ratusan Ton Ikan Mati Mendadak di Cianjur. Liputan6 Online News Portal. 2017. Available from: https://www.liputan6.com/news/read/3111040/ratusan-ton-ikan-mati-mendadak-di-cianjur

[33] Suherlan D. Ratusan Jaring Apung Di Waduk Cirata Ditutup. Jabarnews Online Portal. 2018. Available from: https://jabarnews.com/read/40177

Chapter 9

Virulence Determinants of Non-typhoidal *Salmonellae*

Ruimin Gao, Linru Wang and Dele Ogunremi

Abstract

Non-typhoidal *Salmonellae* (NTS) belong to *Salmonella enterica* subspecies *enterica* and are common causes of foodborne illnesses in humans. Diarrhea is a common symptom but infection occasionally results in life-threatening systemic involvement. One member of the group, *S. enterica* subspecies *enterica* serovar Typhimurium has been extensively studied in live animal models particularly mice and cattle, leading to a better understanding of the pathogenesis of NTS and the development of diarrhea, respectively. This comprehensive review provides an insight into the genetic regulation of over 200 virulence determinants and their involvement in the four steps of *Salmonella* pathogenesis, namely: attachment, invasion, macrophage survival and replication, and systemic dissemination. There is, however, a paucity of information on the functions of some virulence factors present on the *Salmonella* pathogenicity islands (SPIs). The emergence of next generation sequencing (NGS) technology and the availability of more bacterial genomes should provide further insights into the biology of virulence determinants, mechanisms of NTS pathogenesis and host adaptation of *Salmonella*. The new knowledge should translate into improvement and innovations in food safety, and control of salmonellosis as well as better understanding of zoonotic infections in the context of One Health capturing the risks to humans, animals and the environment.

Keywords: non-typhoidal *Salmonellae*, virulence determinants, Typhimurium, attachment, intracellular survival, systemic dissemination, NGS, food safety, *Salmonella* pathogenicity islands, SPI

1. Introduction

Non-typhoidal *Salmonella* (NTS), a major cause of diarrheal disease globally, is estimated to cause 93 million enteric infections and 155,000 diarrheal deaths each year and is a leading cause of foodborne infections worldwide [1]. In Canada, 88,000 people are estimated to fall ill from foodborne NTS each year (90% credible intervals: 58,532–125,525) [2] with a mean hospitalization of about 925 individuals and 17 deaths [3]. An estimated 1 million cases of NTS infections occur annually in the United States alone, resulting in 19,000 hospitalizations and 380 deaths (http://www.cdc.gov/foodborneburden/PDFs/pathogens-complete-list-01-12.pdf). The genus *Salmonella* consists of Gram-negative, facultative intracellular bacteria and belongs to the Enterobacteriaceae family [4]. Historically, *Salmonella* organisms are serologically characterized using the conventional serotyping method known as the White-Kauffmann-Le Minor scheme which is based on the somatic (O), flagellar (H) and capsular (vi) antigens. Over 2600 serotypes are known to be present in a wide range of hosts including humans, cattle, pigs, horses, companion animals, reptiles, fish, avian, and insects [5]. The most commonly encountered pathogenic serovars belong to *S. enterica* subspecies *enterica* [6].

Some pathogenic *Salmonella* serovars are restricted to particular host species and are not found in other species. Examples of host-restricted *Salmonella* are serovars Typhi, Gallinarum, and Abortusovis, and they predictably cause systemic infection in their hosts namely, humans, fowls and ovines, respectively [7]. Another group of serovars are host-adapted including Dublin and Choleraesuis and primarily cause

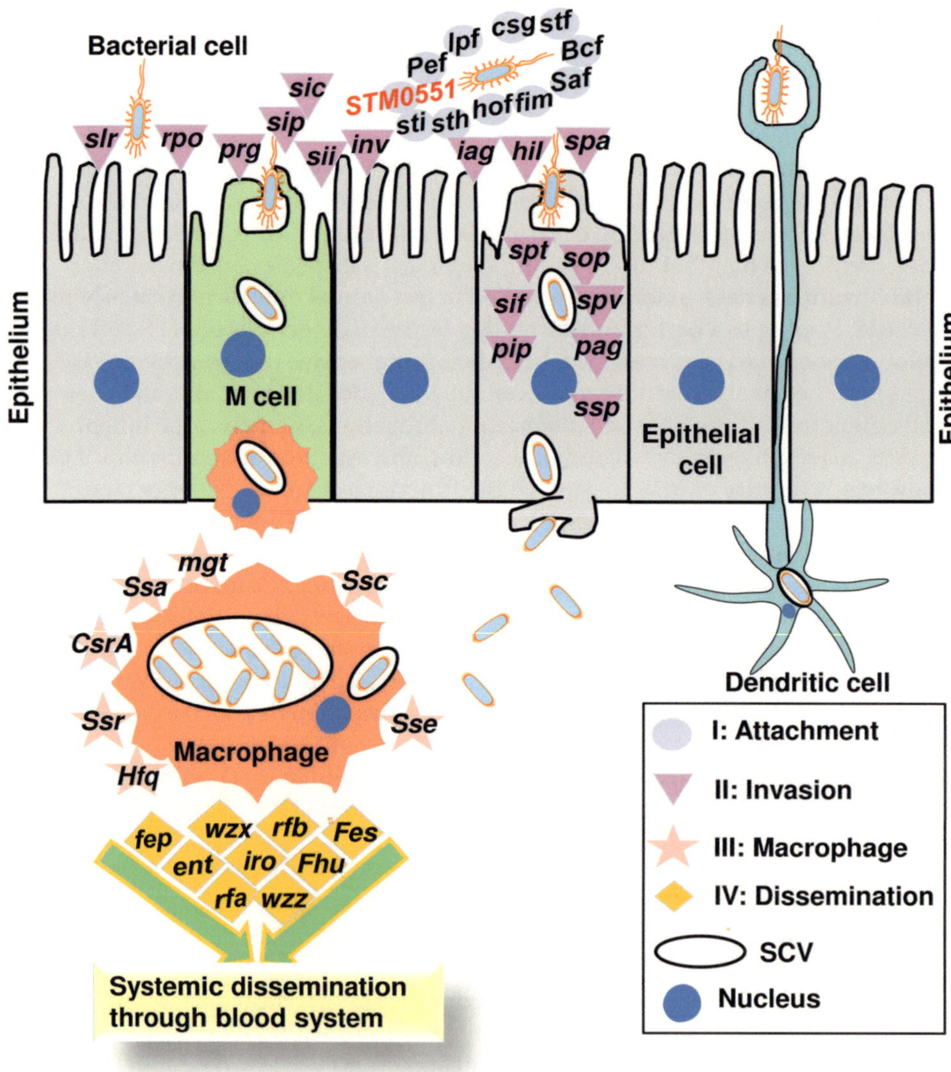

Figure 1.
Pathogenesis of Salmonella following contact with gut epithelium. (I) Salmonella cells attach to the epithelium mainly via adhesins, the representative virulence genes involved are fim, Saf, Bcf, stf, csg, lpf, Pef, sti, sth, hof, as well as a negative regulator of STM0551 (purple circles). (II) Three invasion methods are illustrated: M cells uptake bacteria cells through receptor mediated endocytosis, membrane ruffling and cytoskeletal rearrangement resulting in engulfment; alternatively, bacterial cells can be directly taken up by dendritic cells by phagocytosis. The main virulence factors involved are inv, pip, pag, prg, sap, sip, spa, spv, sop, rop, hil and sii (pink triangles). (III) Salmonella cells taken up by macrophages are localized within a Salmonella containing vacuole (SCV). The representative virulence genes involved in this process are mgt, Ssa, Sse, Ssr, CsrA and Hfq (light red star highlighted). (IV) Phagocyte-mediated systemic dissemination through blood system, mainly to liver, spleen and bone marrow. The virulence genes involved are iro, rfa, rfb, fes, Fhu, fep, ent, wzx and wzz (yellow diamond highlighted).

DOI: http://dx.doi.org/10.5772/intechopen.88904

disease in cattle and pigs respectively, but infrequently cause opportunistic disease in another host species especially humans [7, 8]. The most common non-adapted *Salmonella* are serovars Typhimurium and Enteritidis and they have been studied in live animal models such as mice and cattle, leading to a better understanding of the pathogenesis of NTS and the development of diarrhea [7]. *S. typhimurium* causes a systemic infection in mice that resembles typhoid fever caused by *S. enterica* serovar Typhi in humans [9]. While a vast majority of cases in otherwise healthy, *Salmonella*-infected humans present clinically as a self-limiting gastroenteritis, *S. typhimurium* can cause life-threatening systemic, invasive disease and bacteremia in some patients [10] but the reasons and mechanisms dictating the different disease manifestations in infected humans are not clear.

The advent of microbial whole genome sequencing promises to provide insights to better understand the biology of virulence determinants and mechanisms of NTS pathogenesis. Genomes of *Salmonella* are generated increasingly at a faster rate and deposited in public databases [11]. Further understanding of genome diversity and variation of bacterial pathogens has the potential to improve quantitative risk assessment and assess the evolution of *Salmonella* and emergence of new strains [12]. Mining of the repository of genomes should provide new information expected to complement existing knowledge on virulence genes derived from host infection studies especially involving *Salmonella* mutants. The *Salmonella* Foodborne Syst-OMICS database (SalFoS) was developed as a platform to improve diagnostic accuracy, to develop control methods in the field and to identify prognostic markers in epidemiology and surveillance [13]. Bioinformatics analyses of genomes are expected to reveal the mechanisms of action of virulence genes and help decipher whether there is a dichotomy in the genes contributing to invasive disease compared to restricted pathogenesis in the intestinal tract [14].

This review provides an overview of the genetic regulation of over 200 virulence determinants highlighting their involvement in each of the four steps of *Salmonella* pathogenesis, namely: attachment, invasion, macrophage survival and replication, and systemic dissemination (**Figure 1**). Further analysis of virulence genes will provide us insights in to understanding the mechanisms of invasive disease which appear distinct from gastroenteritis. For instance, the organisms which are responsible for invasive disease have fewer genes because of pseudogenization. Many of these virulence genes have redundant functions; however two *Salmonella* molecules are known to exert a dominant effect in pathogenesis, namely: lipopolysaccharide (LPS) and invasion protein A (invA). Many virulence factors have distinct and unique functions but cooperative crosstalk has been documented at the different steps of infection, e.g., protein products of genes encoded on two *Salmonella* pathogenicity islands (SPI), SPI-2 and SPI-4.

2. Virulence determinants involved in *Salmonella* pathogenesis

2.1 Attachment

In a majority of cases, infection occurs following ingestion of *Salmonella* by the host. Before *Salmonella* can gain entry into the epithelial cell lining the host's gut mucosa, it first needs to attach to the cell. NTS attachment is facilitated by fimbriae, non-fimbriae factors of autotransporter and outer-membrane proteins, which serve as adhesions; up to 20 adhesion molecules have been described so far and it has been demonstrated that the entire adhesiome of *S. enterica* serotype Typhimurium can be expressed [15], which facilitates understanding such a large repertoire of adhesions contributing to colonization of a broad range of host species and adaptation to various environment within the host.

2.1.1 Fimbrial adhesins

Fimbriae, also known as pili, are thin, filamentous appendages protruding on the bacterial surface and consist of polymerized aggregates of small molecular weight monomers of the fimbrin protein [16]. Characteristically, fimbriae mediate the initial attachment of Gram-negative bacterial pathogens to host cells and surfaces [17]. In *Salmonella*, the initial contact results in relatively weak adherence of the bacteria to intestinal epithelial cells but soon induces *de novo* bacterial protein synthesis which increases the strength and intimacy of the attachment [18]. This process is also accompanied by the development and assembly of a unique secretion apparatus called the Type 3 Secretion System (T3SS) which is required for *Salmonella* to invade epithelial cells [19]. The chromosome of *S. typhimurium* contains 13 fimbrial operons, *afg* (*csg*), *bcf*, *fim*, *lpf*, *pef*, *saf*, *stb*, *stc*, *std*, *stf*, *sth*, *sti*, and *stj* [20–22] (**Table 1** and **Figure 1**). Eight types of fimbriae which have been experimentally investigated [23] are outlined below.

2.1.1.1 Mannose-sensitive Type I fimbriae (Fim)

Mannose-sensitive Type I fimbriae (Fim) are encoded by the *fim ACDHIFZYW* operon and bind to D-mannose-containing receptors on host cell surface as well as the glycoprotein laminin of the extracellular matrix [24]. Type I fimbriae promoted bacterial attachment to epithelial cells, facilitated the invasion of HEp-2 cells and HeLa cells and the colonization of the gut mucosa in chicken, mouse, rat and swine [25, 26]. An immunization experiment using purified Fim protein led to the protection of laying hens against egg contamination and colonization of the reproductive organs by *S. enteritidis* [27]. FimA, FimF, and FimH are necessary for the assembly of Type 1 fimbriae on *S. typhimurium* [24]. Differently, *STM0551* gene plays a negative regulatory role in the regulation of type 1 fimbriae in *S. typhimurium* [28].

2.1.1.2 Plasmid-encoded fimbriae (Pef)

Plasmid-encoded fimbriae (Pef) participate in the attachment of bacteria to the surface of murine small intestine and are necessary for fluid production in the infant mouse similar to the observation with the fimbriae of enterotoxigenic *Escherichia coli* and *Vibrio cholerae* [29]. Expression of *pef* gene is regulated by DNA methylation [30]. Purified Pef specifically binds the trisaccharide Galβ1-4(Fucα1-3) GlcNAc (also known as the Lewis X blood group antigen or Le^x), which are preponderant on the surface of human erythrocytes, skin epithelium and mucosal surfaces [31].

2.1.1.3 Long polar fimbriae (Lpf)

Long polar fimbriae (Lpf) encoded by the *lpfABCDE* fimbrial operon is involved in the colonization of murine Peyer's patches by mediating adherence to M cells, a preferred port of entry for *Salmonella* in mice [32]. Mutation of the *lpfC* gene which encodes the fimbrial outer membrane usher attenuated the virulence of *Salmonella typhimurium* in orally exposed mice as shown by a 5-fold increase in the number of organisms needed to kill 50% of test animals (i.e., LD_{50}) when compared to the wild type organism. Lpf is also involved in the early stages of biofilm formation on host epithelial cells [33] and participate in intestinal persistence in mice [34].

Virulence genes	Location*	Functions
BcfABCDEFGH	Chromosome	Contribute to long-term intestinal carriage and bovine colonization
csgABCDEFG	Chromosome	Curlin subunit; assembly and transport component in curli production; DNA-binding transcriptional regulator
fimCDFHIWYZ	Chromosome	Adhesion to epithelial cells; biofilm formation
hofBC	Chromosome	Type IV pilin biogenesis protein
lpfABCDE	Chromosome	Biofilm formation, contribute to long-term intestinal carriage
misL	SPI-3	An extracellular matrix adhesion involved in intestinal colonization
pefA	Plasmid	Adhesion to crypt epithelial cells; induction of proinflammatory response
ppdD	Chromosome	Putative major pilin subunit
SafC	Chromosome	Salmonella atypical fimbria outer membrane usher
ShdA	CS54	Outer membrane
StdB	Chromosome	Contribute to long-term intestinal carriage
stfACDEFG	Chromosome	Not required for long-term intestinal carriage of mice
sthABD	Chromosome	Outer membrane fimbrial usher. Putative fimbrial subunit and chaperone protein
StiABC	Chromosome	Putative fimbrial subunit/usher/chaparone
STM0551	Chromosome	Downregulates fimbrae protein expression and acts as a negative regulator of virulence
STM4595	Chromosome	Unknown function

**SPI-3 and CS54 are genomic islands on Salmonella chromosome.*

Table 1.
Location and function of the major proteins and virulence determinants contributing to Salmonella attachment.

Lpf synthesis is regulated by an on–off switch mechanism (phase variation) to avoid host immune responses [35].

2.1.1.4 Thin aggregative fimbriae

Thin aggregative fimbriae also known as curli [36] with the designation Agf/Csg, are encoded by the *agf/csgBAC* gene cluster [37]. The thin aggregative fimbriae for Enteritidis which is known as SEF 17 is responsible not only for the auto-aggregative phenotype of the bacteria, but for fibronectin binding [38] and has been shown *in vitro* to bind immortalized small intestinal epithelial cells from mice [36]. Mutation in *agfB* resulted in a 3- to 5-fold increase in the oral LD_{50} of Typhimurium for mice [39].

2.1.1.5 Bovine colonization factor (Bcf)

Bovine colonization factor (Bcf) is encoded by genes in the *bcf* gene cluster. The fimbrial usher protein encoded by *bcfC* is required for colonization of bovine but not murine Peyer's patches in oral infection models of calves and mice [40]. The *bcf* gene together with five other fimbrial operons—*lpf*, *stb*, *stc*, *std*, and *sth*—are reported to be required for long-term intestinal carriage of Typhimurium in genetically resistant mice [34].

2.1.1.6 Salmonella atypical fimbriae (Saf)

Salmonella atypical fimbriae (Saf) are encoded by the chromosomal *safABD* operon. A group of BALB/c mice immunized subcutaneously with SafB/D- and recombinant cholera toxin B subunit (rCTB)-conjugated micro-particles had significantly lower CFU counts than the untreated control group [41]. Two additional functions - poly-adhesive and self-associating activities – were attributed to the Saf pili and appear to contribute to host recognition and biofilm formation [42].

2.1.1.7 Typhimurium fimbriae std and stf operons

Std operon is required for adherence to human colonic epithelial cells and for cecal colonization in the mouse by binding to cecal mucosa receptors containing α(1, 2) fucose residues [34, 43]. *Stf* fimbriae share homology with the MR/P fimbriae of *Proteus mirabilis* and *E. coli* Pap fimbriae [44]. *StfA* expression is induced during infection of bovine ileal loops [45].

2.1.1.8 Enteritidis fimbrial SEF14

Enteritidis fimbrial SEF14 contributes to colonization of chicken intestine, liver, spleen and reproductive organs [46, 47]. The fragment encoding genes responsible for SEF14 biosynthesis contain three genes, *sefABC*. The putative adhesion subunit encoded by *sefD* is essential for efficient uptake or survival of Enteritidis in macrophages, as the *sefD* mutants were not readily internalized by peritoneal macrophages compared with the wild-type bacteria soon after intraperitoneal infection of mice [48]. The *sefD* mutant was severely attenuated after both oral and intraperitoneal infection of BALB/c mice (approximate LD50: $>10^4$ (mutant) vs. <10 (wild type)) [48]. In the mouse model, egg-yolk derived anti-SEF14 antibodies afforded passive protection [49].

2.1.2 Non-fimbrial adhesins

Four distinct non-fimbrial intestinal colonization factors have been identified:

2.1.2.1 MisL

MisL encoded within the SPI-3, is an outer membrane fibronectin-binding autotransporter protein which is induced upon bacterial contact with the intestinal epithelial cells, and is required for colonization of the murine cecum and for intestinal persistence. MisL binds fibronectin and collagen IV via its passenger domain [50].

2.1.2.2 ShdA

ShdA gene is located in the 25-kb pathogenicity island called CS54 which is present only in *S. enterica* subspecies *enterica* [51]. ShdA is a large fibronectin/collagen I-binding outer membrane protein which is induced *in vivo* in the murine caecum [52]. It is required for Typhimurium colonization in the murine caecum and Peyer's patches of the terminal ileum [53] and for efficient and prolonged shedding of the organism in feces [51].

2.1.2.3 BapA

BapA is a huge surface-associated protein and secreted via its downstream type I secretion system, BapBCD. BapA contributes to murine intestinal colonization and subsequent organ invasion. Mice orally inoculated with *BapA*-deficient strain survived longer and have a significant reduction in mortality rate than those inoculated with the wild-type strain [54].

2.1.2.4 SiiE

SiiE is a SPI4-encoded protein and works as the substrate protein of the T1SS. SiiE is secreted into the culture medium but mediates contact-dependent adhesion to epithelial cell surfaces. SiiE codes for a giant non-fimbrial adhesion of 600 kDa and consists of 53 repeats of immunoglobulin domains; this is a T1SS-secreted protein that functions as a non-fimbrial adhesion in binding to eukaryotic cells [55].

2.2 Intestinal phase: invasion and intracellular survival

Shortly after adhesion to a host cell, *Salmonella* invasion proceeds as a consequence of the activation of host cell signaling pathways leading to profound cytoskeletal rearrangements [56]. These internal modifications dislocate the normal epithelial brush border and induce the subsequent formation of membrane ruffles that engulf adherent bacteria in barge vesicles called *Salmonella* containing vacuoles (SCVs), which is the only intracellular compartment where *Salmonella* cells survive and replicate [57, 58]. Simultaneously, induction of secretory response in the intestinal epithelium initiates recruitment and transmigration of phagocytes from the submucosal space into the intestinal lumen. Alternatively, *Salmonella* cells may be directly engulfed by dendritic cells from the submucosa. Taken up During SCV maturation, *Salmonella* induces *de novo* formation of an F-actin meshwork around bacterial vacuoles, a process which is termed vacuole-associated action polymerization (VAP) and is important for maintenance of the integrity of the vacuole membrane [59]. Furthermore, intracellular *Salmonella* can induce the formation of long filamentous membrane structure called *Salmonella*-induced filaments (SIFs) [60], which may lead to an increased availability of nutrients within the SCV [61]. A fraction of SCVs transcytose to the basolateral membrane. Once across the intestinal epithelium, *Salmonella* are engulfed by phagocytes and internalized again with SCVs, triggering a response similar to that reported inside epithelial and M cells to ensure bacterial survival and replication [62]. The pathogenic bacterium must at this stage employ many virulence strategies to evade the host defense mechanisms (**Figure 1**).

The majority of the virulence determinants are located within highly conserved SPIs on the chromosome, while others are either on a virulence plasmid (pSLT) or elsewhere in the chromosome. To date, 21 SPIs have been identified in *Salmonella*, and the generalist *S. typhimurium* and the invasive *S. typhi* genomes share 11 (SPIs-1 to 6, 9, 11, 12, 13 and 16). Two SPIs namely SPI-8 and 10 were initially found in *S. typhi* and without counterparts in *S. typhimurium* chromosome; SPI-14 is specific to *S. typhimurium*, while *SPIs*-7, 15, 17 and 18 are specific to *S. typhi*; and SPIs-19, 20 and 21 are absent in both of them [63]. Because of the prominence of the SPIs in pathogenesis, the virulence factors encoded on the major SPIs, SPI-1 to SPI-5 are described below, and their respective functions summarized (**Tables 2** and **3**).

2.2.1 SPI-1 mediates contact-dependent invasion of the intestinal epithelium and enteropathogenesis

SPI-1 codes for several effector proteins that trigger invasion of epithelial cells by mediating actin cytoskeletal rearrangements and hence internalization of the

Virulence genes	Location*	Functions
Crp	Chromosome	cAMP-regulatory protein
hilACD	SPI-1	Promote phop-repressed prgHIJK, sipA, sipC, invF, and orgA; activates the expression of the *hilA* gene
Hnr	SPI-2	SPI-2 regulator (transcriptional and post-transcriptional)
HtrA		Resistance to periplasmic stress
IacP	SPI-1	Posttranslational modification
iagB	SPI-1	Invasion
invABCEFGIJ	SPI-1	Secretion and chaperone; promote sipBCDA, sigD and sicA
msgA	Chromosome	Unknown function
ompR/envZ	SPI-2	Regulates *ssrAB* expression
orgABC	SPI-1	Pathogenesis; secretion
phoR/Q	SPI-2	Regulates *ssrAB* expression; down-regulates the transcription of its master regulator HilA, control *mgtC*
pagACDP	SPI-11	Resistance to AMP, macrophage cytotoxicity
pipABB2CD pipC (*sigE*)	SPI-5	Pathogenesis, effector protein; sif extension; SCV maturation and positioning
prgHIJK	SPI-1	Secretion
Prc		Resistance to periplasmic stress
rpoES rpoS (*katF*)	SPI-2	SPI-2 regulator (transcriptional and post-transcriptional); controls the transcription of the regulatory gene *spvR*; expression of rpoS is induced after entry of *Salmonella* into macrophages or epithelial cells, or in vitro during the stationary growth phase
rtsA	Chromosome	Activates the expression of the *hilA* gene
sapABCDF		Resistance to AMP, macrophage cytotoxicity
sifA	SPI-2	Sif formation in epithelial cells and maintenance of SCV membrane integrity
siiCDEF	SPI-4	Translocation; adhesion to apical side of polarized epithelial cells; involved in T3SS-1 dependent invasion
sicAP	SPI-1	Chaperone for sipBC
sipA (*sspA*)	SPI-1	Stabilization and localization of actin filaments during invasion, stabilization of VAP, correct localization of SifA and PipB2, SCV perinuclear migration and morphology, promote inflammatory response and fluid secretion
sipBCD (*sspBCD*)	SPI-1	Adhesion to epithelial cells, early macrophage pyroptosis, macrophage autophagy; Adhesion to epithelial cells
SpaSRQPO	SPI-1	EscU/YscU/HrcU family type III secretion system export apparatus switch protein; antigen presentation protein SpaO
sptP	SPI-1	Disruption of the actin cytoskeleton rearrangements by antagonizing SopE, SopE2, and SigD, downregulate inflammatory response
sirA	SPI-1	SirA/BarA encoded outside SPI-1 activates HilA
slrP	Chromosome	Adhesion to epithelial cells
slyA	SPI-2	Regulates resistance to oxidative stress
sspH1H2	Phage	Localize to the mammalian nucleus and inhibits NF-κB-dependent gene expression; SCV maturation and positioning

Virulence genes	Location*	Functions
sodABD		Resistance to oxidative stress
SopABDD2EE2 sopB (sigD)	SPI-5	Chloride secretion; promote actin cytoskeletal rearrangements, invasion and inhibition of apoptosis of epithelial cells, induction of proinflammatory response and fluid secretion, SCV size, instability, maturation and positioning, nitrate respiration, outgrowth in the intestine; inhibition of vesicular trafficking; replication inside macrophages; sif formation
spaOPQRS	SPI-1	Secretion
SprB	SPI-1	Regulation of transcription, DNA-templated
spvABCD	Plasmid	Modifies actin and destabilizes the cytoskeleton of infected cells; SCV maturation and positioning; induction of apoptosis; Host cell signaling
SsJ		Resistance to oxidative stress
STM2231	SPI-2	SPI-2 regulator (transcriptional and post-transcriptional)
YejABEF	Chromosome	Resistance to AMP, macrophage cytotoxicity
ymdA	Chromosome	Stress response

**SPI1–5 are genomic islands on Salmonella chromosome.*

Table 2.
Location and function of the major proteins and virulence determinants contributing to Salmonella invasion.

bacteria. These effectors are translocated into host cell by means of a Type III Secretory System or T3SS-1 [64], which is made up of proteins encoded by the SPI-1, such as *inv*, *spa*, *prg* and *org* [65]. Naturally occurring mutants of *Salmonella* have been found in the environment with a deletion of a vast DNA segment of SPI-1 locus and are deficient for *inv*, *spa*, and *hil* hindering their ability to enter cultured epithelial cells [66]. Mutations leading to a defective secretory function of T3SS-1 led to a 50-fold increase in LD_{50} following oral administration of Typhimurium in the mouse model [67]. The *prg/org* and *inv/spa* operons encode the needle complex, whereas the *sic/sip* operons encode the effector proteins and the translocon (SipBCD), a pore-forming structure that embeds in the host cell membrane and delivers these effectors to the host cytosol. In addition, several chaperones are also encoded within SPI-1. For example, SlrP mediate ubiquitination of ubiquitin and thioredoxin [68] and one of the SPI-1 regulons, STM4315 (*rtsA*) interferes with the interactions of *S. typhimurium* and host cells [69]. In general, the expression of SPI-1 genes is subject to control by complex regulatory mechanisms involving local regulators such as HilA, iagB and InvF which are necessary for host invasion by *Salmonella* and induction of gastroenteritis [70, 71]. For example, *prgHIJK*, *invA*, *invJ*, and *orgA* are primarily regulated by HilA [71]. In addition, two major global regulatory networks, SirA/BarA and PhoP/PhoQ, indirectly regulate the expression of the invasion-associated genes via HilA [72, 73].

2.2.2 SPI-2 is essential for survival and replication in macrophage

The SPI-2 is composed of two segments. The smaller portion contains the *ttrRSBCA* operon, which is involved in tetrathionate reduction, and seven open reading frames (ORFs) of unknown function. The expression of these genes may contribute a growth advantage over the microbiota [74]. The larger portion of this island was shown to be critical for the ability of *Salmonella* to survive and replicate

Virulence genes	Location	Functions
CsrA		RNA chaperones
Hfq	SPI-2	SPI-2 regulator (transcriptional and post-transcriptional), RNA chaperones
mgtABCD	SPI-3	A hydrophobic membrane protein; Mg^{2+} transporter (Mg^{2+}-transporting P-type ATPase)
SsaABCDEFGHIJKLMNOPQRSTUV *ssaB* (*spiC*), *ssaC* (*spiA*), *ssaD* (*spiB*), *ssaR* (*yscR*).	SPI-2	Regulate the secretion of translocon proteins under conditions that simulate the vacuolar environment; interferes with vesicular trafficking; intracellular bacterial proliferation; secretion
sscAB	Chromosome	Putative type III secretion system chaperone protein or pathogenicity island effector protein
sseABCDFGIJL	SPI-2	Translocation; sif formation in epithelial cells; SCV maturation and positioning; SCV membrane dynamics; nuclear response-gene expression;
ssrAB (*ssrA/SpiR*)	SPI-2	Regulates SPI-2 gene expression

Table 3.
Location and function of the major proteins and virulence determinants contributing to Salmonella macrophage survival and replication.

inside host cells—both epithelia cells and macrophages—within the SCV [75]. Non-functional SPI-2 mutants are unable to colonize internal target organs such as spleen and liver of mice, although they penetrate the intestinal barrier as efficiently as the wild type strain [76]. These mutants were attenuated by at least five orders of magnitude compared with the wild type strain after either oral or intraperitoneal inoculation of mice [75]. The SPI-2 related events are triggered by the action of effector proteins with its own T3SS known as T3SS-2, which also encodes its proper translocon machinery named SseBCD [77]. Gene sequence similarity to the known components of other T3SS has been used to propose functions for *SsaN*, *SsaR*, *SsaS*, *SsaT*, *SsaU* and *SsaV* as coding for putative proto-channel components, SsaD/SpiB, SsaJ, SsaK and SsaQ appear to code for basal components, whereas SsaC/SpiA may code for an outer ring protein [78]. Generally, SPI-2 contains four types of virulence genes: *ssa* encodes T3SS-2 apparatus; *ssr* encodes regulators; *ssc* encodes the chaperones and *sse* encodes the effectors (**Table 2**) [79, 80].

2.2.3 SPI-3 contributes to intramacrophage proliferation

Unlike SPI-1 and SPI-2, only four ORFs within SPI-3 have been shown to contribute to replication in macrophages via a high-affinity Mg^{2+} uptake system [81]. The *mgtC* gene encoding a 22.5-kDa hydrophobic membrane protein, is the major virulence gene factor found within this locus, and is responsible for growth in Mg^{2+} limiting environment, intramacrophage survival, and systematic virulence in mice [82]. The transcription of *mgtC* is followed by activation of PhoP-PhoQ in response to low Mg^{2+} levels [81].

2.2.4 SPI-4 is involved in colonization

The fourth SPI contributes to *Salmonella* colonization in the intestine of cattle, but not of chicks [83]. Loss of SPI-4 attenuates the oral but not intraperitoneal virulence of serovars Typhimurium and Enteritidis in mice [84]. Three genes namely *SiiC*, *SiiD*, and *SiiF* produce proteins that form the type 1 secretion system (T1SS); the fourth gene, *siiE* codes for a giant non-fimbrial adhesion exported by the T1SS and mediates contact-dependent adhesion to polarized epithelial cells rather than to non-polarized cells. In contrast, SiiA and SiiB are not secreted but represent inner membrane proteins whose function is unknown [55, 85]. Recently, transmembrane mucin MUC1 was shown to be required for *Salmonella* siiE-mediated entry of enterocytes via the apical route [86].

2.2.5 SPI-5 is associated with enteropathogenicity

The SPI-5 locus is well characterized in the serovar Dublin infection in calves. This bovine-adapted serovar primarily causes bacteremia rather than gastroenteritis in humans. This region comprises six genes namely, *pipD*, *orfX*, *sopB* (also known as *sigD*), *pipC* (also known as *sigE*), *pipB*, and *pipA* [87]. Four gene products which include three SPI-5 Pip proteins (PipD, PipB, PipA) and one SPI-1 SopB protein are involved in secretory and inflammatory responses in bovine ligated ileal loops but they do not appear to play a significant role in the development of systemic infection in mice inoculated by the intraperitoneal route [87, 88]. Furthermore, it has been found that SigE serves as a chaperone for the *S. typhimurium* invasion protein, SigD [89].

2.2.6 Crosstalk between SPI-1 and SPI-2 gene products to promote Salmonella survival and virulence

The SPI-2 genes are activated after *Salmonella* gains access into the SCV [76]. T3SS-2 secretes multiple effector proteins into different subcellular fractions where they interfere with various host cellular functions to establish a replication-permissive environment [90]. The identified effectors are encoded within SPI-2 (e.g., SpiC, SseF and SseG) and outside SPI-2 (e.g., SifA, SseI, SseJ and SspH 2) [23]. These SPI-2-encoded effectors together with some of SPI-1-encoded effectors (e.g., SipA, SipD, SopA, SopE, SopB) that persist in the host cytosol after invasion, are distributed in different cellular compartments including the vascular membrane of SCV and Sif, host cytosol, cytoskeleton, Golgi apparatus, and nucleus. These molecules influence distinct intracellular events and collectively contribute to establish a *Salmonella* replicative niche in macrophages [91]. These intracellular events include: inhibition of endocytic trafficking, evasion of NADPH oxidase-dependent killing [92, 93], induction of a delayed apoptosis-like host cell death [94], assembly of a meshwork of F-actin around the SCV [59], accumulation of cholesterol in the SCV [95], and interference with the localization of inducible nitric oxide synthase to the SCV [96]. Efficient replication has been found to be associated with two phenotypes involving host microtubule cytoskeleton and its motor proteins, Golgi apparatus-associated juxtanuclear positioning of SCV [97–99] and Sifs formation which appear as tubular membrane extensions of SCVs enriched in lysosomal glycan proteins [100].

2.2.7 Joint regulation between SPI-1 and SPI-4

The functional relatedness between SPI-1 and SPI-4 is reflected by their co-regulation by the same set of key regulators, for example, a transcriptional activator

SprB encoded within SPI-1 and regulated by HilA under similar environmental conditions; SprB directly activates SPI-4 gene expression and weakly represses SPI-1 gene expression through HilD [101].

2.3 Intramacrophage survival and replication

Similar mechanisms occur inside epithelial cells after intestinal invasion and once bacteria have been internalized by macrophages. Briefly, *Salmonella* cells are localized in the SCV once engulfment is completed. Preserving the SCV membrane integrity plays a crucial role in allowing *Salmonella* replication inside these intracellular niches. These procedures are regulated by T3SS-2 transporting action and its translocon machinery, namely SseBCD complex [77]. Hence, the required effectors which are encoded both inside and outside SPI-2 facilitate the success of *Salmonella* intramacrophage survival. The SPI-2 gene expression is triggered in response to a number of environmental signals mimicking the vacuolar environment of SCV, including stationary growth phase, low osmolarity [102], low concentrations of Mg^{2+}, Ca^{2+} or PO_3 [103, 104], and low pH [76]. The expression of SPI-2 genes is coordinately regulated at both transcriptional and post-transcriptional levels. During the transcription of SPI-2 genes, many two-component regulatory systems are involved, including SsrA-SsrB, OmpR-EnvZ and PhoP-PhoQ as well as transcriptional regulators, namely SlyA and the alternative sigma factor of RNA polymerase RpoE. The main regulatory proteins that act post-transcriptionally are the RNA chaperons, including Hfq, CsrA, and SmpB. The *mgtC* gene located in SPI-3 has been shown to contribute to replication in macrophages. All the mentioned virulence determinants can be found in **Table 3** and **Figure 1**.

2.4 Systemic infection/dissemination

Internalization of the infecting *Salmonella* within SCV is followed by systemic spread through other target organs, such as the spleen and liver. As a prerequisite for spread, the bacterial cells must evade the innate immune system. During this process, serum resistance or resistance to complement-mediated serum killing is a major virulence factor for the development of systemic salmonellosis. It involves three major factors, namely LPS, outer membrane proteins PagC and Rck and siderophores (**Table 4** and **Figure 1**).

2.4.1 LPS constitutes a chemical and physical protective barrier for the cell

LPS of Gram-negative bacteria, a major component of the outer membrane, constitute a chemical and physical protective barrier for the cell. LPS consists of the hydrophobic lipid A, a short non-repeating core oligosaccharide and a long distal repetitive polysaccharide termed O-antigen or O-side chain [105]. Complete LPS is characterized by long O-antigen which confers the smooth (S) phenotype on *Salmonella*. The O-antigen is a major component associated with serum resistance. Incomplete LPS devoid of O-antigen leads to rough (R) phenotype, which is of low virulence [106]. Naturally occurring infections are caused by S-phenotype *Salmonella*, which are resistant to complement killing [107, 108]. There is a correlation between the amount, structure, and chain length of the O-antigen and virulence [109]. The long O-antigen of LPS confers on the organism the ability to resist complement-mediated serum killing by sterically hindering the insertion of the membrane attack complement complex (C5b-9) into the bacterial outer membrane [107, 108].

Virulence genes	Location	Functions
cirA	Chromosome	Colicin I receptor
entABCDEF	Chromosome	Enterobactin synthase
fepABCDEG	Chromosome	Outer membrane receptor; iron-enterobactin transporter binding protein
Fes	Chromosome	Salmochelin secretion/degradation
FhuABCDE	Chromosome	Enterobactin/ferric enterobactin esterase
foxA	Chromosome	Ferrioxamine B receptor precursor
FruR	SPI-2	DNA-binding transcriptional regulator
FUR	Chromosome	Ferric uptake regulator
iroBCDE	Chromosome	Salmochelin glycosylation, transport and processing
MsbA	Chromosome	Lipid transporter ATP-binding/permease protein
rfaBCDFGHIJKLPQYZ	Chromosome	LPS core biosynthesis protein; transcriptional activator; O-antigen ligase
rfbBDFGHIJKMNOPUVX	Chromosome	Glucose biosynthesis pathway; O-chain glycosyltransferase; O-antigen transporter
rfc	Chromosome	O-antigen polymerase
STM0719	Chromosome	Unknown function
wzxCE	Chromosome	Colanic acid exporter; putative LPS biosynthesis protein
wzzBE	Chromosome	LPS chain length regulator and biosynthesis protein
yibR	Chromosome	Unknown function
ybdAB	Chromosome	Enterobactin exporter EntS

Table 4.
Location and function of the major proteins and virulence determinants contributing to Salmonella dissemination.

Surface expression of O-antigen involves multiple steps: O-antigen biosynthesis in the inner membrane (*rfb*), translocation across the inner membrane by Wzx flippase (*wzx*), polymerization (*wzz*, *rfc* and *rfe*) and ligation on to the preformed Core-Lipid A complex by WaaL ligase (*rfaL*). The Core-Lipid A is translocated independently by the ATP-binding cassette (ABC) transporter MsbA [110, 111]. Complete LPS molecules are then transported to the surface across the periplasm and outer membrane by the Lpt (LPS transport) pathway [111]. Defects in any of the above steps would affect the surface display of the O-antigen and its function. The mutants defective in the biosynthesis of LPS core encoded by the *rfa* loci or the O side chain by the *rfb* loci, are significantly attenuated with a LD50 at least 100 times higher than the parental strain in chickens subcutaneously infected with Enteritidis [112].

Typhimurium possesses two functional *wzz* genes responsible for regulating the chain length of the O-antigen [113]. One is *wzzST* encoding a long LPS with 16–35 O-antigen repeat units and the other *fepE* gene coding for a very long LPS estimated to contain more than 100 repeat units [113]. Either gene product is sufficient for complement resistance and virulence in the mouse model of infection, which reflects a degree of functional redundancy of these two *wzz* genes [113]. Double mutation of these two *wzz* genes resulted in relatively short, random-length

O-antigen and the mutant displayed enhanced susceptibility to complement-mediated killing and was highly attenuated in mice [113]. The transcription of *wzzST* gene is independently activated by two-component systems of Typhimurium, PmrA/PmrB (PmrA, sensor; PmrB, response regulator) and RcsC/YojN/RcsB (RcsC, sensor; YojN, intermediate phosphotransfer protein; RcsB, response regulator) [114]. PmrA/PmrB is activated through two pathways: one is directly activated through its cognate sensor PmrB in response to Fe^{3+} and the other is dependent on the PhoP/PhoQ two-component system in response to low Mg^{2+}. The RcsC/YojN/RcsB is activated in the presence of low Mg^{2+} plus Fe^{3+} [114]. In addition, mutants in a number of genes (*rfaG*, *rfaI*, *rfaL*, *rfaQ*, *rfaP*, *rfbC*, *rfbD*, *rfbJ*, *rfbM*, *rfbP*, *yibR*) necessary for LPS biosynthesis/assembly had severely impaired movement on swimming motility agar [115].

2.4.2 PagC and Rck confer resistance to the complement-mediated bacterial activity

In addition to LPS, two outer membrane proteins, the 18-kDa PagC [116] and the 17-kD Rck [117], confer a high level of resistance to the complement-mediated bactericidal activity. These two proteins share homology with virulence-associated outer membrane protein Ail from *Yersinia* that blocks formation of the complement membrane attack complex on the bacterial surface. Similarly, complement resistance mediated by Rck is associated with a failure to form fully polymerized tubular membrane attack complexes [117]. One strain of Typhimurium which contains a single mutation in *pagC* had a virulence defect and decreased survival in cultured murine macrophages and 100-fold reduction in intraperitoneal virulence in mice [118].

2.4.3 Siderophores are important for bacterial growth in serum in the extracellular phase of salmonellosis

Iron is an essential element for the growth of most bacteria through its involvement in a variety of metabolic and regulatory functions [119]. Studies with different iron concentrations in growth media demonstrated an effect on gene expression of the iron acquisition systems encoded both on the chromosome and plasmids at both transcriptional and translational levels [120]. Siderophores which are bacterial molecules that bind and transport iron are important for bacterial growth in serum in the extracellular stage of *Salmonella* systemic infection. They are not required after bacteria reside in SCV where siderophore-independent iron acquisition systems are sufficient for iron uptake during intracellular stage. *Salmonella* produce two major types of siderophores, high-affinity catecholate consisting of salmochelin and enterobactin the latter also known as enterochelin and a low-affinity hydroxamate known as aerobactin which is expressed under iron-restricted conditions [121]. The synthesis, secretion, and uptake of salmochelin requires genes clustered at two genetic loci, the *fepA* gene cluster and *iroBCDEN* operon. The *fepA* gene cluster includes most *ent* genes for synthesis and export [122]. The *iroBCDEN* operon encodes gene products for enterobactin glycosylation (IroB, glycosyltransferase), export (IroC, ABC transporter protein), and utilization (IroD, esterase; IroE, hydrolase; IroN, outer membrane receptor) [122]. Mutants deficient in *iroB* or *iroC* exhibit reduced virulence during systemic infection of mice via intraperitoneal route, as indicated by lower bacterial load in liver and a delayed time of death [122]. Moreover, the enterobactin metabolite, 2, 3-dihydroxybenzoyl serine (DHBS), can also be used by *Salmonella* as sources of iron, albeit at much lower affinities, by recognizing the three catechelate receptors, FepA, IroN and Cir. The three receptors demonstrate a significant degree of functional redundancy. The Typhimurium double mutant Δ*fepA iroN* were similarly virulent to the parental strain after intragastric gavage inoculation of mice, while the triple mutant Δ*fepA iroN cir* was attenuated as indicated by a significantly reduced cecal colonization and no measurable spread to the liver [123, 124].

Furthermore, *Salmonella* also utilize xenosiderophores as iron sources by utilizing the outer membrane receptors, including FhuA, FhuE, and FoxA. For example, utilization of ferrioxamines B, E, and G by Typhimurium is dependent on the FoxA receptor encoded by the Fur repressible *foxA* gene. A strain carrying the foxA mutation exhibited a significantly reduced ability to colonize rabbit ileal loops and was markedly attenuated in mice challenged by either intragastric gavage or intravenously route strain compared to the foxA+ parent [125]. The best characterized regulator for iron uptake is the iron-dependent repressor Fur that acts together with the co[-]repressor ferrous iron (Fe(II)) to regulate genes involved in the iron uptake process in response to iron restriction, including *fhuA*, *fhuB*, *fepA*, *fes*, *fepD*, *entB*, *fur*, *foxA*, *hemP*, and *fhuE* [126, 127].

3. Future directions

The advent of next generation sequencing (NGS) has provided an opportunity to verify or improve on knowledge gained from *in vitro* and *in vivo* analyses of *Salmonella* mutants which were designed for the purpose of understanding gene function and mechanism of action. Recently, Rakov et al. [14] carried out bioinformatics analysis of 500 *Salmonella* genomes and identified 70 allelic variants virulence factors which were associated with different pathogenesis outcomes, i.e. gastrointestinal vs. invasive disease. However, the causative relationship between a putative virulence factor and disease outcome using a genomics based tool is yet to be attained. To that end, we propose the development of a comprehensive genome based tool such as a NGS AmpliSeq assay that can be used to simultaneously interrogate the presence and potential expression of over 200 virulence genes of *Salmonella* identified in this communication. The tool can be used to evaluate differences in strains and correlate the output with virulence phenotype derived from epidemiological or experimental observations which can be developed simultaneously or based on historical documentation. The tool could be used in assessing the potential risk posed by a strain of *Salmonella* given the fact that the serovars obtained from the environment are often distinct with those involved in human diseases. The technology appears suitable for dissecting the complexity associated with the redundancy and pleiotropic nature of some of the currently known virulence genes. In addition, NGS based analysis of virulence genes should provide new insights on *Salmonella* evolution and a better tool for analyzing epidemiological data that could translate to a reduction in the burden on human health posed by this important foodborne and zoonotic pathogen.

4. Conclusions

This review provides an outline of over 200 identified virulence determinants and details of their involvement in the four steps of *Salmonella* pathogenesis, namely: attachment, invasion, intramacrophage survival/replication and systemic dissemination. The genetic regulation of only some of the virulence determinants have been elucidated in live animal models such as mice and cattle, and this has enriched our understanding of the pathogenesis and mechanism of diarrhea and systemic disease. The majority of the current evidence on pathogenesis and virulence determinants of NTS was derived from murine model of serovar Typhimurium infection with and only a few studies focused on NTS infection in humans. For this reason, the relevance of published observations is often called into question. Linking clinical, epidemiological and experimental observations on the nature and severity of diseases caused by *Salmonella* organisms with the presence of a large number of virulence genes currently may not garner enough predictive ability to infer virulence or pathogenetic potential of a strain. Still, the increasing availability of a large number of *Salmonella* genomes in the public databases is proving to be a timely resource. Next generation sequencing and the twin subject of bioinformatics represent an unprecedented opportunity to verify past observations and help improve our understanding of *Salmonella* virulence towards a coherent and comprehensive understanding of the mechanism of *Salmonella* pathogenesis. What is required is a robust laboratory tool that can be used to analyze the large number of virulence genes in an isolate using the tools of whole genome sequencing. We expect that a tool such as an AmpliSeq assay for *Salmonella* virulence could be developed to generate accurate and reliable information that can be fed into a quantitative risk assessment framework. This could usher a new era of risk management customized for a *Salmonella* strain involved in an outbreak and should translate to impactful outcomes in the areas of improved food safety, evaluation of zoonotic diseases and reducing the burden of human salmonellosis.

Acknowledgements

RG is funded by Genome Canada. DO's research program has received funding support from Genome Research and Development Initiative of the Government of Canada, Ontario Ministry of Agriculture, Food and Rural Affairs, Canadian Security and Science Program of the Department of National Defense and the Canadian Food Inspection Agency.

Conflict of interest

The authors declare no conflict of interest.

Acronyms and abbreviations

AMP	antimicrobial peptides
invA	invasion protein A
LPS	lipopolysaccharide
NTS	non-typhoidal *Salmonella*
NGS	next generation sequencing
SalFoS	*Salmonella* Foodborne Syst-OMICS database
SPIs	*Salmonella* pathogenicity islands
SIFs	*Salmonella*-induced filaments
SCV	*Salmonella*-containing vacuole

Author details

Ruimin Gao[1,2], Linru Wang[1,3] and Dele Ogunremi[1]*

1 Ottawa Laboratory Fallowfield, Canadian Food Inspection Agency, Ottawa, Ontario, Canada

2 Department of Food Science and Agricultural Chemistry, McGill University, Sainte Anne de Bellevue, QC, Canada

3 Greater Toronto Area Laboratory, Canadian Food Inspection Agency, Scarborough, Ontario, Canada

*Address all correspondence to: dele.ogunremi@canada.ca

References

[1] Majowicz SE, Musto J, Scallan E, Angulo FJ, Kirk M, O'Brien SJ, et al. International collaboration on Enteric disease 'burden of illness': The global burden of nontyphoidal *Salmonella* gastroenteritis. Clinical Infectious Diseases. 2010;**50**(6):882-889

[2] Thomas MK, Murray R, Flockhart L, Pintar K, Pollari F, Fazil A, et al. Estimates of the burden of foodborne illness in Canada for 30 specified pathogens and unspecified agents, CIRCA 2006. Foodborne Pathogens and Disease. 2013;**10**(7):639-648

[3] Thomas MK, Murray R, Flockhart L, Pintar K, Fazil A, Nesbitt A, et al. Estimates of foodborne illness-related hospitalizations and deaths in Canada for 30 specified pathogens and unspecified agents. Foodborne Pathogens and Disease. 2015;**12**(10):820-827

[4] Hsu HS. Pathogenesis and immunity in murine salmonellosis. Microbiological Reviews. 1989;**53**(4):390-409

[5] Gal-Mor O, Boyle EC, Grassl GA. Same species, different diseases: How and why typhoidal and non-typhoidal *Salmonella enterica* serovars differ. Frontiers in Microbiology. 2014;**5**:391

[6] Abbott SL, Ni FC, Janda JM. Increase in extraintestinal infections caused by *Salmonella enterica* subspecies II-IV. Emerging Infectious Diseases. 2012;**18**(4):637-639

[7] Uzzau S, Brown DJ, Wallis T, Rubino S, Leori G, Bernard S, et al. Host adapted serotypes of *Salmonella enterica*. Epidemiology and Infection. 2000;**125**(2):229-255

[8] Fang FC, Fierer J. Human infection with *Salmonella dublin*. Medicine (Baltimore). 1991;**70**(3):198-207

[9] Chaudhuri D, Roy Chowdhury A, Biswas B, Chakravortty D. *Salmonella typhimurium* infection leads to colonization of the mouse brain and is not completely cured with antibiotics. Frontiers in Microbiology. 2018;**9**:1632

[10] Gordon MA. Invasive nontyphoidal *Salmonella* disease: Epidemiology, pathogenesis and diagnosis. Current Opinion in Infectious Diseases. 2011;**24**(5):484-489

[11] Robertson J, Yoshida C, Kruczkiewicz P, Nadon C, Nichani A, Taboada EN, et al. Comprehensive assessment of the quality of *Salmonella* whole genome sequence data available in public sequence databases using the *Salmonella* in silico Typing Resource (SISTR). Microbial Genomics. 2018;**4**. DOI: 10.1099/mgen.0.000151

[12] Branchu P, Bawn M, Kingsley RA. Genome variation and molecular epidemiology of *Salmonella enterica* serovar Typhimurium pathovariants. Infection and Immunity. 2018;**86**:e00079-18. DOI: 10.1128/IAI.00079-18

[13] Emond-Rheault JG, Jeukens J, Freschi L, Kukavica-Ibrulj I, Boyle B, Dupont MJ, et al. A Syst-OMICS approach to ensuring food safety and reducing the economic burden of Salmonellosis. Frontiers in Microbiology. 2017;**8**:996

[14] Rakov AV, Mastriani E, Liu SL, Schifferli DM. Association of *Salmonella* virulence factor alleles with intestinal and invasive serovars. BMC Genomics. 2019;**20**(1):429

[15] Hansmeier N, Miskiewicz K, Elpers L, Liss V, Hensel M, Sterzenbach T. Functional expression of the entire adhesiome of *Salmonella enterica* serotype Typhimurium. Scientific Reports. 2017;**7**(1):10326

[16] Collinson SK, Liu SL, Clouthier SC, Banser PA, Doran JL, Sanderson KE, et al. The location of four fimbrin-encoding genes, agfA, fimA, sefA and sefD, on the *Salmonella* enteritidis and/or *S. typhimurium* XbaI-BlnI genomic restriction maps. Gene. 1996;**169**(1):75-80

[17] Berne C, Ducret A, Hardy GG, Brun YV. Adhesins involved in attachment to abiotic surfaces by Gram-negative bacteria. Microbiology Spectrum. 2015;**3**(4):MB-0018-2015. DOI:10.1128/microbiolspec.MB-0018-2015

[18] Wagner C, Hensel M. Adhesive mechanisms of *Salmonella enterica*. Advances in Experimental Medicine and Biology. 2011;**715**:17-34

[19] Coburn B, Sekirov I, Finlay BB. Type III secretion systems and disease. Clinical Microbiology Reviews. 2007;**20**(4):535-549

[20] Yue M, Rankin SC, Blanchet RT, Nulton JD, Edwards RA, Schifferli DM. Diversification of the *Salmonella fimbriae*: A model of macro- and microevolution. PLoS One. 2012;**7**(6):e38596

[21] McClelland M, Sanderson KE, Spieth J, Clifton SW, Latreille P, Courtney L, et al. Complete genome sequence of *Salmonella enterica* serovar Typhimurium LT2. Nature. 2001;**413**(6858):852-856

[22] Nuccio SP, Baumler AJ. Evolution of the chaperone/usher assembly pathway: Fimbrial classification goes Greek. Microbiology and Molecular Biology Reviews. 2007;**71**(4):551-575

[23] Fabrega A, Vila J. *Salmonella enterica* serovar Typhimurium skills to succeed in the host: Virulence and regulation. Clinical Microbiology Reviews. 2013;**26**(2):308-341

[24] Zeiner SA, Dwyer BE, Clegg S. FimA, FimF, and FimH are necessary for assembly of type 1 fimbriae on *Salmonella enterica* serovar Typhimurium. Infection and Immunity. 2012;**80**(9):3289-3296

[25] Ernst RK, Dombroski DM, Merrick JM. Anaerobiosis, type 1 fimbriae, and growth phase are factors that affect invasion of HEp-2 cells by *Salmonella typhimurium*. Infection and Immunity. 1990;**58**(6):2014-2016

[26] Horiuchi S, Inagaki Y, Okamura N, Nakaya R, Yamamoto N. Type 1 pili enhance the invasion of *Salmonella braenderup* and *Salmonella typhimurium* to HeLa cells. Microbiology and Immunology. 1992;**36**(6):593-602

[27] De Buck J, Van Immerseel F, Haesebrouck F, Ducatelle R. Protection of laying hens against *Salmonella enteritidis* by immunization with type 1 fimbriae. Veterinary Microbiology. 2005;**105**(2):93-101

[28] Wang KC, Hsu YH, Huang YN, Yeh KS. A previously uncharacterized gene stm0551 plays a repressive role in the regulation of type 1 fimbriae in *Salmonella enterica* serotype Typhimurium. BMC Microbiology. 2012;**12**:111

[29] Baumler AJ, Tsolis RM, Bowe FA, Kusters JG, Hoffmann S, Heffron F. The pef fimbrial operon of *Salmonella typhimurium* mediates adhesion to murine small intestine and is necessary for fluid accumulation in the infant mouse. Infection and Immunity. 1996;**64**(1):61-68

[30] Nicholson B, Low D. DNA methylation-dependent regulation of pef expression in *Salmonella typhimurium*. Molecular Microbiology. 2000;**35**(4):728-742

[31] Chessa D, Dorsey CW, Winter M, Baumler AJ. Binding specificity of *Salmonella* plasmid-encoded fimbriae assessed by glycomics.

The Journal of Biological Chemistry. 2008;**283**(13):8118-8124

[32] Baumler AJ, Tsolis RM, Heffron F. The lpf fimbrial operon mediates adhesion of *Salmonella typhimurium* to murine Peyer's patches. Proceedings of the National Academy of Sciences of the United States of America. 1996;**93**(1):279-283

[33] Ledeboer NA, Frye JG, McClelland M, Jones BD. *Salmonella enterica* serovar Typhimurium requires the Lpf, Pef, and Tafi fimbriae for biofilm formation on HEp-2 tissue culture cells and chicken intestinal epithelium. Infection and Immunity. 2006;**74**(6):3156-3169

[34] Weening EH, Barker JD, Laarakker MC, Humphries AD, Tsolis RM, Baumler AJ. The *Salmonella enterica* serotype Typhimurium lpf, bcf, stb, stc, std, and sth fimbrial operons are required for intestinal persistence in mice. Infection and Immunity. 2005;**73**(6):3358-3366

[35] Norris TL, Kingsley RA, Bumler AJ. Expression and transcriptional control of the *Salmonella typhimurium* Ipf fimbrial operon by phase variation. Molecular Microbiology. 1998;**29**(1):311-320

[36] Sukupolvi S, Lorenz RG, Gordon JI, Bian Z, Pfeifer JD, Normark SJ, et al. Expression of thin aggregative fimbriae promotes interaction of *Salmonella typhimurium* SR-11 with mouse small intestinal epithelial cells. Infection and Immunity. 1997;**65**(12):5320-5325

[37] Collinson SK, Emody L, Muller KH, Trust TJ, Kay WW. Purification and characterization of thin, aggregative fimbriae from *Salmonella enteritidis*. Journal of Bacteriology. 1991;**173**(15):4773-4781

[38] Collinson SK, Doig PC, Doran JL, Clouthier S, Trust TJ, Kay WW. Thin, aggregative fimbriae mediate binding of *Salmonella enteritidis* to fibronectin. Journal of Bacteriology. 1993;**175**(1):12-18

[39] van der Velden AW, Baumler AJ, Tsolis RM, Heffron F. Multiple fimbrial adhesins are required for full virulence of *Salmonella typhimurium* in mice. Infection and Immunity. 1998;**66**(6):2803-2808

[40] Tsolis RM, Townsend SM, Miao EA, Miller SI, Ficht TA, Adams LG, et al. Identification of a putative *Salmonella enterica* serotype Typhimurium host range factor with homology to IpaH and YopM by signature-tagged mutagenesis. Infection and Immunity. 1999;**67**(12):6385-6393

[41] Strindelius L, Folkesson A, Normark S, Sjoholm I. Immunogenic properties of the *Salmonella* atypical fimbriae in BALB/c mice. Vaccine. 2004;**22**(11-12):1448-1456

[42] Zeng L, Zhang L, Wang P, Meng G. Structural basis of host recognition and biofilm formation by *Salmonella* Saf pili. eLife. 2017;**6**:e28619. DOI: 10.7554/eLife.28619

[43] Chessa D, Winter MG, Jakomin M, Baumler AJ. *Salmonella enterica* serotype Typhimurium Std fimbriae bind terminal alpha(1,2)fucose residues in the cecal mucosa. Molecular Microbiology. 2009;**71**(4):864-875

[44] Morrow BJ, Graham JE, Curtiss R 3rd. Genomic subtractive hybridization and selective capture of transcribed sequences identify a novel *Salmonella typhimurium* fimbrial operon and putative transcriptional regulator that are absent from the *Salmonella typhi* genome. Infection and Immunity. 1999;**67**(10):5106-5116

[45] Humphries AD, Raffatellu M, Winter S, Weening EH, Kingsley RA, Droleskey R, et al. The use of flow

cytometry to detect expression of subunits encoded by 11 *Salmonella enterica* serotype Typhimurium fimbrial operons. Molecular Microbiology. 2003;**48**(5):1357-1376

[46] de Louvois J. *Salmonella* contamination of eggs. Lancet. 1993;**342**(8867):366-367

[47] Thiagarajan D, Thacker HL, Saeed AM. Experimental infection of laying hens with *Salmonella enteritidis* strains that express different types of fimbriae. Poultry Science. 1996;**75**(11):1365-1372

[48] Edwards RA, Schifferli DM, Maloy SR. A role for *Salmonella fimbriae* in intraperitoneal infections. Proceedings of the National Academy of Sciences of the United States of America. 2000;**97**(3):1258-1262

[49] Peralta RC, Yokoyama H, Ikemori Y, Kuroki M, Kodama Y. Passive immunisation against experimental salmonellosis in mice by orally administered hen egg-yolk antibodies specific for 14-kDa fimbriae of *Salmonella enteritidis*. Journal of Medical Microbiology. 1994;**41**(1):29-35

[50] Dorsey CW, Laarakker MC, Humphries AD, Weening EH, Baumler AJ. *Salmonella enterica* serotype Typhimurium MisL is an intestinal colonization factor that binds fibronectin. Molecular Microbiology. 2005;**57**(1):196-211

[51] Kingsley RA, van Amsterdam K, Kramer N, Baumler AJ. The shdA gene is restricted to serotypes of *Salmonella enterica* subspecies I and contributes to efficient and prolonged fecal shedding. Infection and Immunity. 2000;**68**(5):2720-2727

[52] Kingsley RA, Abi Ghanem D, Puebla-Osorio N, Keestra AM, Berghman L, Baumler AJ. Fibronectin binding to the *Salmonella enterica* serotype Typhimurium ShdA autotransporter protein is inhibited by a monoclonal antibody recognizing the A3 repeat. Journal of Bacteriology. 2004;**186**(15):4931-4939

[53] Kingsley RA, Humphries AD, Weening EH, De Zoete MR, Winter S, Papaconstantinopoulou A, et al. Molecular and phenotypic analysis of the CS54 island of *Salmonella enterica* serotype Typhimurium: Identification of intestinal colonization and persistence determinants. Infection and Immunity. 2003;**71**(2):629-640

[54] Latasa C, Roux A, Toledo-Arana A, Ghigo JM, Gamazo C, Penades JR, et al. BapA, a large secreted protein required for biofilm formation and host colonization of *Salmonella enterica* serovar Enteritidis. Molecular Microbiology. 2005;**58**(5):1322-1339

[55] Gerlach RG, Jackel D, Stecher B, Wagner C, Lupas A, Hardt WD, et al. *Salmonella* pathogenicity island 4 encodes a giant non-fimbrial adhesin and the cognate type 1 secretion system. Cellular Microbiology. 2007;**9**(7):1834-1850

[56] Finlay BB, Ruschkowski S, Dedhar S. Cytoskeletal rearrangements accompanying *Salmonella* entry into epithelial cells. Journal of Cell Science. 1991;**99**(Pt 2):283-296

[57] Francis CL, Ryan TA, Jones BD, Smith SJ, Falkow S. Ruffles induced by *Salmonella* and other stimuli direct macropinocytosis of bacteria. Nature. 1993;**364**(6438):639-642

[58] Garcia-del Portillo F, Finlay BB. *Salmonella* invasion of nonphagocytic cells induces formation of macropinosomes in the host cell. Infection and Immunity. 1994;**62**(10):4641-4645

[59] Meresse S, Unsworth KE, Habermann A, Griffiths G, Fang F,

Martinez-Lorenzo MJ, et al. Remodelling of the actin cytoskeleton is essential for replication of intravacuolar *Salmonella*. Cellular Microbiology. 2001;**3**(8):567-577

[60] Garcia-del Portillo F, Zwick MB, Leung KY, Finlay BB. *Salmonella* induces the formation of filamentous structures containing lysosomal membrane glycoproteins in epithelial cells. Proceedings of the National Academy of Sciences of the United States of America. 1993;**90**(22):10544-10548

[61] Rajashekar R, Liebl D, Seitz A, Hensel M. Dynamic remodeling of the endosomal system during formation of *Salmonella*-induced filaments by intracellular *Salmonella enterica*. Traffic. 2008;**9**(12):2100-2116

[62] Ohl ME, Miller SI. *Salmonella*: A model for bacterial pathogenesis. Annual Review of Medicine. 2001;**52**:259-274

[63] Sabbagh SC, Forest CG, Lepage C, Leclerc JM, Daigle F. So similar, yet so different: Uncovering distinctive features in the genomes of *Salmonella enterica* serovars Typhimurium and Typhi. FEMS Microbiology Letters. 2010;**305**(1):1-13

[64] Salmond GP, Reeves PJ. Membrane traffic wardens and protein secretion in Gram-negative bacteria. Trends in Biochemical Sciences. 1993;**18**(1):7-12

[65] Nieto PA, Pardo-Roa C, Salazar-Echegarai FJ, Tobar HE, Coronado-Arrazola I, Riedel CA, et al. New insights about excisable pathogenicity islands in *Salmonella* and their contribution to virulence. Microbes and Infection. 2016;**18**(5):302-309

[66] Ginocchio CC, Rahn K, Clarke RC, Galan JE. Naturally occurring deletions in the centisome 63 pathogenicity island of environmental isolates of *Salmonella* spp. Infection and Immunity. 1997;**65**(4):1267-1272

[67] Penheiter KL, Mathur N, Giles D, Fahlen T, Jones BD. Non-invasive *Salmonella typhimurium* mutants are avirulent because of an inability to enter and destroy M cells of ileal Peyer's patches. Molecular Microbiology. 1997;**24**(4):697-709

[68] Bernal-Bayard J, Ramos-Morales F. *Salmonella* type III secretion effector SlrP is an E3 ubiquitin ligase for mammalian thioredoxin. The Journal of Biological Chemistry. 2009;**284**(40):27587-27595

[69] Ellermeier CD, Slauch JM. RtsA and RtsB coordinately regulate expression of the invasion and flagellar genes in *Salmonella enterica* serovar Typhimurium. Journal of Bacteriology. 2003;**185**(17):5096-5108

[70] Lim S, Choi J, Kim D, Seo HS. Transcriptional analysis of the iagB within *Salmonella* pathogenicity island 1 (SPI1). Journal of Bacteriology and Virology. 2016;**46**(3):128-134

[71] Bajaj V, Hwang C, Lee CA. hilA is a novel ompR/toxR family member that activates the expression of *Salmonella typhimurium* invasion genes. Molecular Microbiology. 1995;**18**(4):715-727

[72] Teplitski M, Goodier RI, Ahmer BM. Pathways leading from BarA/SirA to motility and virulence gene expression in *Salmonella*. Journal of Bacteriology. 2003;**185**(24):7257-7265

[73] Pegues DA, Hantman MJ, Behlau I, Miller SI. PhoP/PhoQ transcriptional repression of *Salmonella typhimurium* invasion genes: Evidence for a role in protein secretion. Molecular Microbiology. 1995;**17**(1):169-181

[74] Winter SE, Thiennimitr P, Winter MG, Butler BP, Huseby DL,

Crawford RW, et al. Gut inflammation provides a respiratory electron acceptor for *Salmonella*. Nature. 2010;**467**(7314):426-429

[75] Shea JE, Hensel M, Gleeson C, Holden DW. Identification of a virulence locus encoding a second type III secretion system in *Salmonella typhimurium*. Proceedings of the National Academy of Sciences of the United States of America. 1996;**93**(6):2593-2597

[76] Cirillo DM, Valdivia RH, Monack DM, Falkow S. Macrophage-dependent induction of the *Salmonella* pathogenicity island 2 type III secretion system and its role in intracellular survival. Molecular Microbiology. 1998;**30**(1):175-188

[77] Hensel M, Shea JE, Waterman SR, Mundy R, Nikolaus T, Banks G, et al. Genes encoding putative effector proteins of the type III secretion system of *Salmonella* pathogenicity island 2 are required for bacterial virulence and proliferation in macrophages. Molecular Microbiology. 1998;**30**(1):163-174

[78] Makishima S, Komoriya K, Yamaguchi S, Aizawa SI. Length of the flagellar hook and the capacity of the type III export apparatus. Science. 2001;**291**(5512):2411-2413

[79] Marcus SL, Brumell JH, Pfeifer CG, Finlay BB. *Salmonella* pathogenicity islands: Big virulence in small packages. Microbes and Infection. 2000;**2**(2):145-156

[80] Kuhle V, Hensel M. Cellular microbiology of intracellular *Salmonella enterica*: Functions of the type III secretion system encoded by *Salmonella* pathogenicity island 2. Cellular and Molecular Life Sciences. 2004;**61**(22):2812-2826

[81] Blanc-Potard AB, Groisman EA. The *Salmonella* selC locus contains a pathogenicity island mediating intramacrophage survival. The EMBO Journal. 1997;**16**(17):5376-5385

[82] Moncrief MB, Maguire ME. Magnesium and the role of MgtC in growth of *Salmonella typhimurium*. Infection and Immunity. 1998;**66**(8):3802-3809

[83] Morgan E, Campbell JD, Rowe SC, Bispham J, Stevens MP, Bowen AJ, et al. Identification of host-specific colonization factors of *Salmonella enterica* serovar Typhimurium. Molecular Microbiology. 2004;**54**(4):994-1010

[84] Kiss T, Morgan E, Nagy G. Contribution of SPI-4 genes to the virulence of *Salmonella enterica*. FEMS Microbiology Letters. 2007;**275**(1):153-159

[85] Morgan E, Bowen AJ, Carnell SC, Wallis TS, Stevens MP. SiiE is secreted by the *Salmonella enterica* serovar Typhimurium pathogenicity island 4-encoded secretion system and contributes to intestinal colonization in cattle. Infection and Immunity. 2007;**75**(3):1524-1533

[86] Li X, Bleumink-Pluym NMC, Luijkx Y, Wubbolts RW, van Putten JPM, Strijbis K. MUC1 is a receptor for the *Salmonella* SiiE adhesin that enables apical invasion into enterocytes. PLoS Pathogens. 2019;**15**(2):e1007566

[87] Wood MW, Jones MA, Watson PR, Hedges S, Wallis TS, Galyov EE. Identification of a pathogenicity island required for *Salmonella* enteropathogenicity. Molecular Microbiology. 1998;**29**(3):883-891

[88] Galyov EE, Wood MW, Rosqvist R, Mullan PB, Watson PR, Hedges S, et al. A secreted effector protein of *Salmonella dublin* is translocated into eukaryotic cells and mediates inflammation and fluid secretion in infected ileal

mucosa. Molecular Microbiology. 1997;**25**(5):903-912

[89] Darwin KH, Robinson LS, Miller VL. SigE is a chaperone for the *Salmonella enterica* serovar Typhimurium invasion protein SigD. Journal of Bacteriology. 2001;**183**(4):1452-1454

[90] Fields PI, Swanson RV, Haidaris CG, Heffron F. Mutants of *Salmonella typhimurium* that cannot survive within the macrophage are avirulent. Proceedings of the National Academy of Sciences of the United States of America. 1986;**83**(14):5189-5193

[91] Steele-Mortimer O, Brumell JH, Knodler LA, Meresse S, Lopez A, Finlay BB. The invasion-associated type III secretion system of *Salmonella enterica* serovar Typhimurium is necessary for intracellular proliferation and vacuole biogenesis in epithelial cells. Cellular Microbiology. 2002;**4**(1):43-54

[92] Vazquez-Torres A, Xu Y, Jones-Carson J, Holden DW, Lucia SM, Dinauer MC, et al. *Salmonella* pathogenicity island 2-dependent evasion of the phagocyte NADPH oxidase. Science. 2000;**287**(5458):1655-1658

[93] Gallois A, Klein JR, Allen LA, Jones BD, Nauseef WM. *Salmonella* pathogenicity island 2-encoded type III secretion system mediates exclusion of NADPH oxidase assembly from the phagosomal membrane. Journal of Immunology. 2001;**166**(9):5741-5748

[94] van der Velden AW, Lindgren SW, Worley MJ, Heffron F. *Salmonella* pathogenicity island 1-independent induction of apoptosis in infected macrophages by *Salmonella enterica* serotype Typhimurium. Infection and Immunity. 2000;**68**(10):5702-5709

[95] Catron DM, Sylvester MD, Lange Y, Kadekoppala M, Jones BD, Monack DM, et al. The *Salmonella*-containing vacuole is a major site of intracellular cholesterol accumulation and recruits the GPI-anchored protein CD55. Cellular Microbiology. 2002;**4**(6):315-328

[96] Waterman SR, Holden DW. Functions and effectors of the *Salmonella* pathogenicity island 2 type III secretion system. Cellular Microbiology. 2003;**5**(8):501-511

[97] Knodler LA, Steele-Mortimer O. Taking possession: Biogenesis of the *Salmonella*-containing vacuole. Traffic. 2003;**4**(9):587-599

[98] Abrahams GL, Muller P, Hensel M. Functional dissection of SseF, a type III effector protein involved in positioning the *Salmonella*-containing vacuole. Traffic. 2006;**7**(8):950-965

[99] Freeman JA, Ohl ME, Miller SI. The *Salmonella enterica* serovar Typhimurium translocated effectors SseJ and SifB are targeted to the *Salmonella*-containing vacuole. Infection and Immunity. 2003;**71**(1):418-427

[100] Stein MA, Leung KY, Zwick M, Garcia-del Portillo F, Finlay BB. Identification of a *Salmonella* virulence gene required for formation of filamentous structures containing lysosomal membrane glycoproteins within epithelial cells. Molecular Microbiology. 1996;**20**(1):151-164

[101] Saini S, Rao CV. SprB is the molecular link between *Salmonella* pathogenicity island 1 (SPI1) and SPI4. Journal of Bacteriology. 2010;**192**(9):2459-2462

[102] Lee AK, Detweiler CS, Falkow S. OmpR regulates the two-component system SsrA-ssrB in *Salmonella* pathogenicity island 2. Journal of Bacteriology. 2000;**182**(3):771-781

[103] Deiwick J, Hensel M. Regulation of virulence genes by environmental signals in *Salmonella typhimurium*. Electrophoresis. 1999;**20**(4-5):813-817

[104] Deiwick J, Nikolaus T, Erdogan S, Hensel M. Environmental regulation of *Salmonella* pathogenicity island 2 gene expression. Molecular Microbiology. 1999;**31**(6):1759-1773

[105] Raetz CR, Whitfield C. Lipopolysaccharide endotoxins. Annual Review of Biochemistry. 2002;**71**:635-700

[106] Nakano M, Saito K. Chemical components in the cell wall of *Salmonella typhimurium* affecting its virulence and immunogenicity in mice. Nature. 1969;**222**(5198):1085-1086

[107] Joiner KA, Hammer CH, Brown EJ, Frank MM. Studies on the mechanism of bacterial resistance to complement-mediated killing. II. C8 and C9 release C5b67 from the surface of *Salmonella minnesota* S218 because the terminal complex does not insert into the bacterial outer membrane. The Journal of Experimental Medicine. 1982;**155**(3):809-819

[108] Joiner KA, Hammer CH, Brown EJ, Cole RJ, Frank MM. Studies on the mechanism of bacterial resistance to complement-mediated killing. I. Terminal complement components are deposited and released from *Salmonella minnesota* S218 without causing bacterial death. The Journal of Experimental Medicine. 1982;**155**(3):797-808

[109] Lerouge I, Vanderleyden J. O-antigen structural variation: Mechanisms and possible roles in animal/plant-microbe interactions. FEMS Microbiology Reviews. 2002;**26**(1):17-47

[110] Wang X, Quinn PJ. Lipopolysaccharide: Biosynthetic pathway and structure modification. Progress in Lipid Research. 2010;**49**(2):97-107

[111] Ruiz N, Kahne D, Silhavy TJ. Transport of lipopolysaccharide across the cell envelope: The long road of discovery. Nature Reviews. Microbiology. 2009;**7**(9):677-683

[112] Chang J, Pang E, He H, Kwang J. Identification of novel attenuated *Salmonella enteritidis* mutants. FEMS Immunology and Medical Microbiology. 2008;**53**(1):26-34

[113] Murray GL, Attridge SR, Morona R. Regulation of *Salmonella typhimurium* lipopolysaccharide O antigen chain length is required for virulence: Identification of FepE as a second Wzz. Molecular Microbiology. 2003;**47**(5):1395-1406

[114] Delgado MA, Mouslim C, Groisman EA. The PmrA/PmrB and RcsC/YojN/RcsB systems control expression of the *Salmonella* O-antigen chain length determinant. Molecular Microbiology. 2006;**60**(1):39-50

[115] Bogomolnaya LM, Aldrich L, Ragoza Y, Talamantes M, Andrews KD, McClelland M, et al. Identification of novel factors involved in modulating motility of *Salmonella enterica* serotype Typhimurium. PLoS One. 2014;**9**(11):e111513

[116] Nishio M, Okada N, Miki T, Haneda T, Danbara H. Identification of the outer-membrane protein PagC required for the serum resistance phenotype in *Salmonella enterica* serovar Choleraesuis. Microbiology. 2005;**151**(Pt 3):863-873

[117] Heffernan EJ, Reed S, Hackett J, Fierer J, Roudier C, Guiney D. Mechanism of resistance to complement-mediated killing of bacteria encoded by the *Salmonella typhimurium* virulence plasmid gene rck. The Journal of Clinical Investigation. 1992;**90**(3):953-964

[118] Miller SI, Kukral AM, Mekalanos JJ. A two-component regulatory system (phoP phoQ) controls *Salmonella typhimurium* virulence. Proceedings of the National Academy of Sciences of the United States of America. 1989;**86**(13):5054-5058

[119] Miethke M, Marahiel MA. Siderophore-based iron acquisition and pathogen control. Microbiology and Molecular Biology Reviews. 2007;**71**(3):413-451

[120] Khajanchi BK, Xu J, Grim CJ, Ottesen AR, Ramachandran P, Foley SL. Global transcriptomic analyses of *Salmonella enterica* in iron-depleted and iron-rich growth conditions. BMC Genomics. 2019;**20**(1):490

[121] Fischbach MA, Lin H, Liu DR, Walsh CT. How pathogenic bacteria evade mammalian sabotage in the battle for iron. Nature Chemical Biology. 2006;**2**(3):132-138

[122] Crouch ML, Castor M, Karlinsey JE, Kalhorn T, Fang FC. Biosynthesis and IroC-dependent export of the siderophore salmochelin are essential for virulence of *Salmonella enterica* serovar Typhimurium. Molecular Microbiology. 2008;**67**(5):971-983

[123] Rabsch W, Methner U, Voigt W, Tschape H, Reissbrodt R, Williams PH. Role of receptor proteins for enterobactin and 2,3-dihydroxybenzoylserine in virulence of *Salmonella enterica*. Infection and Immunity. 2003;**71**(12):6953-6961

[124] Williams PH, Rabsch W, Methner U, Voigt W, Tschape H, Reissbrodt R. Catecholate receptor proteins in *Salmonella enterica*: Role in virulence and implications for vaccine development. Vaccine. 2006;**24**(18):3840-3844

[125] Kingsley RA, Reissbrodt R, Rabsch W, Ketley JM, Tsolis RM, Everest P, et al. Ferrioxamine-mediated iron(III) utilization by *Salmonella enterica*. Applied and Environmental Microbiology. 1999;**65**(4):1610-1618

[126] Ernst JF, Bennett RL, Rothfield LI. Constitutive expression of the iron-enterochelin and ferrichrome uptake systems in a mutant strain of *Salmonella typhimurium*. Journal of Bacteriology. 1978;**135**(3):928-934

[127] Tsolis RM, Baumler AJ, Stojiljkovic I, Heffron F. Fur regulon of *Salmonella typhimurium*: Identification of new iron-regulated genes. Journal of Bacteriology. 1995;**177**(16):4628-4637

Chapter 10

New Approaches for Competing Microbial Resistance and Virulence

Mohammed El-Mowafy, Abdelaziz Elgaml and Mona Shaaban

Abstract

The spread of multidrug-resistant pathogens together with the development of fatal cases of infectious microorganisms is on the rise. Therefore, there must be new approaches for combating pathogenic microorganisms, either by overcoming antibiotic resistance or via inhibiting their virulence factors. Several virulence factors extremely increase the antimicrobial resistance of various species of pathogens; as a result, the screening of antivirulence agents has gained more and more attention recently. In this aspect, non-traditional strategies that are considered promising in overcoming virulence and pathogenicity of microorganisms will be discussed including; quorum sensing inhibition, antibiofilm, control of the global regulators, bacteriocins and bacteriophages. Applying these methods could provide innovative approaches for competing microbial resistance and virulence.

Keywords: bacterial virulence, resistance, quorum sensing inhibition, global regulators, phage therapy, inhibition of biofilm formation, bacteriocins

1. Introduction

The high incidence of microbial resistance and the spread of multidrug-resistant and pan drug-resistant pathogens have been developed to threaten human mankind. Fortunately, there are upcoming alternative therapeutic approach for eliminating bacterial virulence and host-pathogen interaction [1, 2]. Quorum sensing signals [3, 4] and global regulators represent the main players to control virulence circuits and coordinate host-pathogen interaction [5]. Thus, targeting these regulators provide a promising trend to overcome microbial pathogenicity. Bacterial cells have the ability to grow in matrices of polysaccharides, proteins and DNA forming biofilm [6]. The cell communities inside the biofilm matrices are highly resistant to antibiotics [7]. In this chapter, we will focus on the agents that are known to exhibit antibiofilm assembly including bacteriocins.

Moreover, bacteriophages have specific ability to infect and lyse bacteria [8]. Hence, phage therapy has many potential applications in the treatment of infectious diseases, with high therapeutic index and diminished adverse effects [9, 10]. Inhibitors of quorum sensing signaling, control of the global regulators, and the development of antibiofilm agents will be discussed in detail in this chapter. Additionally, the use of bacteriophages either for eradication of bacterial infections or as an efficient delivery system for antimicrobial agents will be described in this part.

2. Control of microbial virulence and resistance

2.1 Quorum sensing inhibition

Quorum sensing (QS) is a cellular signaling system, which is developed in response to population cell density [3, 4]. QS cascade relays on the release of signaling molecules called QS autoinducers/signals. The QS signals are produced at low levels with the start of microbial growth and accumulate upon increase in the cell density. Quorum sensing signals coordinate the microbial virulence behaviors such as secretion of toxins, secretion of exoenzymes, microbial motility, adhesion and biofilm assembly [11]. Furthermore, microbial communication systems have been assigned in fungi [12] and viruses [13]. Studies of QS provide significant insights into different mechanisms that control the interactions in microbial communities and how these interactions affect microbial pathogenesis. Several QS systems are well understood including Gram-negative bacteria that produce acyl-homoserine lactone (AHL) signals, including *Pseudomonas aeruginosa*, *Vibrio* sp., *Acinetobacter baumannii* and *Serratia marcescens* [5, 14, 15]. Alternatively, Gram-positive species such as *Staphylococcus aureus* utilize autoinducer peptide (AIP)-based QS systems [16].

Various strategies for quorum sensing inhibition have been explored. The quorum sensing inhibition approaches could be accomplished via interference with the synthesis of QS signals, elimination of the signal accumulation and disruption of signal-receptor interaction [17–19].

2.1.1 Interference with the synthesis of the autoinducing signals

One of the main quorum sensing inhibiting approaches is the interference with the synthesis of the autoinducing signals [20]. AI-2 compounds are considered as "universal" signal molecules of Gram-negative and Gram-positive bacteria [14, 21]. Moreover, they are encountered in species communications. The biosynthesis of AI-2 requires two main enzymes: methylthioadenosine/S-adenosylhomocysteine nucleosidase (MTA/SAH nucleosidase) and LuxS. AI-2 molecules contribute in various virulence behaviors, biofilm formation and host-pathogen interaction. Therefore, targeting AI-2 elaborates broad spectrum quorum sensing inhibition [22, 23]. In this instance, Gutierrez group have identified the transition analogs, 5′-methylthio- (MT-), 5′-ethylthio-(EtT-) and 5′-butylthio- (BuT) DADMe-immucillin, which specifically bind and inhibit MTA enzymes in *Escherichia coli* O157:H7. Also, 4,5-dihydroxy-2,3-pentanedioneS-ribosyl-homocysteine analogs have been developed as competitive inhibitor of LuxS [24–26].

On other instance, inhibiting AHL-synthesis has been extensively studied, for instance, triclosan inhibited both N-3-oxo-dodecanoyl-L-homoserine lactone and N-butyryl-L-homoserine lactone [27, 28], anthranilate derivatives are a *Pseudomonas* quinolone signal inhibitors [28], and proanthocyanidins have been approved as inhibitor of LasI/RhlI AHL synthases expression [29]. Furthermore, precursors of *Pseudomonas* quinolone signals (PQS) such anthranilatic acid derivatives reduced the pathogenicity of *P. aeruginosa* in lung-infected mice [15].

2.1.2 Elimination of the QS signals accumulation

Other common strategy is eliminating the accumulation of the QS signals, which have been attained by degrading the QS signal using enzymes or through sequestering the signal by synthetic polymers [30, 31] or utilizing antibodies that bind with the signals. Synthesized monoclonal antibodies (AP4-24H11) by Park group provoke high binding affinity for sequestering AIP-IV and decrease α-hemolysin production in *S. aureus* with relief of abscess formation in the infected murine model [32]. Kaufmann and coauthors inhibited the *P. aeruginosa* QS cascade via development of AHL-specific monoclonal antibodies. Synthetic polymers such as itaconic acid sequester the signaling molecules AHL and attenuate QS in *V. fischeri* [31, 33].

Moreover, disturbing enzymes responsible for biosynthesis of QS signals is a chief method, which affects both production and accumulation of different signals and perturb quorum sensing circuit [30]. Acylases, lactonases and oxidoreductases are the widely identified enzymes that target AHLs. AHL lactonases are broad AHL degrading enzymes, which produce its effect via hydrolyzing the ester bond of the AHL ring [34]. Lactonases have been isolated from various *Bacillus* sp., which harbor *aiiA* (autoinducer inactivation gene) [35, 36]. Ulrich study showed that, the heterologous expression of *aiiA* in *Burkholderia thailandensis* and *P. aeruginosa* lowered the levels of AHL and QS-related virulence factors [37]. Other important AHL lactonases are AttM and AiiB, which have been isolated from *Agrobacterium* sp. [38], AhlD from *Arthrobacterium*, AhlK from *Klebsiella* [39] and AidC from *Chryseobacterium* [40], QsdA from *Rhodococcus erythropolis* strain W2 [41], AiiM of *Microbacterium testaceum* [42], AidH of *Ochrobactrum* sp. T63 [43] and QsdH of *Pseudoalteromonas yunnanensis* [44]. Furthermore, paraoxonases 1, 2 and 3 (PON1 to −3) are mammalian lactonases were identified in the airway epithelia and mammalian sera [45].

AHL acylases enzymes (*aiiD)* and homologs were found in *Ralstonia* [46], *Actinoplanes utahensis* and *Pseudomonas* sp. The purified AiiD protein has the ability to degrade 3OC10HSL into HSL and 3-oxodecanoic acid. In addition, PvdQ, QuiP and HacB are specific AHL acylases of *P. aeruginosa*, in addition, HacA and HacB acylases of *Pseudomonas syringae* [47, 48]. Furthermore, the broader substrate specificity of AHL acylase (AhlM) was detected in *Streptomyces* sp. strain M664 with activity towards medium- and long-chain AHLs [49].

Oxidoreductases from *Rhodococcus erythropolis* inactivates AHLs (oxidation or reduction) with subsequent elimination of bacterial virulence *in vivo. Rhizobium* strain NGR234 possess diverse AHL-inactivation loci: *dhlR*, *qsdR1* and *qsdR2*, with lactonases activity, *aldR*, and *hydR-hitR* [50]. Enzymatic degradation of other QS autoinducers have been described: *carA* and *carB* from *Bacillus*, *E. coli* DH10B, *Staphylococcus* and *Pseudomonas* as the genes responsible for inhibition of DSF signaling [51]. Hod (3-hydroxy-2-methyl-4(1*H*)-quinolone 2,4-dioxygenase) stimulates the cleavage of PQS and attenuates PQS-regulated virulence factors. Roy and coauthors elicit the AI-2 activation activity of endogenous LsrK in *E. coli*, however, exogenously phosphorylation of AI-2 by LsrK eliminates its intracellular transport and hinders subsequent activation of AI-2 [52].

2.1.3 Elimination of the QS signal-receptor interaction

Interference with signal detection through eliminating the QS signal-receptor binding represents a successful approach [53, 54]. Various synthetic and natural AHL analogs have been reported to block the binding of the signal with specific receptors in *P. aeruginosa* and *Vibrio* sp. The prototype signal inhibitors, halogenated furanones, which are produced from *Delisea pulchra* represent a good example [55, 56]. Natural analogs have been also isolated with signal-receptor interference including ajoene [57], eugenol [58], flavonoids [59], iberin [60], furocoumarins [61], ellagic acid, penicillanic acid and patulin [62], phenethyl amide [63] and 1H-pyrrole-2-carboxylic acid [64].

The synthetic furanone derivative C-30 interferes and hinders the interaction of AHLs with the receptors [65]. Other furanone analogs have been developed including S-phenyl-L-cysteine sulfoxide and diphenyl disulfide [66] and tetrazole derivatives [67]. Furthermore, synthetic LasR derivatives have been developed such as indole derivatives, non-AHL-like antagonists [68], the synthesized azines derivatives, 4-(alkyloxy)-6-methyl-2H-pyran-2-one [69] and aspirin [70]. Triphenyl hybridγ-butyrolactones and cyclopentanones derivatives are potent inhibitors of LuxR [71]. Putative LasI inhibitors have been identified using molecular docking methods including the trans-cinnamaldehyde [72], (z)-5-octylidenethiazolidine-2, 4-dione [73] and fatty acyl purified from marine *Streptomyces* sp. [74]. Additionally, meta-bromo-thiolactone is a potent inhibitor of RhiI and subsequent PQS cascade [11].

In *S. aureus*, the interference with agr system has been accomplished using solonamide A and B that are cyclodepsipeptides derivatives, which purified from marine *Photobacterium* and reduced the expression of *hla* and RNAIII. Solonamide can act through competitive inhibition of agr system such as *S. aureus* agr system via structure similarity to the AIPs [75]. Other *S. aureus* quorum-sensing inhibitors have been identified including linear peptidomimetics as competitive inhibitors to AgrC [76], savirin as potent inhibitor of AgrA [77] and the polyhydroxy anthraquinone ω-hydroxyemodin as inhibitor of AgrA [78].

2.2 Control of the global regulators

Beside the QS regulons, other global regulators exhibit crucial functions in dominating the expression of various genes in assortment style as a response to environmental stimuli and changes, most notably the temperature change [5]. These so-called global regulators enable the bacterial communities to survive different environmental stresses including starvations, pH changes and temperature fluctuations, through the quick conformation of bacterial physiology and structure [79].

Among many regulators that coordinate gene expression in bacteria, in Gram-negative bacteria, the global regulator termed histone-like nucleoid-structuring (H-NS) protein is relatively significant and of paramount importance [80]. H-NS has been considered as the main model of studying how global regulators can affect bacterial structure and physiology. The H-NS protein is incorporated in the regulation of many genes responsible for controlling the physiological functions of Gram-negative bacterial cells involving cellular functions, survival under different environmental conditions and production of various virulence factors [81, 82]. Moreover, in Gram-positive bacteria, there are several global regulatory loci [83]. Among them in the *S. aureus*, SarA, a regulatory DNA binding protein involved in controlling the virulence genes expression, is well documented [84]. During regulation of the expression of various genes, these regulators have been demonstrated to act either as a positive regulators through enhancing the stability of the mRNA of expressed genes, resulting in excessive translation, or as a silencer protein that alter and decrease the gene expression by hindering binding of RNA polymerases to the promoters of target genes [85, 86].

This would open up novel approaches for the treatment and eradication of pathogenic bacteria utilizing inhibitors or modulators of these global loci to vanquish the global concerns of antimicrobial resistance and immune evasion of microbial pathogens. Among these approaches, the interesting inhibitor of SarA (SarABI), 4-[(2,4-diflurobenzyl)amino] cyclohexanol, was confirmed as SarA-based new curative medicament against *S. aureus*-related infections [87]. This might encourage research groups for screening other compounds that might affect global regulators in bacteria to give a new therapy for multi-drug resistant (MDR) bacterial strains.

2.3 Biofilm inhibition and eradication

Biofilm is a sessile community of microbial cells that is found to be attached to animate or inanimate surface, and usually surrounded by a matrix of polysaccharides, proteins and DNA [6]. The cells in these sessile communities differs phenotypically form those present in planktonic communities [88]. Bacterial cells in planktonic forms are almost one thousand times more sensitive to antibiotics than their biofilm counterparts [7]. Additionally, biofilms act as a defense mechanism against different stress conditions or immune cells attack [89].

In this part, we will focus on the agents that are known to exhibit antibiofilm activity.

2.3.1 Antimicrobial peptides

Antimicrobial peptides (AMPs) that are crucial players of innate immunity are reported to prevent biofilm formation in different pathogens. AMPs with antibiofilm activity are either natural or synthetic. The human cathelicidin peptide, LL-37, has been demonstrated to have antibiofilm activity in case of *P. aeruginosa* (at a concentration of 0.5 μg/mL), while the minimum inhibitory concentration for planktonic cells was 64 μg/mL [90]. In this study, it was reported that LL-37 was able to interfere with the adherence of microbial cells, enhancing twitching motility and downregulation of genes required for biofilm formation via affecting quorum sensing systems (Las and Rhl) [90]. Furthermore, such peptide was shown to prevent biofilm formation in *E. coli* and *S. aureus* [91]. The mouse cathelicidin-derived peptide AS10 was reported to exhibit antibiofilm activity in *Candida albicans* [92]. The synthetic cathelicidin-derived peptides; peptide 1018, DJK5 and DJK6, were reported to prevent biofilm formation in addition to enhancement of biofilm dispersion via prompting the hydrolysis of nucleotide signaling systems, and therefore, leads to its depletion in bacteria [93].

Another synthetic peptide, S4(1–16) M4Ka, has been found to inhibit biofilm formation and detach bacterial cells in *P. aeruginosa* [94]. The human β-defensin 3 (hBD-3) was found to inhibit the expression of *icaA*, *icaD* and *icaR* genes of *Staphylococcus epidermidis,* thus interfering with biofilm formation, where biofilm formation in *Staphylococci* is dependent on the synthesis of the polysaccharide intercellular adhesin PIA encoded by *icaADBC* locus [95]. Another example of human AMP with antibiofilm activity in *S. epidermidis*, is the liver-derived hepcidin 20. This peptide can inhibit extracellular matrix formation of biofilms via targeting PIA [95].

The natural AMP piscidin-3, obtained from fish, exhibits nucleosidase activity and can degrade extracellular DNA of *P. aeruginosa* [96]. Another example of natural AMP, that possesses antibiofilm activity, is esculentin, which is obtained from frog's skin. It acts by permeabilization of the cellular membrane of *P. aeruginosa* PAO1 cells in the biofilm [97]. A synthetic peptide P1, derived from a tick antifreeze protein, significantly inhibited biofilm formation in *Streptococcus mutans*. Such peptide reduced biofilm biomass by about 75% in microtiter plates and *in vitro* tooth models [98].

2.3.2 Surfactants

The anionic surfactant, sodium dodecyl sulfate, has been reported to destruct biofilm via enhancing the formation of central cavity within biofilm [99]. Cetyltrimethylammonium bromide (Catanionic surfactant), together with application of high shear stress, increased the detachment of *Pseudomonas fluorescens* biofilms [100]. The non-ionic surfactants, polyoxy ethylene sorbitan monolaurate

(Tween-20) and ethoxylated p-tert-octyl phenol (Triton X-100), were demonstrated to cause biofilm detachment [100]. Certain biosurfactants, which are surface active molecules formed by microorganisms, were reported to have antibiofilm activity. For example, surfactin, obtained from *Bacillus subtilis*, was found to have antibiofilm activity in case of *Salmonella enterica* in polyvinyl chloride microtiter wells and urethral catheters [101]. Another example is Rhamnolipids, that are produced principally, by *P. aeruginosa*, were found to promote the dispersal of bacterial biofilm [99]. Additionally, biosurfactants from *P. fluorescens* prevent the attachment of *Listeria monocytogenes* to stainless steel surfaces [102].

2.3.3 Free fatty acids

Free fatty acids obtained via hydrolysis of lipids by enzymes [103]. Certain members of free fatty acids are reported to exhibit antibiofilm activity [104]. For example, cis-2-decenoic acid from *P. aeruginosa* enhanced the dispersal of biofilms and inhibited its formation in different pathogens, such as *Klebsiella pneumoniae*, *E. coli*, *Proteus mirabilis*, *Streptococcus pyogenes*, *B. subtilis* and *S. aureus*, in addition to *C. albicans* [105]. Another example is *cis*-9-octadecenoic acid (oleic acid) that was reported to repress biofilm formation in *S. aureus* by interference with the initial attachment of bacterial cells [106]. The diffusible signal factor; *cis*-11-methyl-2-dodecenoic acid, from *Xanthomonas campestris* inhibits biofilm formation in case of *Bacillus cereus* [107]. This study showed also that diffusible signal factor or its structural analogs increased the antibiotic susceptibility of numerous bacterial pathogens, by inhibition of biofilm formation [107].

2.3.4 Metal chelators

Removal of metals from the microbial environment via metal chelators renders bacteria more susceptible to antimicrobial agents, as metals are essential for different cellular processes [108]. Ethylenediaminetetraacetic acid (EDTA), the most-known metal chelator, has been reported to exhibit antibiofilm activity against *S. aureus*, and to eradicate the *in vivo* biofilm models on catheters [109]. Combination of EDTA with minocycline has effectively reduced the colonization of *S. epidermidis*, *S. aureus* and *C. albicans* on catheters [110]. Similarly, the combination of EDTA and fluconazole remarkably inhibited biofilm assembly in *C. albicans* [111].

2.3.5 Enzymes

Based on their target, the antibiofilm enzymes are classified into three types: polysaccharide-degrading enzymes, nucleases and proteases.

2.3.5.1 Polysaccharide-degrading enzymes

Alpha amylase enzyme was found to inhibit biofilm formation by *S. aureus* through the detachment of biofilm and interfering with aggregation of cells [112]. Dispersion B, a bacterial glycoside hydrolase, degrades poly-N-acetylglucosamine (PNAG), a main matrix exopolysaccharide of *S. aureus* and *E. coli* [113]. Such polysaccharide is produced by many bacteria and fungi and plays an important role in surface adhesion, and biofilm formation. Furthermore, PNAG was reported to successfully disrupt the biofilm matrix of *S. epidermidis* [114]. Moreover, the combination of dispersion B and triclosan was reported to significantly reduce biofilm formation of *E. coli*, *S. aureus* and *S. epidermidis* [115].

2.3.5.2 Nucleases enzymes

Deoxyribonuclease I (DNase I) degrades DNA in biofilm matrix [104]. Moreover, it was shown to have antibiofilm activity and to detach the biofilms produced by different bacterial species [116]. Such nuclease can prevent the initial adherence of microbial cells to surfaces via the degradation of cell surface-associated nucleic acids that act as surface adhesins [117]. Furthermore, DNase I has been found to increase the sensitivity of bacterial cells in biofilm matrix to antibiotics, resulting in reduction of biofilm mass [118].

2.3.5.3 Proteases

Proteases act as antibiofilm agents because they are able to inhibit cell-cell communication, in biofilms, via hydrolysis of extracellular protein fibers and surface adhesins [104]. Subtilisins, a class of serine proteases produced by Bacillus species, were reported to prevent the adherence of microorganisms to surfaces [119]. The coating of silicone surfaces with multiple layers of amylase or acylase has been found to inhibit biofilm formation in case of *P. aeruginosa* and *S. aureus* [120]. Another example is lysostaphin, a metalloprotease produced by *Staphylococcus simulans*, was shown to prevent the adherence of *S. aureus* to lysostaphin-coated catheters [121].

2.3.6 Amino acids

D-Amino acids have been shown to inhibit biofilm formation in *B. subtilis*, via activating the release of amyloid fibers [122]. Such inhibitory effect was reversed by their cognate L-amino acids [123]. Furthermore, D-amino acids were shown to have antibiofilm activity in case of *P. aeruginosa* and *S. aureus* [122].

2.3.7 Nitric oxide generators

Exogenous generation of nitric oxide (NO) by agents, for example, sodium nitroprusside has been shown to trigger the bacterial growth from the biofilm form to the planktonic form via the reduction of the level of cyclic di-GMP inside the bacterial cells [104]. Further NO-generators, for example, S-nitroso-N-acetyl penicillamine and S-nitroso-L-glutathione were found also to induce the dispersion of *P. aeruginosa* biofilm [124]. The dispersion of biofilm by NO-generators was also demonstrated in *B. subtilis* [125]. Recently, it has been reported that catheters charged with NO prevented the adherence and the colonization of *P. aeruginosa*, *E. coli* and *C. albicans* on their surfaces [126].

2.3.8 Natural agents

Alkaloids are a group of natural organic compounds that contain a nitrogen atom and are present in different species of plants. The alkaloid berberine has been reported to inhibit biofilm formation in *S. epidermidis* biofilm at a concentration of 30 μg/mL, possibly via binding to the amyloid proteins in the biofilm matrix [127]. Reserpine has been shown to effectively prevent biofilm formation in *K. pneumoniae* at a concentration of 0.0156 mg/mL, which was 64-fold lower than its minimum inhibitory concentration [128]. Tetrandrine inhibited biofilm formation of *C. albicans* at a concentration of 32 mg/L, which is the MIC_{50} of that alkaloid against *C. albicans* SC5314 [129].

Guaijaverin, a flavonoid obtained from the leaves of *Psidium guajava*, has been shown to prevent the attachment of *S. mutans* to smooth surfaces by 83.7% at a concentration of 500 μg/mL. Eembelin, which is isolated from *Embelia ribes*, has been shown to inhibit biofilm formation in *S. mutans* [130]. Macelignan, isolated from the nutmegs of *Myristica fragrans*, was shown to reduce more than 50% of *S. mutans* biofilm at a concentration of 10 μg/mL [131].

Terpenes are a large class of natural hydrocarbons that are synthesized in microorganisms, plants and animals. Bakuchiol, isolated from the seeds of *Psoralea corylifolia*, has been shown to inhibit the adherence of *S. mutans* [132]. Other examples for terpenes that inhibit biofilm formation in *S. mutans*, are Xanthorrhizol (in combination with chlorhexidine gluconate) and casbane diterpene [133, 134].

2.4 Bacteriocins

Bacteriocins are proteins or peptides that are produced by bacteria or archaea, and are usually active against strains of bacteria that are related or unrelated to the producer strain [135]. Several bacteriocins are reported to exhibit antibiofilm activity and/or antimicrobial activity. The results of some these reports are summarized in **Table 1**.

2.5 Phage therapy

Phage therapy, which is also termed viral phage therapy, is the utilization of bacteriophages as medicaments for controlling and treating diseases brought by pathogenic bacterial infections [145]. Bacteriophages, like other viruses, are obligate intracellular parasites that utilize the enzymatic machinery of their hosts for establishing their physiological functions and replication [131]. The hosts for bacteriophages are bacteria, and phages have unique ability to specifically infect bacterial hosts resulting in their lysis [8].

Bacteriocin	Source	Antimicrobial activity	Antibiofilm activity
Mutacin 1140	*Streptococcus mutans*		Oral biofilm-associated with *Streptococcus sobrinus*, *Streptococcus oralis* [136]
Nisin A	*Lactococcus lactis* subsp. lactis	*Enterococcus faecalis* and *Streptococcus gordonii* [137]	*Listeria monocytogenes* [138]
Gallidermin	*Staphylococcus gallinarum*		*Staphylococcus aureus* and *Staphylococcus epidermidis* [139, 140]
Sonorensin	*Bacillus sonorensis* MT93	*Listeria monocytogenes* and *Staphylococcus aureus* [141]	*Staphylococcus aureus* [141]
Epidermicin NI0	*Staphylococcus epidermidis*	MRSA, *Enterococci* [142]	*Staphylococcus epidermidis* [142]
Amylolysin	*Bacillus amyloliquefaciens* GA1	*Listeria monocytogenes*, *Staphylococcus aureus* and *Staphylococcus epidermidis* [143]	
Philipimycin	*Actinoplanes philippinensis* MA7347	MRSA [144]	

Table 1.
Bacteriocins produced from different sources and exhibit antimicrobial and antibiofilm activity.

There are many conceivable usages for phage therapy in the treatment of crucial diseases in plants, animals as well as human [8, 145]. An outstanding advantage of utilizing bacteriophages over commonly used antibiotics, during treating infectious diseases, is their selectivity and specificity to infect and lyse infectious bacteria only without harming the host [9]. Besides, bacteriophages cause no harm to other organisms that live in a commensalism within hosts, such as the normal flora in human, which decreases significantly the incidence of superinfections or other opportunistic infections [10]. Moreover, due to their mode of action that phages replicate *in vivo* within their bacterial hosts, they can be used in modicum concentrations, which results in decreasing any side effects may rise during therapy and giving them a high therapeutic index [9, 10]. In addition, the capability of bacteriophages to penetrate bacterial biofilms that act as shields during the conventional antibiotic therapy, gives phages a superiority in controlling and treating diseases brought by pathogenic bacterial infections [146]. As living organisms, the capability of bacteriophages of continuous evolution, gives them the ability to overcome any resistance that can be developed by the evolution of pathogenic bacteria [146, 147]. All these tremendous advantages put the bacteriophage treatment as a superior alternative for treating diseases brought about multidrug resistant MDR bacterial pathogens [132]. On the other hand, the high bacterial host specificity of bacteriophages is encountered as a disadvantage during therapy, where, a phage can kill only its specific bacterial strain. However, this drawback can be solved by utilizing mixtures of bacteriophages, which is termed phage cocktails that have different pathogenic specific bacterial hosts as targets, to enhance the opportunities of unguis complete treatment [148]. Attention must be given, during the preparation of these cocktails, to the fact of continuous evolution of new MDR strains, so the cocktails must be updated periodically to be sufficient enough to treat infections brought by these strains [148, 149].

Historically, the first trials for the utilization of bacteriophages as medicaments for treating bacterial pathogens was reported in the Eastern world before the discovery of marvelous medicaments so-called antibiotics; however, there was any report of their usage in the Western world [150, 151]. The ability of bacteriophages to infect and lyse pathogenic bacteria was discovered by the scientists Frederick Twort and Felix D'Hérelle, who worked on *Shigella dysenteriae* [152]. They found that the cultures of stool specimens recovered from convalescent patients who were suffering from *Shigella* dysentery always depicting a high titer of phages [153]. Subsequently, they recorded that phages are the most abundant organisms in the environment and there are many sources where they can be found combined with their bacterial hosts; including gut and feces of convalescent patients as well as sewages [153]. Thereafter, due to their ubiquity especially in sewages, bacteriophages were widely utilized as medicaments for controlling and eradication of diseases brought by pathogenic bacteria [8].

It has been estimated that there are more than 100 different phage species and at least 10 phages for each bacterium. The International Committee for the Taxonomy of Viruses (ICTV) was affirmed at 1971 with the objective to always bring to date the taxonomic guidelines of viruses. The ICTV classified tailed bacteriophages (bacterial infecting phages) under the order of viruses which is termed *Caudovirales*. In this respect, three main families are involved within this order named *Siphoviridae*, *Myoviridae* and *Podoviridae*. The main difference between bacteriophages belonging to each of these families is the characteristics of the tail. Phages under the *Siphoviridae* family have long and non-contractile tails, and those belong to *Myoviridae* family have long and contractile tails, while those belong to the *Podoviridae* family have short, stubbed tails and a striking lack of features. Each of these three families can also be divided into different genera [8].

Compared with antibiotics and other therapeutic regimens, the steps and cost of production of bacteriophages are much easier and cheaper, respectively [10]. The easiest process for capturing of bacteriophages is done through collecting samples that seem to involve high titers of phages like sewage water samples. The collected samples are inoculated with the host bacterium, which seems to be infected by phages, on suitable growth medium. The successful isolation of certain lytic phage is depicted by the presence of clear inhibition zones in which bacteria cannot grow termed plaques; which indicates the lytic power of the isolated phage. Thereafter, the titer of isolated phage is increased by passing the phage in its specific bacterial strain several times to increase its concentration. Then, the pure supernatants containing phages are gained by centrifugation of bacterial/phage mixture, filtered through bacterial filters to remove any bacterial debris and pure phages are participated using special solutions containing NaCl and polyethylene glycol 8000 (PEG8000) [154].

Caution must be given during isolation of phages as a type called lysogenic bacteriophage may be isolated rather than the required bacterial pathogen killing type, which is called lytic bacteriophage. Lysogenic bacteriophages do not lyse bacterial cells, but they perform as tools for transfer of genetic elements of the nucleic acid between bacteria; including the genes responsible for antibiotic resistance. Fortunately, the most abundant phages are of the lytic type not the lysogenic [8, 145, 150].

Practically, bacteriophages can be dispensed and used through many routes including; less commonly oral or systemic route and most commonly topical route as sprays, liquid solutions or their application on surgical dressings for the treatment of wound infections [154]. The possibility of their clearance during the presence in blood stream by immune system or presence of any trace hazards of chemicals or parts of the bacterial host used during their production, made bacteriophage usage as intravenous injections uncommon and very rare [148, 149]. Lyophilization of bacteriophages and their production as solid dosage forms as pills or tablets do not decrease their potency and increase their shelf life as oral dosage forms [155, 156]. The supplementation of oral forms of phages, either solid or liquid, with antacid increases its stability, as it protect them from the high acidity during their bypassing in the stomach [155, 156].

The application of bacteriophages as therapeutic medicaments has been extensively reported. For example, in the field of human health promotion and food protection, different bacteriophages have been employed to eradicate common bacterial pathogens that may cause food spoilage as *Listeria* sp. and *Campylobacter* sp. [157, 158]. In the fields of veterinary medicine and agriculture different bacteriophages were employed to control and eradicate bacterial pathogens like *Xanthomonas*, *Escherichia*, *Campylobacter* and *Salmonella* [159]. Moreover, in the field of fish production and aquacultures, different bacteriophages were employed to control and eradicate bacterial pathogens like *Vibrio* sp. [160]. In the field of human medicine, different bacteriophages were employed to control and eradicate bacterial pathogens including *P. aeruginosa*, *Staphylococci*, *Streptococci*, *E. coli*, *Vibrio* and *Shigella* and *Mycobacterium* sp. [161, 162]. Most recent application of bacteriophages in human medicine is their utilization as drug delivery system, which is very interesting as they can be used for the delivery of common antibiotics [163, 164] or antitumor agents [165].

A more recent policy, termed enzybiotic, for using phages as therapeutic agents is the utilization of their enzymes only, which are produced by recombinant technology, combined with other antibacterial agents or as a separate antibacterial agents [166].

As other therapeutic regimens for controlling bacterial pathogens, the patients may develop extensive fever and shock, when the bacteria are lysed due to the release of what is called pyrogens or endotoxins within the patient [167]. This problem can be coped during phage therapy through the utilization of genetically modified phages that harbor enzymes having the ability to lyse these endotoxins and the other bacterial structures into harmless products [168].

Examples of therapeutic approaches of bacteriophages and their enzymes are illustrated in **Table 2**.

Infection/ disease	Model	Causative agent	Route of administration of phages/enzymes	Treatment outcomes	Reference
Chronic otitis	Human	*Pseudomonas aeruginosa*	Oral administration of phages	Successful treatment	[169]
Typhoid	Human	*Salmonella typhi*	Oral administration of phages	Successful treatment	[170]
Diabetic foot ulcer	Human	*Staphylococcus aureus*	Topical application of phages	Successful treatment	[171]
Sepsis	Murine	*Vibrio parahaemolyticus*	Intraperitoneal and oral administration of phages	Successful treatment	[172]
Pneumonia	Murine	*Pseudomonas aeruginosa*	Intranasal administration of phages	Successful treatment	[154]
Ulcers and wounds	Human	*Proteus vulgaris*	Topical application of phages	Successful treatment	[173]
Meningitis	Murine	*Escherichia coli*	Intraperitoneal or subcutaneous administration of phages.	Successful treatment	[174]
Sepsis	Murine	*Acinetobacter baumannii*	Intraperitoneal administration of phages	Successful treatment	[175]
Bacteremia	Murine	*Enterococcus faecium*	Intraperitoneal administration of phages	Successful treatment	[176]
Ileocecitis	Hamster	*Clostridium difficile*	Oral administration of phages	Successful treatment	[177]
Dysentery	Human	*Shigella dysenteriae*	Oral administration of phages	Successful treatment	[178]
Cholera	Human	*Vibrio cholerae*	Oral administration of phages	Successful treatment	[178]
Pneumonia	Murine	*Streptococcus pneumoniae*	Intraperitoneal administration of Cpl-1 lysin enzyme	Successful treatment	[179]
Bacteremia	Murine	*Streptococcus pyogenes*	Intraperitoneal administration of PlySs2 lysin enzyme	Successful treatment	[179]
In vitro	*In vitro*	*Bacillus anthracis*	Application of PlyG lysin enzyme	Significant reduction in bacterial density	[180]
Endophthalmitis	Murine	*Staphylococcus aureus*	Application of Ply187 lysin as eye drops	Successful treatment	[181]
Bacteremia	Murine	*Acinetobacter baumannii*	Administration of PlyF307 lysin enzyme	Successful treatment	[182]
In vitro	*In vitro*	*Pseudomonas aeruginosa* and *Salmonella typhimurium*	Application of ABgp46 lysin enzyme	Significant reduction in bacterial density	[183]

Table 2.
Therapeutic approaches of bacteriophages and their enzymes.

3. Conclusion

Various approaches have been developed for competing microbial virulence and resistance. Quorum sensing signals and global regulators play an essential role in controlling the gene expression of virulence factors, and the expression of proteins required for adaptation to environmental and stress condition. Therefore, control of these regulators will stop the microbial pathogenicity. In addition, biofilms act as a defense mechanism against host immunity and antimicrobial therapy. Natural and synthetic compounds have approved activities in eradication of biofilm formation. Besides, phage therapy, which is currently successful in destruction of bacterial pathogens that do not respond to conventional antimicrobials. These methods would open up new perspectives for management the up growing problem of microbial resistance. Further, *in vivo* studies are required for real applications of these trends in eradication of microbial infections.

Author details

Mohammed El-Mowafy, Abdelaziz Elgaml and Mona Shaaban*
Department of Microbiology and Immunology, Faculty of Pharmacy, Mansoura University, Mansoura, Egypt

*Address all correspondence to: mona_ibrahem@mans.edu.eg

References

[1] Reuter K, Steinbach A, Helms V. Interfering with bacterial quorum sensing. Perspectives in Medicinal Chemistry. 2016;**8**:1-15

[2] Finch RG, Pritchard DI, Bycroft BW, Williams P, Stewart GS. Quorum sensing: A novel target for anti-infective therapy. The Journal of Antimicrobial Chemotherapy. 1998;**42**(5):569-571

[3] Miller MB, Bassler BL. Quorum sensing in bacteria. Annual Review of Microbiology. 2001;**55**:165-199

[4] Geske GD, O'Neill JC, Miller DM, Mattmann ME, Blackwell HE. Modulation of bacterial quorum sensing with synthetic ligands: Systematic evaluation of N-acylated homoserine lactones in multiple species and new insights into their mechanisms of action. Journal of the American Chemical Society. 2007;**129**(44):13613-13625

[5] Elgaml A, Miyoshi SI. Regulation systems of protease and hemolysin production in *Vibrio vulnificus*. Microbiology and Immunology. 2017;**61**(1):1-11

[6] Bogino PC, Oliva Mde L, Sorroche FG, Giordano W. The role of bacterial biofilms and surface components in plant-bacterial associations. International Journal of Molecular Sciences. 2013;**14**(8):15838-15859

[7] Hengzhuang W, Wu H, Ciofu O, Song Z, Høiby N. Pharmacokinetics/pharmacodynamics of colistin and imipenem on mucoid and nonmucoid *Pseudomonas aeruginosa* biofilms. Antimicrobial Agents and Chemotherapy. 2011;**55**(9):4469-4474

[8] Abedon ST. Phage therapy: Various perspectives on how to improve the art. Methods in Molecular Biology. 1734;**2018**:113-127

[9] Grasis JA. Host-associated bacteriophage isolation and preparation for viral metagenomics. Methods in Molecular Biology. 1746;**2018**:1-25

[10] Nobrega FL, Costa AR, Kluskens LD, Azeredo J. Revisiting phage therapy: New applications for old resources. Trends in Microbiology. 2015;**23**(4):185-191

[11] O'Loughlin CT, Miller LC, Siryaporn A, Drescher K, Semmelhack MF, Bassler BL. A quorum-sensing inhibitor blocks *Pseudomonas aeruginosa* virulence and biofilm formation. Proceedings of the National Academy of Sciences of the United States of America. 2013;**110**(44):17981-17986

[12] Hornby JM, Jensen EC, Lisec AD, Tasto JJ, Jahnke B, Shoemaker R, et al. Quorum sensing in the dimorphic fungus *Candida albicans* is mediated by farnesol. Applied and Environmental Microbiology. 2001;**67**(7):2982-2992

[13] Erez Z, Steinberger-Levy I, Shamir M, Doron S, Stokar-Avihail A, Peleg Y, et al. Communication between viruses guides lysis-lysogeny decisions. Nature. 2017;**541**(7638):488-493

[14] Schaefer AL, Val DL, Hanzelka BL, Cronan JE Jr, Greenberg EP. Generation of cell-to-cell signals in quorum sensing: Acyl homoserine lactone synthase activity of a purified *Vibrio fischeri* LuxI protein. Proceedings of the National Academy of Sciences of the United States of America. 1996;**93**(18):9505-9509

[15] Calfee MW, Coleman JP, Pesci EC. Interference with pseudomonas quinolone signal synthesis inhibits virulence factor expression by *Pseudomonas aeruginosa*. Proceedings of the National Academy of Sciences of the United States of America. 2001;**98**(20):11633-11637

[16] Xiong YQ, Willard J, Yeaman MR, Cheung AL, Bayer AS. Regulation of *Staphylococcus aureus* alpha-toxin gene (hla) expression by agr, sarA, and sae in vitro and in experimental infective endocarditis. The Journal of Infectious Diseases. 2006;**194**(9):1267-1275

[17] Kong C, Neoh HM, Nathan S. Targeting *Staphylococcus aureus* toxins: A potential form of anti-virulence therapy. Toxins. 2016;**8**(3): pii: E72; 1-21

[18] Remy B, Mion S, Plener L, Elias M, Chabriere E, Daude D. Interference in bacterial quorum sensing: A biopharmaceutical perspective. Frontiers in Pharmacology. 2018;**9**:203

[19] Shaaban M, Elgaml A, Habib EE. Biotechnological applications of quorum sensing inhibition as novel therapeutic strategies for multidrug resistant pathogens. Microbial Pathogenesis. 2019;**127**:138-143

[20] Sedlmayer F, Hell D, Muller M, Auslander D, Fussenegger M. Designer cells programming quorum-sensing interference with microbes. Nature Communications. 2018;**9**(1):1822

[21] Xavier KB, Bassler BL. LuxS quorum sensing: More than just a numbers game. Current Opinion in Microbiology. 2003;**6**(2):191-197

[22] Guo M, Gamby S, Nakayama S, Smith J, Sintim HO. A pro-drug approach for selective modulation of AI-2-mediated bacterial cell-to-cell communication. Sensors. 2012;**12**(3):3762-3772

[23] Ren D, Li C, Qin Y, Yin R, Li X, Tian M, et al. Inhibition of *Staphylococcus aureus* adherence to Caco-2 cells by lactobacilli and cell surface properties that influence attachment. Anaerobe. 2012;**18**(5):508-515

[24] Gutierrez JA, Crowder T, Rinaldo-Matthis A, Ho MC, Almo SC, Schramm VL. Transition state analogs of 5′-methylthioadenosine nucleosidase disrupt quorum sensing. Nature Chemical Biology. 2009;**5**(4):251-257

[25] Malladi VL, Sobczak AJ, Meyer TM, Pei D, Wnuk SF. Inhibition of LuxS by S-ribosylhomocysteine analogues containing a [4-aza]ribose ring. Bioorganic & Medicinal Chemistry. 2011;**19**(18):5507-5519

[26] Wnuk SF, Lalama J, Garmendia CA, Robert J, Zhu J, Pei D. S-Ribosylhomocysteine analogues with the carbon-5 and sulfur atoms replaced by a vinyl or (fluoro)vinyl unit. Bioorganic & Medicinal Chemistry. 2008;**16**(9):5090-5102

[27] Hoang TT, Schweizer HP. Characterization of *Pseudomonas aeruginosa* enoyl-acyl carrier protein reductase (FabI): A target for the antimicrobial triclosan and its role in acylated homoserine lactone synthesis. Journal of Bacteriology. 1999;**181**(17):5489-5497

[28] Coleman JP, Hudson LL, McKnight SL, Farrow JM 3rd, Calfee MW, Lindsey CA, et al. *Pseudomonas aeruginosa* PqsA is an anthranilate-coenzyme A ligase. Journal of Bacteriology. 2008;**190**(4):1247-1255

[29] Maisuria VB, Los Santos YL, Tufenkji N, Deziel E. Cranberry-derived proanthocyanidins impair virulence and inhibit quorum sensing of *Pseudomonas aeruginosa*. Scientific Reports. 2016;**6**:30169

[30] LaSarre B, Federle MJ. Exploiting quorum sensing to confuse bacterial pathogens. Microbiology and Molecular Biology Reviews. 2013;**77**(1):73-111

[31] Piletska EV, Stavroulakis G, Karim K, Whitcombe MJ, Chianella I, Sharma A, et al. Attenuation of *Vibrio fischeri* quorum sensing using rationally

designed polymers. Biomacromolecules. 2010;**11**(4):975-980

[32] Park J, Jagasia R, Kaufmann GF, Mathison JC, Ruiz DI, Moss JA, et al. Infection control by antibody disruption of bacterial quorum sensing signaling. Chemistry & Biology. 2007;**14**(10):1119-1127

[33] Cavaleiro E, Duarte AS, Esteves AC, Correia A, Whitcombe MJ, Piletska EV, et al. Novel linear polymers able to inhibit bacterial quorum sensing. Macromolecular Bioscience. 2015;**15**(5):647-656

[34] Dong YH, Xu JL, Li XZ, Zhang LH. AiiA, an enzyme that inactivates the acylhomoserine lactone quorum-sensing signal and attenuates the virulence of *Erwinia carotovora*. Proceedings of the National Academy of Sciences of the United States of America. 2000;**97**(7):3526-3531

[35] Dong YH, Gusti AR, Zhang Q, Xu JL, Zhang LH. Identification of quorum-quenching N-acyl homoserine lactonases from Bacillus species. Applied and Environmental Microbiology. 2002;**68**(4):1754-1759

[36] Liu D, Momb J, Thomas PW, Moulin A, Petsko GA, Fast W, et al. Mechanism of the quorum-quenching lactonase (AiiA) from *Bacillus thuringiensis*. 1. Product-bound structures. Biochemistry. 2008;**47**(29):7706-7714

[37] Ulrich RL. Quorum quenching: Enzymatic disruption of N-acylhomoserine lactone-mediated bacterial communication in *Burkholderia thailandensis*. Applied and Environmental Microbiology. 2004;**70**(10):6173-6180

[38] Zhang HB, Wang LH, Zhang LH. Genetic control of quorum-sensing signal turnover in *Agrobacterium tumefaciens*. Proceedings of the National Academy of Sciences of the United States of America. 2002;**99**(7):4638-4643

[39] Park SY, Lee SJ, Oh TK, Oh JW, Koo BT, Yum DY, et al. AhlD, an N-acylhomoserine lactonase in *Arthrobacter* sp., and predicted homologues in other bacteria. Microbiology. 2003;**149**(Pt 6):1541-1550

[40] Wang WZ, Morohoshi T, Someya N, Ikeda T. AidC, a novel N-acylhomoserine lactonase from the potato root-associated cytophaga-flavobacteria-bacteroides (CFB) group bacterium *Chryseobacterium* sp. strain StRB126. Applied and Environmental Microbiology. 2012;**78**(22):7985-7992

[41] Uroz S, Oger PM, Chapelle E, Adeline MT, Faure D, Dessaux Y. A Rhodococcus qsdA-encoded enzyme defines a novel class of large-spectrum quorum-quenching lactonases. Applied and Environmental Microbiology. 2008;**74**(5):1357-1366

[42] Wang WZ, Morohoshi T, Ikenoya M, Someya N, Ikeda T. AiiM, a novel class of N-acylhomoserine lactonase from the leaf-associated bacterium Microbacterium testaceum. Applied and Environmental Microbiology. 2010;**76**(8):2524-2530

[43] Mei GY, Yan XX, Turak A, Luo ZQ, Zhang LQ. AidH, an alpha/beta-hydrolase fold family member from an Ochrobactrum sp. strain, is a novel N-acylhomoserine lactonase. Applied and Environmental Microbiology. 2010;**76**(15):4933-4942

[44] Huang W, Lin Y, Yi S, Liu P, Shen J, Shao Z, et al. QsdH, a novel AHL lactonase in the RND-type inner membrane of marine *Pseudoalteromonas byunsanensis* strain 1A01261. PLoS One. 2012;**7**(10):e46587

[45] Yang F, Wang LH, Wang J, Dong YH, Hu JY, Zhang LH. Quorum

quenching enzyme activity is widely conserved in the sera of mammalian species. FEBS Letters. 2005;**579**(17):3713-3717

[46] Lin YH, Xu JL, Hu J, Wang LH, Ong SL, Leadbetter JR, et al. Acyl-homoserine lactone acylase from Ralstonia strain XJ12B represents a novel and potent class of quorum-quenching enzymes. Molecular Microbiology. 2003;**47**(3):849-860

[47] Bokhove M, Nadal Jimenez P, Quax WJ, Dijkstra BW. The quorum-quenching N-acyl homoserine lactone acylase PvdQ is an Ntn-hydrolase with an unusual substrate-binding pocket. Proceedings of the National Academy of Sciences of the United States of America. 2010;**107**(2):686-691

[48] Huang JJ, Petersen A, Whiteley M, Leadbetter JR. Identification of QuiP, the product of gene PA1032, as the second acyl-homoserine lactone acylase of *Pseudomonas aeruginosa* PAO1. Applied and Environmental Microbiology. 2006;**72**(2):1190-1197

[49] Park SY, Kang HO, Jang HS, Lee JK, Koo BT, Yum DY. Identification of extracellular N-acylhomoserine lactone acylase from a Streptomyces sp. and its application to quorum quenching. Applied and Environmental Microbiology. 2005;**71**(5):2632-2641

[50] Krysciak D, Schmeisser C, Preuss S, Riethausen J, Quitschau M, Grond S, et al. Involvement of multiple loci in quorum quenching of autoinducer I molecules in the nitrogen-fixing symbiont Rhizobium (Sinorhizobium) sp. strain NGR234. Applied and Environmental Microbiology. 2011;**77**(15):5089-5099

[51] Llamas I, Suarez A, Quesada E, Bejar V, del Moral A. Identification and characterization of the carAB genes responsible for encoding carbamoylphosphate synthetase in *Halomonas eurihalina*. Extremophiles. 2003;**7**(3):205-211

[52] Roy V, Fernandes R, Tsao CY, Bentley WE. Cross species quorum quenching using a native AI-2 processing enzyme. ACS Chemical Biology. 2010;**5**(2):223-232

[53] Koch B, Liljefors T, Persson T, Nielsen J, Kjelleberg S, Givskov M. The LuxR receptor: The sites of interaction with quorum-sensing signals and inhibitors. Microbiology. 2005; **151**(Pt 11):3589-3602

[54] Singh RP, Desouky SE, Nakayama J. Quorum quenching strategy targeting Gram-positive pathogenic bacteria. Advances in Experimental Medicine and Biology. 2016;**901**:109-130

[55] Manefield M, de Nys R, Kumar N, Read R, Givskov M, Steinberg P, et al. Evidence that halogenated furanones from Delisea pulchra inhibit acylated homoserine lactone (AHL)-mediated gene expression by displacing the AHL signal from its receptor protein. Microbiology. 1999;**145** (Pt 2):283-291

[56] Givskov M, de Nys R, Manefield M, Gram L, Maximilien R, Eberl L, et al. Eukaryotic interference with homoserine lactone-mediated prokaryotic signalling. Journal of Bacteriology. 1996;**178**(22):6618-6622

[57] Fong J, Yuan M, Jakobsen TH, Mortensen KT, Delos Santos MM, Chua SL, et al. Disulfide bond-containing Ajoene analogues As novel quorum sensing inhibitors of *Pseudomonas aeruginosa*. Journal of Medicinal Chemistry. 2017;**60**(1):215-227

[58] Zhou L, Zheng H, Tang Y, Yu W, Gong Q. Eugenol inhibits quorum sensing at sub-inhibitory concentrations. Biotechnology Letters. 2013;**35**(4):631-637

[59] Paczkowski JE, Mukherjee S, McCready AR, Cong JP, Aquino CJ, Kim H, et al. Flavonoids suppress *Pseudomonas aeruginosa* virulence through allosteric inhibition of quorum-sensing receptors. The Journal of Biological Chemistry. 2017;**292**(10):4064-4076

[60] Jakobsen TH, Bragason SK, Phipps RK, Christensen LD, van Gennip M, Alhede M, et al. Food as a source for quorum sensing inhibitors: Iberin from horseradish revealed as a quorum sensing inhibitor of *Pseudomonas aeruginosa*. Applied and Environmental Microbiology. 2012;**78**(7):2410-2421

[61] Girennavar B, Cepeda ML, Soni KA, Vikram A, Jesudhasan P, Jayaprakasha GK, et al. Grapefruit juice and its furocoumarins inhibits autoinducer signaling and biofilm formation in bacteria. International Journal of Food Microbiology. 2008;**125**(2):204-208

[62] Rasmussen TB, Skindersoe ME, Bjarnsholt T, Phipps RK, Christensen KB, Jensen PO, et al. Identity and effects of quorum-sensing inhibitors produced by Penicillium species. Microbiology. 2005;**151** (Pt 5):1325-1340

[63] Teasdale ME, Liu J, Wallace J, Akhlaghi F, Rowley DC. Secondary metabolites produced by the marine bacterium *Halobacillus salinus* that inhibit quorum sensing-controlled phenotypes in gram-negative bacteria. Applied and Environmental Microbiology. 2009;**75**(3):567-572

[64] Hassan R, Shaaban MI, Abdel Bar FM, El-Mahdy AM, Shokralla S. Quorum sensing inhibiting activity of *Streptomyces coelicoflavus* isolated from soil. Frontiers in Microbiology. 2016;7:659

[65] Wu H, Song Z, Hentzer M, Andersen JB, Molin S, Givskov M, et al. Synthetic furanones inhibit quorum-sensing and enhance bacterial clearance in *Pseudomonas aeruginosa* lung infection in mice. The Journal of Antimicrobial Chemotherapy. 2004;**53**(6):1054-1061

[66] Cady NC, McKean KA, Behnke J, Kubec R, Mosier AP, Kasper SH, et al. Inhibition of biofilm formation, quorum sensing and infection in *Pseudomonas aeruginosa* by natural products-inspired organosulfur compounds. PLoS One. 2012;**7**(6):e38492

[67] Muh U, Schuster M, Heim R, Singh A, Olson ER, Greenberg EP. Novel *Pseudomonas aeruginosa* quorum-sensing inhibitors identified in an ultra-high-throughput screen. Antimicrobial Agents and Chemotherapy. 2006;**50**(11):3674-3679

[68] Biswas NN, Kutty SK, Barraud N, Iskander GM, Griffith R, Rice SA, et al. Indole-based novel small molecules for the modulation of bacterial signalling pathways. Organic & Biomolecular Chemistry. 2015;**13**(3):925-937

[69] Park S, Kim HS, Ok K, Kim Y, Park HD, Byun Y. Design, synthesis and biological evaluation of 4-(alkyloxy)-6-methyl-2H-pyran-2-one derivatives as quorum sensing inhibitors. Bioorganic & Medicinal Chemistry Letters. 2015;**25**(15):2913-2917

[70] El-Mowafy SA, Abd El Galil KH, El-Messery SM, Shaaban MI. Aspirin is an efficient inhibitor of quorum sensing, virulence and toxins in *Pseudomonas aeruginosa*. Microbial Pathogenesis. 2014;**74**:25-32

[71] O'Reilly MC, Blackwell HE. Structure-based design and biological evaluation of triphenyl scaffold-based hybrid compounds as hydrolytically stable modulators of a LuxR-type quorum sensing receptor. ACS Infectious Diseases. 2016;**2**(1):32-38

[72] Chang CY, Krishnan T, Wang H, Chen Y, Yin WF, Chong YM, et al.

Non-antibiotic quorum sensing inhibitors acting against N-acyl homoserine lactone synthase as druggable target. Scientific Reports. 2014;**4**:7245

[73] Lidor O, Al-Quntar A, Pesci EC, Steinberg D. Mechanistic analysis of a synthetic inhibitor of the *Pseudomonas aeruginosa* LasI quorum-sensing signal synthase. Scientific Reports. 2015;**5**:16569

[74] Kamarudheen N, Rao KVB. Fatty acyl compounds from marine *Streptomyces griseoincarnatus* strain HK12 against two major bio-film forming nosocomial pathogens; an in vitro and in silico approach. Microbial Pathogenesis. 2019;**127**:121-130

[75] Mansson M, Nielsen A, Kjaerulff L, Gotfredsen CH, Wietz M, Ingmer H, et al. Inhibition of virulence gene expression in *Staphylococcus aureus* by novel depsipeptides from a marine photobacterium. Marine Drugs. 2011;**9**(12):2537-2552

[76] Karathanasi G, Bojer MS, Baldry M, Johannessen BA, Wolff S, Greco I, et al. Linear peptidomimetics as potent antagonists of *Staphylococcus aureus* agr quorum sensing. Scientific Reports. 2018;**8**(1):3562

[77] Sully EK, Malachowa N, Elmore BO, Alexander SM, Femling JK, Gray BM, et al. Selective chemical inhibition of agr quorum sensing in *Staphylococcus aureus* promotes host defense with minimal impact on resistance. PLoS Pathogens. 2014;**10**(6):e1004174

[78] Daly SM, Elmore BO, Kavanaugh JS, Triplett KD, Figueroa M, Raja HA, et al. Omega-hydroxyemodin limits *Staphylococcus aureus* quorum sensing-mediated pathogenesis and inflammation. Antimicrobial Agents and Chemotherapy. 2015;**59**(4):2223-2235

[79] Lee SE, Kim SY, Kim CM, Kim MK, Kim YR, Jeong K, et al. The pyrH gene of *Vibrio vulnificus* is an essential in vivo survival factor. Infection and Immunity. 2007;**75**(6):2795-2801

[80] Elgaml A, Miyoshi S. Role of the histone-like nucleoid structuring protein (H-NS) in the regulation of virulence factor expression and stress response in *Vibrio vulnificus*. Biocontrol Science. 2015; **20**(4):263-274

[81] Dorman CJ. H-NS: A universal regulator for a dynamic genome. Nature Reviews. Microbiology. 2004;**2**(5):391-400

[82] Dorman CJ, Deighan P. Regulation of gene expression by histone-like proteins in bacteria. Current Opinion in Genetics & Development. 2003;**13**(2):179-184

[83] Zheng W, Liang Y, Zhao H, Zhang J, Li Z. 5,5′-methylenedisalicylic acid (MDSA) modulates SarA/MgrA phosphorylation by targeting Ser/Thr phosphatase Stp1. Chembiochem. 2015;**16**(7):1035-1040

[84] Cheung AL, Nishina KA, Trotonda MP, Tamber S. The SarA protein family of *Staphylococcus aureus*. The International Journal of Biochemistry & Cell Biology. 2008;**40**(3):355-361

[85] Ono S, Goldberg MD, Olsson T, Esposito D, Hinton JC, Ladbury JE. H-NS is a part of a thermally controlled mechanism for bacterial gene regulation. The Biochemical Journal. 2005;**391**(Pt 2):203-213

[86] Brescia CC, Kaw MK, Sledjeski DD. The DNA binding protein H-NS binds to and alters the stability of RNA in vitro and in vivo. Journal of Molecular Biology. 2004;**339**(3): 505-514

[87] Arya R, Ravikumar R, Santhosh RS, Princy SA. SarA based novel therapeutic candidate against *Staphylococcus aureus* associated with vascular graft infections. Frontiers in Microbiology. 2015;**6**:416

[88] Donlan RM, Costerton JW. Biofilms: Survival mechanisms of clinically relevant microorganisms. Clinical Microbiology Reviews. 2002;**15**(2):167-193

[89] Roy R, Tiwari M, Donelli G, Tiwari V. Strategies for combating bacterial biofilms: A focus on anti-biofilm agents and their mechanisms of action. Virulence. 2018;**9**(1):522-554

[90] Overhage J, Campisano A, Bains M, Torfs EC, Rehm BH, Hancock RE. Human host defense peptide LL-37 prevents bacterial biofilm formation. Infection and Immunity. 2008;**76**(9):4176-4182

[91] Aka ST. Killing efficacy and anti-biofilm activity of synthetic human cationic antimicrobial peptide cathelicidin hCAP-18/LL37 against urinary tract pathogens. Journal of Microbiology and Infectious Diseases. 2015;**5**:15-20

[92] De Brucker K, Delattin N, Robijns S, Steenackers H, Verstraeten N, Landuyt B, et al. Derivatives of the mouse cathelicidin-related antimicrobial peptide (CRAMP) inhibit fungal and bacterial biofilm formation. Antimicrobial Agents and Chemotherapy. 2014;**58**(9):5395-5404

[93] De la Fuente-Núñez C, Reffuveille F, Haney EF, Straus SK, Hancock RE. Broad-spectrum anti-biofilm peptide that targets a cellular stress response. PLoS Pathogens. 2014;**10**(5):e1004152

[94] Quiles F, Saadi S, Francius G, Bacharouche J, Humbert F. In situ and real time investigation of the evolution of a *Pseudomonas fluorescens* nascent biofilm in the presence of an antimicrobial peptide. Biochimica et Biophysica Acta. 2016;**1858**(1):75-84

[95] Rohde H, Frankenberger S, Zahringer U, Mack D. Structure, function and contribution of polysaccharide intercellular adhesin (PIA) to Staphylococcus epidermidis biofilm formation and pathogenesis of biomaterial-associated infections. European Journal of Cell Biology. 2010;**89**(1):103-111

[96] Libardo MDJ, Bahar AA, Ma B, Fu R, McCormick LE, Zhao J, et al. Nuclease activity gives an edge to host-defense peptide piscidin 3 over piscidin 1, rendering it more effective against persisters and biofilms. The FEBS Journal. 2017;**284**(21):3662-3683

[97] Luca V, Stringaro A, Colone M, Pini A, Mangoni ML. Esculentin(1-21), an amphibian skin membrane-active peptide with potent activity on both planktonic and biofilm cells of the bacterial pathogen *Pseudomonas aeruginosa*. Cellular and Molecular Life Sciences: CMLS. 2013;**70**(15):2773-2786

[98] Ansari JM, Abraham NM, Massaro J, Murphy K, Smith-Carpenter J, Fikrig E. Anti-biofilm activity of a self-aggregating peptide against *Streptococcus mutans*. Frontiers in Microbiology. 2017;**8**:488

[99] Boles BR, Thoendel M, Singh PK. Rhamnolipids mediate detachment of *Pseudomonas aeruginosa* from biofilms. Molecular Microbiology. 2005;**57**(5):1210-1223

[100] Simoes M, Pereira MO, Vieira MJ. Action of a cationic surfactant on the activity and removal of bacterial biofilms formed under different flow regimes. Water Research. 2005;**39**(2-3):478-486

[101] Mireles JR 2nd, Toguchi A, Harshey RM. *Salmonella enterica* serovar typhimurium swarming mutants with altered biofilm-forming abilities: Surfactin inhibits biofilm formation. Journal of Bacteriology. 2001;**183**(20):5848-5854

[102] Meylheuc T, van Oss CJ, Bellon-Fontaine MN. Adsorption of biosurfactant on solid surfaces and consequences regarding the bioadhesion of Listeria monocytogenes LO28. Journal of Applied Microbiology. 2001;**91**(5):822-832

[103] Desbois AP, Smith VJ. Antibacterial free fatty acids: Activities, mechanisms of action and biotechnological potential. Applied Microbiology and Biotechnology. 2010;**85**(6):1629-1642

[104] Li X-H, Lee J-H. Antibiofilm agents: A new perspective for antimicrobial strategy. Journal of Microbiology. 2017;**55**(10):753-766

[105] Davies DG, Marques CN. A fatty acid messenger is responsible for inducing dispersion in microbial biofilms. Journal of Bacteriology. 2009;**191**(5):1393-1403

[106] Stenz L, Francois P, Fischer A, Huyghe A, Tangomo M, Hernandez D, et al. Impact of oleic acid (cis-9-octadecenoic acid) on bacterial viability and biofilm production in *Staphylococcus aureus*. FEMS Microbiology Letters. 2008;**287**(2):149-155

[107] Deng Y, Lim A, Lee J, Chen S, An S, Dong YH, et al. Diffusible signal factor (DSF) quorum sensing signal and structurally related molecules enhance the antimicrobial efficacy of antibiotics against some bacterial pathogens. BMC Microbiology. 2014;**14**:51

[108] Gadd GM. Metals, minerals and microbes: Geomicrobiology and bioremediation. Microbiology (Reading, England). 2010;**156**(Pt 3):609-643

[109] Kite P, Eastwood K, Sugden S, Percival SL. Use of in vivo-generated biofilms from hemodialysis catheters to test the efficacy of a novel antimicrobial catheter lock for biofilm eradication in vitro. Journal of Clinical Microbiology. 2004;**42**(7):3073-3076

[110] Raad I, Chatzinikolaou I, Chaiban G, Hanna H, Hachem R, Dvorak T, et al. In vitro and ex vivo activities of minocycline and EDTA against microorganisms embedded in biofilm on catheter surfaces. Antimicrobial Agents and Chemotherapy. 2003;**47**(11):3580-3585

[111] Casalinuovo IA, Sorge R, Bonelli G, Di Francesco P. Evaluation of the antifungal effect of EDTA, a metal chelator agent, on *Candida albicans* biofilm. European Review for Medical and Pharmacological Sciences. 2017;**21**(6):1413-1420

[112] Craigen B, Dashiff A, Kadouri DE. The use of commercially available alpha-amylase compounds to inhibit and remove *Staphylococcus aureus* biofilms. Open Microbiology Journal. 2011;**5**:21-31

[113] Ramasubbu N, Thomas LM, Ragunath C, Kaplan JB. Structural analysis of dispersin B, a biofilm-releasing glycoside hydrolase from the periodontopathogen *Actinobacillus actinomycetemcomitans*. Journal of Molecular Biology. 2005;**349**(3):475-486

[114] Chaignon P, Sadovskaya I, Ragunah C, Ramasubbu N, Kaplan JB, Jabbouri S. Susceptibility of staphylococcal biofilms to enzymatic treatments depends on their chemical composition. Applied Microbiology and Biotechnology. 2007;**75**(1):125-132

[115] Darouiche RO, Mansouri MD, Gawande PV, Madhyastha S. Antimicrobial and antibiofilm efficacy of triclosan and DispersinB

combination. Journal of Antimicrobial Chemotherapy. 2009;**64**(1):88-93

[116] Kaplan JB. Therapeutic potential of biofilm-dispersing enzymes. The International Journal of Artificial Organs. 2009;**32**(9):545-554

[117] Qin Z, Ou Y, Yang L, Zhu Y, Tolker-Nielsen T, Molin S, et al. Role of autolysin-mediated DNA release in biofilm formation of *Staphylococcus epidermidis*. Microbiology (Reading, England). 2007;**153**(Pt 7):2083-2092

[118] Tetz GV, Artemenko NK, Tetz VV. Effect of DNase and antibiotics on biofilm characteristics. Antimicrobial Agents and Chemotherapy. 2009;**53**(3):1204-1209

[119] Leroy C, Delbarre C, Ghillebaert F, Compere C, Combes D. Effects of commercial enzymes on the adhesion of a marine biofilm-forming bacterium. Biofouling. 2008;**24**(1):11-22

[120] Ivanova K, Fernandes MM, Francesko A, Mendoza E, Guezguez J, Burnet M, et al. Quorum-quenching and matrix-degrading enzymes in multilayer coatings synergistically prevent bacterial biofilm formation on urinary catheters. ACS Applied Materials & Interfaces. 2015;**7**(49):27066-27077

[121] Shah A, Mond J, Walsh S. Lysostaphin-coated catheters eradicate *Staphylococccus aureus* challenge and block surface colonization. Antimicrobial Agents and Chemotherapy. 2004;**48**(7):2704-2707

[122] Kolodkin-Gal I, Romero D, Cao S, Clardy J, Kolter R, Losick R. D-amino acids trigger biofilm disassembly. Science. 2010;**328**(5978):627-629

[123] Leiman SA, May JM, Lebar MD, Kahne D, Kolter R, Losick R. D-amino acids indirectly inhibit biofilm formation in *Bacillus subtilis* by interfering with protein synthesis. Journal of Bacteriology. 2013;**195**(23):5391-5395

[124] Barraud N, Schleheck D, Klebensberger J, Webb JS, Hassett DJ, Rice SA, et al. Nitric oxide signaling in *Pseudomonas aeruginosa* biofilms mediates phosphodiesterase activity, decreased cyclic di-GMP levels, and enhanced dispersal. Journal of Bacteriology. 2009;**191**(23):7333-7342

[125] Schreiber F, Beutler M, Enning D, Lamprecht-Grandio M, Zafra O, Gonzalez-Pastor JE, et al. The role of nitric-oxide-synthase-derived nitric oxide in multicellular traits of *Bacillus subtilis* 3610: Biofilm formation, swarming, and dispersal. BMC Microbiology. 2011;**11**:111

[126] Margel D, Mizrahi M, Regev-Shoshani G, Ko M, Moshe M, Ozalvo R, et al. Nitric oxide charged catheters as a potential strategy for prevention of hospital acquired infections. PLoS One. 2017;**12**(4):e0174443

[127] Wang X, Yao X, Zhu Z, Tang T, Dai K, Sadovskaya I, et al. Effect of berberine on *Staphylococcus epidermidis* biofilm formation. International Journal of Antimicrobial Agents. 2009;**34**(1):60-66

[128] Magesh H, Kumar A, Alam A, Priyam SU, Sumantran VN, et al. Identification of natural compounds which inhibit biofilm formation in clinical isolates of *Klebsiella pneumoniae*. Indian Journal of Experimental Biology. 2013;**51**(9):764-772

[129] Zhao L-X, Li D-D, Hu D-D, Hu G-H, Yan L, Wang Y, et al. Effect of tetrandrine against *Candida albicans* biofilms. PLoS One. 2013;**8**(11):e79671

[130] Dwivedi D, Singh V. Effects of the natural compounds embelin and piperine on the biofilm-producing property of *Streptococcus mutans*. Journal

of Traditional and Complementary Medicine. 2015;**6**(1):57-61

[131] Rukayadi Y, Kim KH, Hwang JK. In vitro anti-biofilm activity of macelignan isolated from *Myristica fragrans* Houtt. against oral primary colonizer bacteria. Phytotherapy Research. 2008;**22**(3):308-312

[132] Katsura H, Tsukiyama RI, Suzuki A, Kobayashi M. In vitro antimicrobial activities of bakuchiol against oral microorganisms. Antimicrobial Agents and Chemotherapy. 2001;**45**(11):3009-3013

[133] Rukayadi Y, Hwang JK. In vitro activity of xanthorrhizol against *Streptococcus mutans* biofilms. Letters in Applied Microbiology. 2006;**42**(4):400-404

[134] Sá NC, Cavalcante TTA, Araújo AX, Santos HS, Albuquerque MRJR, Bandeira PN, et al. Antimicrobial and antibiofilm action of Casbane Diterpene from *Croton nepetaefolius* against oral bacteria. Archives of Oral Biology. 2012;**57**(5):550-555

[135] Santos V, Nardi R, Dias-Souza M. Bacteriocins as antimicrobial and antibiofilm agents. In: Current Developments in Biotechnology and Bioengineering: Human and Animal Health Applications. 2017. pp. 403-436

[136] Hillman JD. Genetically modified *Streptococcus mutans* for the prevention of dental caries. Antonie Van Leeuwenhoek. 2002;**82**(1-4):361-366

[137] Turner SR, Love RM, Lyons KM. An in-vitro investigation of the antibacterial effect of nisin in root canals and canal wall radicular dentine. International Endodontic Journal. 2004;**37**(10):664-671

[138] Gonzalez-Toledo SY, Dominguez-Dominguez J, Garcia-Almendarez BE, Prado-Barragan LA, Regalado-Gonzalez C. Optimization of nisin production by *Lactococcus lactis* UQ2 using supplemented whey as alternative culture medium. Journal of Food Science. 2010;**75**(6):M347-M353

[139] Gerke C, Kraft A, Sussmuth R, Schweitzer O, Gotz F. Characterization of the N-acetylglucosaminyltransferase activity involved in the biosynthesis of the *Staphylococcus epidermidis* polysaccharide intercellular adhesin. The Journal of Biological Chemistry. 1998;**273**(29):18586-18593

[140] Zoll S, Schlag M, Shkumatov AV, Rautenberg M, Svergun DI, Gotz F, et al. Ligand-binding properties and conformational dynamics of autolysin repeat domains in staphylococcal cell wall recognition. Journal of Bacteriology. 2012;**194**(15):3789-3802

[141] Chopra L, Singh G, Choudhary V, Sahoo DK. Sonorensin: An antimicrobial peptide, belonging to the heterocycloanthracin subfamily of bacteriocins, from a new marine isolate, *Bacillus sonorensis* MT93. Applied and Environmental Microbiology. 2014;**80**(10):2981-2990

[142] Sandiford S, Upton M. Identification, characterization, and recombinant expression of epidermicin NI01, a novel unmodified bacteriocin produced by *Staphylococcus epidermidis* that displays potent activity against Staphylococci. Antimicrobial Agents and Chemotherapy. 2012;**56**(3):1539-1547

[143] Arguelles Arias A, Ongena M, Devreese B, Terrak M, Joris B, Fickers P. Characterization of amylolysin, a novel lantibiotic from *Bacillus amyloliquefaciens* GA1. PLoS One. 2013;**8**(12):e83037

[144] Zhang C, Occi J, Masurekar P, Barrett JF, Zink DL, Smith S, et al. Isolation, structure, and antibacterial activity of philipimycin, a thiazolyl

peptide discovered from Actinoplanes philippinensis MA7347. Journal of the American Chemical Society. 2008;**130**(36):12102-12110

[145] Gu Y, Xu Y, Xu J, Yu X, Huang X, Liu G, et al. Identification of novel bacteriophage vB_EcoP-EG1 with lytic activity against planktonic and biofilm forms of uropathogenic *Escherichia coli*. Applied Microbiology and Biotechnology. 2019;**103**(1):315-326

[146] Pires DP, Melo L, Vilas Boas D, Sillankorva S, Azeredo J. Phage therapy as an alternative or complementary strategy to prevent and control biofilm-related infections. Current Opinion in Microbiology. 2017;**39**:48-56

[147] Colavecchio A, Goodridge LD. Phage therapy approaches to reducing pathogen persistence and transmission in animal production environments: Opportunities and challenges. Microbiology Spectrum. 2017;**5**(3):1-14

[148] Villarroel J, Larsen MV, Kilstrup M, Nielsen M. Metagenomic analysis of therapeutic PYO phage cocktails from 1997 to 2014. Viruses. 2017;**9**(11): pii: E328; 1-22

[149] McCallin S, Alam Sarker S, Barretto C, Sultana S, Berger B, Huq S, et al. Safety analysis of a Russian phage cocktail: From metagenomic analysis to oral application in healthy human subjects. Virology. 2013;**443**(2):187-196

[150] Pallavali RR, Degati VL, Lomada D, Reddy MC, Durbaka VRP. Isolation and in vitro evaluation of bacteriophages against MDR-bacterial isolates from septic wound infections. PLoS One. 2017;**12**(7):e0179245

[151] Lin DM, Koskella B, Lin HC. Phage therapy: An alternative to antibiotics in the age of multi-drug resistance. World Journal of Gastrointestinal Pharmacology and Therapeutics. 2017;**8**(3):162-173

[152] Twort FW. An investigation on the nature of ultra-microscopic viruses. The Lancet. 1915;**186**(4814):1241-1243

[153] Schofield DA, Wray DJ, Molineux IJ. Isolation and development of bioluminescent reporter phages for bacterial dysentery. European Journal of Clinical Microbiology & Infectious Diseases. 2015;**34**(2):395-403

[154] Abd, El-Aziz AM, Elgaml A, Ali YM. Bacteriophage therapy increases complement-mediated lysis of bacteria and enhances bacterial clearance after acute lung infection with multidrug-resistant *Pseudomonas aeruginosa*. The Journal of Infectious Diseases. 2019;**219**(9):1439-1447

[155] Orlova ZN, Garnova NA. Use of tablet-form polyvalent bacteriophage with acid resistant coating in the treatment of dysentery in children. Voprosy Okhrany Materinstva i Detstva. 1970;**15**(3):25-29

[156] Vinner GK, Rezaie-Yazdi Z, Leppanen M. Microencapsulation of salmonella-specific bacteriophage Felix O1 using spray-drying in a pH-responsive formulation and direct compression tableting of powders into a solid oral dosage form. 2019;**12**(1): pii: E43; 1-14

[157] Endersen L, O'Mahony J, Hill C, Ross RP, McAuliffe O, Coffey A. Phage therapy in the food industry. Annual Review of Food Science and Technology. 2014;**5**:327-349

[158] Janez N, Loc-Carrillo C. Use of phages to control Campylobacter spp. Journal of Microbiological Methods. 2013;**95**(1):68-75

[159] Svircev A, Roach D, Castle A. Framing the future with bacteriophages in agriculture. Viruses. 2018;**10**(5): pii: E218; 1-13

[160] Doss J, Culbertson K, Hahn D, Camacho J, Barekzi N. A review of

phage therapy against bacterial pathogens of aquatic and terrestrial organisms. Viruses. 2017;**9**(3): pii: E50; 1-10

[161] Eyer L, Pantucek R, Ruzickova V, Doskar J. New perspectives of the phage therapy. Klinická Mikrobiologie a Infekční Lékařství. 2007;**13**(6):231-235

[162] Kutateladze M, Adamia R. Bacteriophages as potential new therapeutics to replace or supplement antibiotics. Trends in Biotechnology. 2010;**28**(12):591-595

[163] Yacoby I, Bar H, Benhar I. Targeted drug-carrying bacteriophages as antibacterial nanomedicines. Antimicrobial Agents and Chemotherapy. 2007;**51**(6):2156-2163

[164] Yacoby I, Shamis M, Bar H, Shabat D, Benhar I. Targeting antibacterial agents by using drug-carrying filamentous bacteriophages. Antimicrobial Agents and Chemotherapy. 2006;**50**(6):2087-2097

[165] Bar H, Yacoby I, Benhar I. Killing cancer cells by targeted drug-carrying phage nanomedicines. BMC Biotechnology. 2008;**8**:37

[166] Borysowski J, Weber-Dabrowska B, Gorski A. Bacteriophage endolysins as a novel class of antibacterial agents. Experimental Biology and Medicine (Maywood, NJ). 2006;**231**(4):366-377

[167] Pohane AA, Jain V. Insights into the regulation of bacteriophage endolysin: Multiple means to the same end. Microbiology (Reading, England). 2015;**161**(12):2269-2276

[168] Schmelcher M, Loessner MJ. Bacteriophage endolysins: Applications for food safety. Current Opinion in Biotechnology. 2016;**37**:76-87

[169] Wright A, Hawkins CH, Anggard EE, Harper DR. A controlled clinical trial of a therapeutic bacteriophage preparation in chronic otitis due to antibiotic-resistant *Pseudomonas aeruginosa*; a preliminary report of efficacy. Clinical Otolaryngology. 2009;**34**(4):349-357

[170] Kutateladze Á, Adamia R. Phage therapy experience at the Eliava Institute. Médecine et Maladies Infectieuses. 2008;**38**(8):426-430

[171] Fish R, Kutter E, Wheat G, Blasdel B, Kutateladze M, Kuhl S. Bacteriophage treatment of intransigent diabetic toe ulcers: A case series. Journal of Wound Care. 2016;**25**(Sup7):S27-S33

[172] Jun JW, Shin TH, Kim JH, Shin SP, Han JE, Heo GJ, et al. Bacteriophage therapy of a *Vibrio parahaemolyticus* infection caused by a multiple-antibiotic–resistant O3: K6 pandemic clinical strain. The Journal of Infectious Diseases. 2014;**210**(1):72-78

[173] Markoishvili K, Tsitlanadze G, Katsarava R, Glenn J, Morris M Jr, Sulakvelidze A. A novel sustained-release matrix based on biodegradable poly (ester amide) s and impregnated with bacteriophages and an antibiotic shows promise in management of infected venous stasis ulcers and other poorly healing wounds. International Journal of Dermatology. 2002;**41**(7):453-458

[174] Pouillot F, Chomton M, Blois H, Courroux C, Noelig J, Bidet P, et al. Efficacy of bacteriophage therapy in experimental sepsis and meningitis caused by a clone O25b: H4-ST131 *Escherichia coli* strain producing CTX-M-15. Antimicrobial Agents and Chemotherapy. 2012;**56**(7):3568-3575

[175] Soothill J. Treatment of experimental infections of mice with bacteriophages. Journal of Medical Microbiology. 1992;**37**(4):258-261

[176] Biswas B, Adhya S, Washart P, Paul B, Trostel AN, Powell B, et al. Bacteriophage therapy rescues mice bacteremic from a clinical isolate of vancomycin-resistant *Enterococcus faecium*. Infection and Immunity. 2002;**70**(1):204-210

[177] Ramesh V, Fralick JA, Rolfe RD. Prevention of *Clostridium difficile*-induced ileocecitis with bacteriophage. Anaerobe. 1999;**5**(2):69-78

[178] Chanishvili N. Phage therapy—History from Twort and d'Herelle through Soviet experience to current approaches. Advances in Virus Research. 2012;**83**:3-40

[179] Gilmer DB, Schmitz JE, Euler CW, Fischetti VA. Novel bacteriophage lysin with broad lytic activity protects against mixed infection by *Streptococcus pyogenes* and methicillin-resistant *Staphylococcus aureus*. Antimicrobial Agents and Chemotherapy. 2013;**57**(6):2743-2750

[180] Yang H, Wang D-B, Dong Q, Zhang Z, Cui Z, Deng J, et al. Existence of separate domains in lysin PlyG for recognizing *Bacillus anthracis* spores and vegetative cells. Antimicrobial Agents and Chemotherapy. 2012;**56**(10):5031-5039

[181] Singh PK, Donovan DM, Kumar A. Intravitreal injection of the chimeric phage endolysin Ply187 protects mice from *Staphylococcus aureus* endophthalmitis. Antimicrobial Agents and Chemotherapy. 2014;**58**(8):4621-4629

[182] Lood R, Winer BY, Pelzek AJ, Diez-Martinez R, Thandar M, Euler CW, et al. Novel phage lysin capable of killing the multidrug-resistant gram-negative bacterium *Acinetobacter baumannii* in a mouse bacteremia model. Antimicrobial Agents and Chemotherapy. 2015;**59**(4):1983-1991

[183] Oliveira H, Vilas Boas D, Mesnage S, Kluskens LD, Lavigne R, Sillankorva S, et al. Structural and enzymatic characterization of ABgp46, a novel phage endolysin with broad anti-gram-negative bacterial activity. Frontiers in Microbiology. 2016;**7**:208

Chapter 11

Interplay between Human Intestinal Microbiota and Gut-to-Brain Axis: Relationship with Autism Spectrum Disorders

Francisco Javier Díaz-García, Saúl Flores-Medina and Diana Mercedes Soriano-Becerril

Abstract

A growing body of scientific reports suggests a relevant key role of human intestinal microbiota (HIM) in maintaining the host's physiological and mental balance; thus any disturbance in the microbiota diversity and/or concentrations may result in impaired stimulation of the gastrointestinal (GI) system-central nervous system (CNS) bidirectional pathway, termed gut-to-brain axis. Recent data show that HIM composition is significantly unbalanced among a subset of autism spectrum disorder (ASD) subjects, as compared with non-ASD siblings or age-matched control subjects. Several authors claim that specific changes in HIM (diet-based alteration of Bacteroidetes/Firmicutes ratio and death of predominant microbiota after antibiotic treatments, among others) could either trigger or be highly associated events with persistent ASD signs and behaviors. Whether HIM plays a causative or a circumstantial role in ASD severity, then HIM manipulation might be applied as a therapeutic alternative to improve ASD clinical manifestations and behaviors.

Keywords: human intestinal microbiota, dysbiosis, gut-to-brain axis, autism spectrum disorders, short-chain fatty acids

1. Introduction

Communication between the gastrointestinal (GI) system and the central nervous system (CNS) occurs constantly and plays a critical role in maintaining the healthy status. That communication engages a bidirectional stimulation system, involving not only the brain and gut cells but endocrine-, immune-, and microbiota-derived components as well, the so-called gut-to-brain axis (GBA) [1, 2]. Consequently, impaired communication between both ends of the GBA, associated with or as consequence of disturbance of the GI microbial diversity, has been associated with a negative health outcome later in life [1, 3].

Increasing evidence points out that there is a link between alterations of gut microbiota and several disorders of the central nervous system including autism spectrum disorders (ASD) depression, anxiety, irritable bowel syndrome, attention deficit and hyperactivity disorder (ADHD), Parkinson's disease, disorders of mood and affect, and chronic pain [3–5].

The abovementioned psychiatric disorders frequently co-occur with each other, but interestingly they also occur in comorbidity with metabolic disorders, such as diabetes, cardiovascular disease, and metabolic syndrome [6–8], and are associated with adverse outcomes including higher mortality [6]. The insights of how those disorders are linked remain unclear. One likely explanation is that gut microbiota can trigger and guide the communication network of GBA and subsequently alter metabolic and psychological equilibrium [3, 5, 7, 8].

ASD are a heterogeneous set of lifelong neurodevelopmental diseases, whose incidence increased significantly over the past decades [9]. No unique etiology of ASD has been identified, though both genetic and environmental factors have been suggested [9, 10]. However, findings of candidate genes do not conclusively explain the etiopathology of ASD; thus, scientific research has been redirected to GI comorbidities of ASD, under the premise that the high frequency of gut microbiota alterations seen in these patients may be associated with autism symptoms severity [10]. Indeed, the independent observations of Rodakis [10] and Sandler et al. [11] about improvements in autism clinical manifestations after antibiotic treatments prompted intense research around the issue, including therapeutic interventions such as diet modification, supplementation with biotics (prebiotics, probiotics, synbiotics, and/or postbiotics), alternative antibiotic treatments, and fecal microbiota transplantation, among others, with variable outcomes [12].

2. Human intestinal microbiota

Colonization of the human body occurs after birth, and possibly before birth, with a diverse microbial community of archaea, bacteria, fungi, viruses, and protozoa. This diverse community is referred to as the HIM. The prokaryote organisms colonizing the human body encompass nearly 90% of all HIM [13, 14]. Resident microbiota of the human GI tract, the one that colonizes permanently, is one of the most densely populated communities, even more so than the soil, the subsoil, and the oceans [15].

Colonization of GI tract is influenced by many factors like mode of birth delivery, infant feeding method, and the environment (stress, frequency of exercise, hygiene habits, infections, pharmaceuticals use, and type of feeding) [16, 17]. Within the human intestinal microbiota, there are both types of microorganisms: those who are essential, and even indispensable, for the survival of the host, and those who are potentially pathogenic. The vast majority have beneficial rather than detrimental effects on the host's health [15].

The importance of the GI microbiota was overlooked for a long time, and efforts to determine its composition and functions were unsuccessful; on the one hand, cultures from stool samples are unproductive, and on the other hand, according to estimations, 80% of the GI microbiota are anaerobe uncultivable organisms [10]. Anaerobic bacteria outnumber aerobic and facultative anaerobic bacteria by 100- to 1000-fold [16, 17]. Calculations of microbial counts in the colon of adult humans reach a mean of 10^{11} organisms/gram of wet stool, a quantity updated that is similar to the total number of human cells [18]. Estimated HIM composition comprise up to 1800 genera representing 7000–40,000 bacterial strains belonging to 500–1000 resident species [17, 18].

Taking into account the presence of gene content and metabolic products, along with the microbiota organisms contained within a particular body site, we must refer to it as a microbiome [15]. Studies on composition and function of uncultured microbial communities, more specifically by sequencing-based assays, are referred to as metagenomics. First, community DNA is extracted from a sample containing multiple microbial members. Second, bacterial taxa present in the community are then defined by amplification of the 16S rRNA gene followed by sequencing. Highly similar sequences are grouped into operational taxonomic units (OTUs) or phylotypes, which can be compared to 16S rRNA databases to identify them as accurately as possible. An alternate method identifies community taxa after the total DNA is metagenomically sequenced and compared to reference genomes or gene catalogs. The OTUs can be described in terms of their relative abundance and/or their phylogenetic relationships, while sequenced genomes can be described as relative abundances of its genes and pathways [19].

The human intestinal microbiome is mainly defined by the high abundance of two bacterial phylotypes: Bacteroidetes and Firmicutes. Other phylotypes present at lesser amounts are Proteobacteria, *Actinomyces*, Fusobacterium, and Verrucomicrobia [14]. The gut microbiome is conformed with nearly 470 phylotypes, more than 1000 bacterial species representing more than 5000 strains, which in turn encode between 5 and 10 million of nonredundant genes (150-fold the number of genes identified in the human genome) [16, 17]. Studies on intestinal microbiome in health and disease revealed two microbiome subpopulations, one with high-gene counts and the other with low-gene counts; the first one seems to be associated with a healthy digestive status [20, 21].

Every person has a unique microbiome profile; still there is a reduced number of species shared between persons. The aforesaid feature allowed to classify individuals into one of three enterotypes, each one based on the proportions of the three predominant intestinal genera, based on their abundance, *Bacteroides*, *Prevotella*, and *Ruminococcus*. The first two genera represent the Bacteroidetes, and the last one represents the Firmicutes. Enterotype 1 shows predominance of *Bacteroides*, while enterotypes 2 and 3 were defined by predominance of *Prevotella* and *Ruminococcus*, respectively [22].

Alterations of the typical GI microbiota, in number and abundance distribution of distinctive types of microorganisms, and the host's adverse response to such changes have been called as dysbiosis. Thus, dysbiosis with low diversity has been linked particularly with obesity, inflammatory bowel disease, and ASD [16].

There are two ongoing multi-group projects on human microbiome, the Europe-based Metagenomics of the Human Intestinal Tract (MetaHIT) and the US-based Human Microbiome Project (HMP). Both of them will allow to define to its finest details the microbiome diversity, at least to species level, their genetic load, and how microbiota interacts with the host [23, 24].

2.1 Biological role of GI microbiota

The intestinal microbiota maintains a symbiotic relationship with the host. Studies in both humans and mammals have implicated the intestinal microbiome in several physiological processes that are pivotal to the host health, from food digestion and energy homeostasis to immune and neurobehavioral development [25].

The single layer of intestinal epithelial cells, connected by tight junctions, constitutes itself a physical and biochemical barrier that segregates the commensal microbiota organisms to maintain intestinal homeostasis. This occurs through regulation of nutrients, electrolytes, and water absorption, as well as through release of mucins, antimicrobial peptides, and IgA for the prevention of the entry of pathogenic microorganisms [26–28]. The interaction between the microbiota and intestinal epithelial cells also promotes tissue restoration in the setting of injury or acute inflammation, thus supporting epithelium integrity. Besides the above statement, the microbiota provides protection against exogenous pathogenic organisms, either through competition for common nutrients and niches or by prompting development and functional maturation of the gut immune system, including gut-associated lymphoid tissue, T-helper 17 cells, inducible regulatory T cells, IgA-producing B cells, and innate lymphoid cells [13].

The HIM microbiota has a considerable input on the metabolomic profile, the complete set of intestinal metabolites, of the host [29]. Specifically, the microbiota is a major source of both circulating organic acids and tryptophan metabolites, which have beneficial effects on the host health (**Table 1**) [30–53].

Fermentative processes of nondigestible complex carbohydrates, from dietary fiber, by Firmicutes and Bacteroidetes, result in the production of various short-chain fatty acids (SCFAs), such as acetate, propionate, butyrate, isobutyrate, valerate, and isovalerate. These bacteria-derived SCFAs, in a physiological context, may serve as an energy source for enterocytes, stimulate water and sodium absorption, decrease colonic pH, etc. [30].

Hyperproduction or deficiency of SCFAs may also affect the pathogenesis of a diverse range of diseases, from allergies and asthma to neurological diseases [17, 29, 30]. For example, a diet high in fat and digestible saccharides provokes that majority of nutrients be absorbed in the duodenum, leaving very few substrates for the colonic bacteria, leading to dysbiosis. Higher levels of SCFAs can also alter the intercellular spaces between the cells, resulting in a leaky gut that allows for more metabolites and bacteria to pass through the epithelial barrier, where bacterial endotoxins and other microbial-derived metabolites can gain entry into the bloodstream [17] Furthermore, dysbiosis can affect host immunity and neurobehavioral responses [17, 29]. Among SCFAs, butyrate is a promoter of colonic functionality and physical integrity, via cholesterol-rich membrane microdomain, as well as the preferred metabolic substrate for the colonocytes' energy requirements [54].

Essential vitamins such as folate, vitamin K, and vitamin B12, for the host's growth, are synthesized by gut microbiota, which in turn may affect DNA and histone protein methylation [55–57]. Certain hormones and vitamins also participate in drug and poison removal [58].

Apart from carbohydrates, GI bacteria also metabolize complex lipids and proteins that are indigestible by the host [59–61]. Expression of colipase, a critical protein factor for lipid metabolism, and subsequent stimulation of the release of pancreatic lipases appear to be regulated by *Bacteroides thetaiotaomicron* [59].

The metabolism of tryptophan by the HIM and/or gut and immune cells follows three alternative pathways: (a) the transformation to ligands of the aryl hydrocarbon receptor (AhR), (b) the kynurenine pathway (via indoleamine 2,3-dioxygenase 1), and (c) the serotonin (5-HT) production pathway. These pathways are performed by HIM, enterocytes/immune cells, and enterochromaffin cells, respectively. The HIM pathway yields several molecules, indole-3-aldehyde, indole-3-acid-acetic, indole-3-propionic acid, indole-3-acetaldehyde, and indoleacrylic acid. AhR signaling is crucial for gut epithelium renewal and barrier integrity and acts over many immune cell types for responsiveness [62].

Microbial metabolism of tryptophan is very important for intestinal AhR activity, since the absence or imbalance of tryptophan-metabolizing organisms generate deficiency of AhR agonists [46]. The production of AhR ligands have been determined among few HIM species, *Peptostreptococcus russellii* and *Lactobacillus* ssp. Many GI and neuropsychiatric diseases have been related to dysbiotic impairment of tryptophan metabolism or to accumulation of the end products [62].

Several HIM species not only can synthesize but respond as well to hormones and neurotransmitters of bacterial and human origin, which impact their growth and virulence. Beneficial *Lactobacillus* spp. are able to synthesize acetylcholine and gamma-aminobutyric acid (GABA), while *Bifidobacterium* spp. produce GABA. *Escherichia* spp. produce norepinephrine, serotonin, and dopamine; other Firmicutes species

Pathway	Metabolite	Microbial agent	Health benefits	Refs.
Carbohydrate metabolism	Butyrate	Clostridia (clusters IV and IVa)	Increased intestinal barrier function	[30, 31]
		Faecalibacterium. Prausnitzii	Modulate intestinal macrophage function	[32]
		Eubacterium spp.	Regulation of colonic regulatory T cell homeostasis	[33, 34]
		Roseburia spp.	Induction of tolerogenic dendritic cells that polarize naive CD4+ T cells toward IL-10–producing regulatory T cells	[35]
		Coprococcus catus	Suppression of colonic inflammation	[36, 37]
		Anaerostipes hadrus	Improvements in insulin sensitivity	[38]
	Propionate	*Bacteroides* spp.	Regulation of colonic regulatory T cell homeostasis	[33, 34]
		Blautia obeum	Suppression of colonic inflammation	[39]
		C. catus	Decreased innate immune responses to microbial stimulation	[40]
		Roseburia inulinivorans	Protection from allergic airway inflammation	[41]
		P. copri	Improvements in insulin sensitivity and weight control in obese mice	[42]
Tryptophan metabolism	Indole	Various tryptophanase-producing bacteria such as *Lactobacillus* spp.	Maintenance of host–microbe homeostasis at mucosal surfaces via IL-22	[43]
		B. longum	Increased barrier function	[44]
		B. fragilis	Modulation of host metabolism	[45]
	I3A	*Lactobacillus* spp.	Maintenance of mucosal homeostasis and intestinal barrier function Protection against mouse intestinal inflammation.	[43, 46]
	IPA	*Clostridium sporogenes*	• Maintenance of intestinal barrier function and mucosal homeostasis • Increased production of antioxidant and neuroprotectant molecules	[47, 48]
Llipid metabolism	HYA	*Lactobacillus* spp.	• Maintenance of intestinal barrier function • Decreased inflammation • Increased intestinal IgA production	[49, 50]
	CLA	*Lactobacillus* spp.	Decreased inflammation	[51]
		Bifidobacterium spp.	Reduced adiposity	[52]
		F. prausnitzii	Improved insulin sensitivity	[53]

I3A, indole-3-aldehyde; IPA, indole-3-propionate; HYA, 10-hydroxy-cis-12-octadecoate (linoleic acid derivative); CLA, conjugated linoleic acid.
Modified from [29].

Table 1.
Examples of intestinal microbiota-derived metabolites and their beneficial effects on human health.

belonging to *Streptococcus* and *Enterococcus* produce serotonin, and *Bacillus* produce norepinephrine and dopamine [17]. Those bacteria-derived neurotransmitters released directly to the intestinal lumen may either induce epithelial cells to in turn release molecules that modulate neural signaling within the enteric nervous system (ENS) or, after passing through the gut wall, gain entry into the portal circulation to exert direct effects on afferent axons [63, 64]. Indeed, several reports documented elevated levels of noradrenaline and adrenaline in the plasma of subjects coursing with systemic infections by gram-negative Proteobacteria, like *Escherichia coli*. (Reviewed in [64]).

2.2 Stability of HIM

Bacterial colonization of the human gut likely occurs at the time of birth, when infants born via vaginal delivery are inoculated with a complex mixture of maternal vaginal microorganisms. According to Dominguez-Bello et al. [65], those infants had colonizing *Lactobacillus*, *Prevotella*, or *Sneathia* species in their skin and mucosae, which resembled their own mother's vaginal microbiota. In contrast, infants delivered by cesarean section had predominantly *Staphylococcus*, *Corynebacterium*, and *Propionibacterium* species, akin to their mothers' skin microbiota. Thus, there is concern that babies delivered via cesarean section may receive an insufficient maternal bacterial load [17].

After birth, breastfeeding is the main factor defining the composition of newborn's GI microbiota, since breast milk provides a variety of specific antibodies and immediate immunity molecules that neutralize pathogenic bacteria. Breast milk also contains more than 200 oligosaccharides (prebiotics) that favor the growth of bifidobacteria [66, 67], which have been reported to prevent gastrointestinal infections by competitive exclusion of pathogens based on common binding sites on epithelial cells [67]. Therefore, in breastfed children, bifidobacteria reaches up to 90% of GI microbiota, followed by lactobacilli, *Bacteroides*, coliforms, and clostridia. In contrast, infants fed with infant formula have predominance of *Bacteroides*, enterococci, coliforms, and clostridia, with much lesser bifidobacteria, resembling the more diverse GI microbiota of adults [66, 67].

The initial breastfeeding-driven colonization is essential for induction of adaptive immunity and for early metabolic programming. After the introduction of complementary feeding, the microbiota differences between breastfed children and those fed with formula tend to disappear. It is assumed that the predominant bacteria in the intestinal microbiome of 3-year-olds are similar to those of adults and remain relatively stable lifelong [66, 67].

Daily variability of the HIM composition has been assessed in controlled feeding studies, specifically short-term administration of extreme amount of fat and fiber intake, which revealed disturbance of the intestinal microbiome, but this effect was of low-scale and transient that not changed the individual's enterotype designation [66, 67].

3. Gut-to-brain axis

The basis of the GBA cross-communication includes an array of multichannel sensing and trafficking pathways (neural, endocrine, immune, and metabolic) to transfer the enteric signals to the brain (**Figure 1**), which ultimate results in keeping proper maintenance of GI homeostasis, although its multiple effects likely impacts on brain performance and higher cognitive functions [1–3, 68].

The GBA comprises highly interconnected body systems. Those systems are the CNS, the autonomic nervous system (vagal and spinal nerves), and the ENS

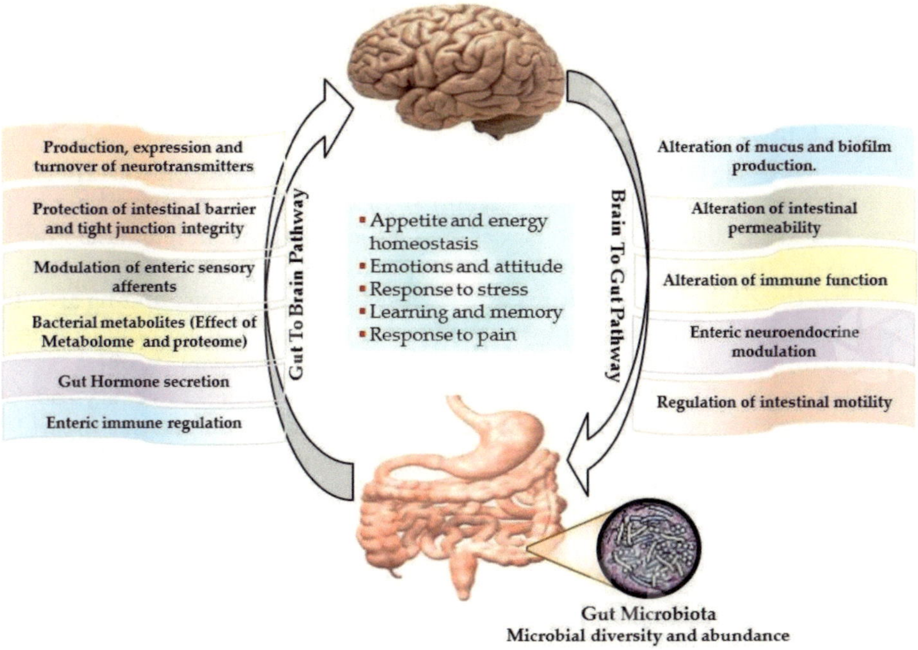

Figure 1.
The bidirectional pathways of the gut-to-brain axis and their effects. Modulation of the CNS by the gut microbiome (through microbial-derived molecules such as SCFAs, neurotransmitters, hormones and tryptophan metabolites) occurs primarily via neuro-immune and neuroendocrine mechanisms. Those microbial molecules reach brain sites directly or only induce central responses through long-distance neural signaling by vagal and/or spinal afferents. The autonomic nervous system regulates gut functions (motility, secretion, intestinal permeability, and mucosal immune response), which ultimately affect the microbial habitat, thereby modulating microbiota composition and activity.

(the arrangement of neurons and supporting cells throughout and embedded within GI tract, from the esophagus to the anus). Other critical components of GBA include the hypothalamic pituitary adrenal axis (HPA; release of gut hormones), the immune system (release of multiple cytokines), and bacteria-derived metabolites (SCFAs and free amino acids). In fact, gut microbes have evolved alongside their host, through complex relationships, so influencing their own genotypic and phenotypic features [1–3]. However, failures in the GBA cross talk may lead to a number of health disorders, from inflammatory to metabolic and neurodevelopmental conditions, including ASD [1].

The following pathways may explain the influence of the gut microbiota on neurologic disorders through GBA: (a) production of neurotransmitters, (b) triggering release of gut hormones from entero-endocrine cells, (c) stimulation of the ENS and signaling to the brain via ascending neural pathways, and (d) activation of the immune system via cytokine release by the mucosa-associated immune cells.

At physiological conditions, GBA modulates the digestive processes like motility and secretion, immune function, and perception and emotional response to visceral stimuli [17]. The high comorbidity of stress-related neurologic disorders with GI disorders proves the impact of altered function of GBA [3].

4. Autism spectrum disorders

ASD is a group of neurodevelopmental abnormalities whose clinical manifestations begin in early childhood (although their diagnosis may delay months to years later in life). Clinically ASDs show complex and heterogeneous features but generally are defined by a core symptomatology including impaired social communication (oral and nonverbal languages, eye contact), behavioral problems (fixated interests in the daily routine, engagement in repetitive manners, exacerbated responses to external stimuli), and self-isolation, with or without impairment of cognitive abilities and competences [9, 69].

According to the latest American Psychiatric Association's diagnostic criteria [69], ASDs include conditions known as autism disorder (AD), Asperger's syndrome, childhood disintegrative disorder, and pervasive developmental disorder not otherwise specified (PDD-NOS).

Noteworthy ASD clinical features show extensive heterogeneity among affected subjects, according to the developmental stage, to chronological age, and to specific disorder within the spectrum (and even within the same disorder) [9, 69].

Until of April 2018, ASD were estimated to affect, in average, 1 in every 160 children worldwide, with a yearly rising incidence, and an estimated boy to girl ratio of 5:1 [70]. Data from the USA reveal that prevalence of ASDs has dramatically increased from 4.5 in 10,000 children in 1966 to 1 in 68 in 2010 and finally to 1 in 59 children in 2014 [71].

This recent outburst in frequency may be partly attributed to increased public awareness and or to better diagnosis; however, the occurrence of other factors, such as exposure to environmental chemicals, diet alterations, metabolic status, and changes in microbiota composition, cannot be excluded [17].

Despite the alarming rise trend in frequency of diagnosed cases in developed countries, the etiopathogenesis of ASD is still unknown; thus, there are no consensus in medical, neurologic, or psychiatric treatments [10]. Moreover, a diversity of comorbidities also affect ASD individuals, including one or more of the following: anxiety, intellectual disability, epilepsy/seizures, attention deficit and hyperactivity disorder, GI disorders, sleep disorders, obesity, depression, bipolar disorder, and Tourette's syndrome, among others (**Figure 2**) [6, 7, 9].

Among the most frequent GI comorbidities in ASD subjects are exacerbated flatulence (60%), bloating (38%), abdominal pain (37%), diarrhea (28%), burping/belching (25%), gastroesophageal reflux symptoms (16%), and constipation (10%) [8].

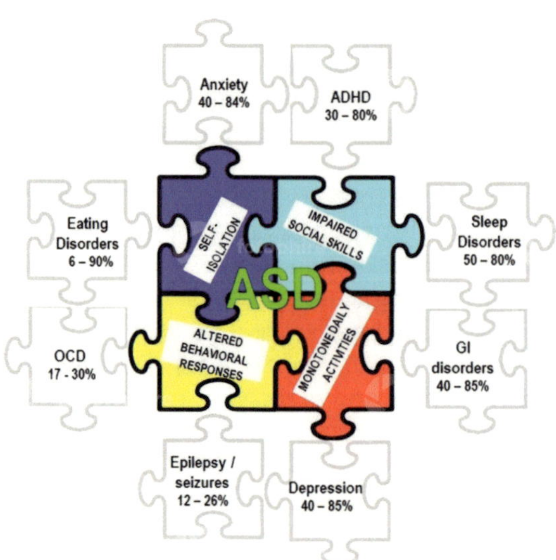

Figure 2.
Relevant features of ASD and their most frequent comorbidities. The colored figures represent typical features defining ASD, while colorless figures represent the most prevalent of its comorbidities. ADHD, attention deficit and hyperactivity disorder; GI, gastrointestinal, OCD, obsesive –compulsive disorder.

Research on ASD was primarily focused on genetic associations, but recent evidence has suggested that other environmental factors, including pre- or postnatal exposure to chemicals and drugs, air pollution, stress, maternal infection, the HIM, and dietary factors, may play a role in the clinical manifestations of the ASD [17].

5. Interplay between HIM and GBA in the context of ASD

About 40–60% of ASD children suffer from gastrointestinal comorbidities [8], although due to their social and communicative impairments, the real prevalence of gastrointestinal issues among ASD patients may be higher. Such intestinal dysfunction in this group of patients may be caused by disturbances in the pathways underlying the GBA, with a central role of the HIM and including an immune component.

Several studies have demonstrated HIM dysbiosis in ASD subjects; however, little or null correlation between studies has been obtained, mainly due to variations in study groups, control groups, and the use of diverse methods for microbiota/microbiome determinations and analysis (**Table 2**) [67, 72]. In short, 13 of the 15 studies showed some degree of dysbiosis among ASD patients as compared with controls (total combined sample of 585 individuals, 339 ASD, 61 control siblings, and 185 unrelated neurotypical controls), whereas 2 of the 15 studies found no significant differences among ASD subjects as compared with siblings controls (no neurotypical controls were included).

Altogether the microbiome data from the studies showed in **Table 2** suggests some important features among stool samples of ASD subjects: (a) levels of clostridia, *Desulfovibrio*, and *Sutterella* seem consistently elevated; (b) on the opposite, levels of *Prevotella* and bifidobacteria appears to be reduced; (c) the Bacteroidetes/Firmicutes ratio showed inconsistent results over different cohorts. There are significant, but not consistent, distinctive different microbiome compositions in ASD patients, regardless of gastrointestinal problems, compared to controls [73–90]. Moreover, the presence of HIM dysbiosis may correlate with ASD phenotype [91].

Dysbiosis in ASD is also associated with increased permeability of the GI tract, the leaky gut, which leads to the entry of endotoxins, and other bacterial products into the bloodstream [92]. Bacterial lipopolysaccharide (LPS) can alter neuronal as well as microglial activity in brain regions involved in emotional control [93–95]. In fact, serum levels of LPS were significantly higher among ASD subjects compared to healthy individuals and correlated with impaired social behavioral scores [96].

Serotonin synthesis in the gut and the brain depends on the availability of dietary tryptophan. High levels of blood serotonin were found in children with ASD [97–99], which contrasts with finding of decreased brain serotonin synthesis in ASD subjects [100]. A significant correlation between whole-blood serotonin levels and low-grade intestinal inflammation in ASD was demonstrated [101]. Regarding these findings, a likely explanation was proposed by de Theije et al. (2011) [91]: After GI inflammation, the intestinal serotonin release provokes changes in motility, secretion, vasodilation, and permeability, leading to functional intestinal dysmotility, stool inconsistency, and abdominal pain. Since the majority of dietary tryptophan is transformed in serotonin by HIM during inflammation, less tryptophan (and serotonin) will be available for the brain resulting in mood and cognitive dysfunction in ASD and increased autistic behavior [102].

Propionic acid, a major SCFA produced by clostridia, *Bacteroides*, and *Desulfovibrio,* has been associated with ASD, since it can induce ASD-like behavioral deficits in rats [103, 104]. Detrimental effects of propionic acid are suggested to be through mitochondrial and epigenetic modulation of ASD-associated genes. In fact, elevated levels of SCFAs are described in the stool of ASD children [82, 105].

Country (Year)	Study Group			Specimen type	Analytical method	Changes in fecal microbiome in ASD	Refs.
	ASD (GI+/GI−)	SIB (GI+/GI−)	NTC (GI+GI−)				
USA (2002)	13	–	8	Stool	Bacterial cultures	↑ Nine species of *Clostridium*	[73]
USA (2004)	15	–	8	Stool	16S rRNA gene sequencing	↑ *C. bolteae* and cluster I/IX	[74]
United Kingdom (2005)	58	12	10	Stool	FISH analysis	↑ *C. histolyticum* and cluster I/II. Siblings show intermediate levels.	[75]
USA (2010)	33 (33/0)	7 (0/7)	8 (0/8)	Stool	16S rRNA gene sequencing	↑ Bacteroidetes and Proteobacteria: *Desulfovibrio, B. Alkaliflexus, Acetanaerobacterium, Parabacteroides* ↓ Firmicutes and Actinobacteria: *Clostridium*, *Weissella*, *Turicibacter*, *Anaerofilum*, *Ruminococcus*, *Streptococcus, Pseudoramibacter*,	[76]
USA (2011)	58 (58/0)	–	39 (0/39)	Stool	Bacterial cultures	↓ *Bifidobacterium* and *Enterococcus* ↑ *Bacillus* spp. *(Lactobacillus)*	[77]
Poland (2011)	41	–	10	Stool	Bacterial cultures	↑ *Clostridium perfringens*	[78]
USA (2011, 2012)	23 (23/0)	–	9 (9/0)	Intestinal biopsies	16S rRNA gene sequencing	↓ Bacteroidetes ↑ Firmicutes, Proteobacteria, *Sutterella*	[79, 80]
Australia (2011, 2012 2013)	23 (9/14)	22 (6/16)	9 (1/8)	Stool	Targeted qPCR GC HPLC	↓*Bifidobacterium* spp., *Akkermansia muciniphilia* ↑ *Sutterella* spp. (↑ Relative abundance of *Clostridium difficile* in ASD, NS) ↑ *Ruminococcus torques* (only in ASD-GI+). ↑Ammonia and SCFA (acetic, butyric, isobutyric, valeric, isovaleric acids), likely microbial-derived. No differences in phenol and p-cresol levels	[81–83]

Country (Year)	Study Group			Specimen type	Analytical method	Changes in fecal microbiome in ASD	Refs.
	ASD (GI+/GI−)	SIB (GI+/GI−)	NTC (GI+GI−)				
Australia (2012)	51 (28/23)	53 (4/49)	–	Stool	16S rRNA gene sequencing	No differences	[84]
USA (2013)	20 (20/0)	–	20 (0/20)	Stool	16S rRNA gene sequencing	↓ *Prevotella, Coprocuccus, Veillonellaceae*	[85]
Italy (2013)	10	10	10	Stool	16S rRNA gene sequencing GC-MS/SPME	↓ *Caloramator, Sarcina, Clostridium, Sutterellaceae* ↓ *Eubacterium, Bifidobacterium* ↓SCFA (except PPA) ↑ Phenol, 4-(1,1-dimethylethyl)-phenol, and p-cresol. ↑ Free amino acids (Proteolytic bacteria)	[86]
USA (2015)	59 (25/34)	44 (13/31)	-	Stool	16S rRNA gene sequencing	No differences	[87]
Slovakia (2015)	10	10	10	Stool	Targeted qPCR	↓Bacteroidetes/Firmicutes ↑ *Lactobacillus* spp. (↑Clostridia cluster I and *Desulfovibrio*, NS.) *Desulfovibrio* spp.: strong positive association with ASD severity	[88]
USA (2017)	14 (14/0)	-	21 (15/6)	Rectal biopsies	16S rDNA PCR and sequencing HPLC of mucosal supernatant.	↑ Clostridiales (*C lituseburense, Lachnoclostridum bolteae, L hathewayi, C aldenense*, and *Flavonifractor plautii*) ↓ *Dorea formicigenerans, Blautia luti, Sutterella* spp. ↓Tryptophan (correlation with ↑*Erysipelotrichaceae*, *C. lituseburense*, and *Terrisporobacter*) ↑ Serotonergic metabolites, including 5-HIAA (associated with abdominal pain and with the following: ↓ *Akkermansia muciniphila*, *Coprococcus catus*, *Odoribacter splanchnicus*, *C. lactatifermentans*, and *Ruminococcus lactaris;* ↑ *L. bolteae, L. hathewayi*, and *F. plautii*).	[89]

Country (Year)	Study Group			Specimen type	Analytical method	Changes in fecal microbiome in ASD	Refs.
	ASD (GI+/GI−)	SIB (GI+/GI−)	NTC (GI+GI−)				
USA (2018)	21	-	23	Stool	^{1}H NMR spectroscopy 16S rRNA gene sequencing	↑ Isopropanol, p-cresol. ↓ GABA (associated to ↓ *Streptococcus thermophiles)* ↓Phylotypes closely related to *Prevotella copri* , ↓ *Feacalibacterium prausnitzii and Haemophilus parainfluenzae*.	[90]

Data presented here include microbial phylotypes or species and/or relevant metabolites pertaining ASD-associated alterations, compared to non-ASD siblings or unrelated healthy controls.
ASD, children with autism spectrum disorder; SIB, siblings without ASD; NTC, neuro typical controls; GI+, with gastrointestinal comorbidities; GI−, without gastrointestinal comorbidities, ↑, increased level(s); ↓, decreased level(s); NS, non statistically significant; FISH, fluorescent in situ hybridization; GC, gas chromatography; HPLC, high performance liquid chromtography; SCFA, short-chain fatty acids; PPA, propionic acid MS, mass spectroscopy; SPME, solid phase microextraction; 5-HIAA: 5-hydoxy-indoleacetic acid; H-NMR, proton nuclear magnetic resonance; GABA, gamma-amino butyric acid.

Table 2.
Studies on gut microbiome in ASD.

Scientific literature supports the notion that the HIM plays a crucial role in the pathogenesis of ASD, so scientists are now targeting gut microbiome as a therapeutic approach for such disorder (reviewed in [106]). First, modification of high lipid and sugar diet for a fiber- and protein-containing one showed improved skills while ameliorated ASD behavioral deficits. Second, supplementation with prebiotics (inulin, fructo-oligosaccharides, galacto-oligosaccharides, and lactulose) allows specific changes, both in the composition and/or activity of the gut microflora, mainly inducing the growth of indigenous lactobacilli and bifidobacteria. Third, probiotics administration, either *Bacteroides fragilis* or *Lactobacillus reuteri*, there were improvements in ASD-associated behaviors, counteract effect of harmful infections and stimulation of the host's immune system. Fourth, fecal microbiota transplant, usually applied for treating recurrent *Clostridium difficile* infection and other GI disorders, consists of a sample containing about a thousand indigenous bacterial species of the GI from a neurotypical donor, treatment showed sustained improvement of both GI- and ASD related symptoms (up to 8 weeks posttreatment).

6. Conclusion

After the complete sequencing of the human genome was achieved, the scientific community began, in the second half of the past decade, the task of mapping the human microbiota, mainly the intestinal microbiota. In parallel, the notion that the ENS interplay with the intestinal microbiota, generating responses in the CNS, through the GBA and HPA axis, has opened an avenue for the study of gastrointestinal, metabolic, and/or neuropsychiatric disorders.

In this landscape, an increasing body of evidence suggests that HIM has a key role in gut and brain development and functionality but also in pathogenesis of mental disorders, including ASD. Studies on ASD have showed that HIM dysbiosis, with altered Bacteroidetes/Firmicutes ratio, presence of detrimental key species, and dysregulation of bacterial metabolite release, appears to correlate with severity of ASD symptoms. In this regard, intervention measures to restore HIM homeostasis are likely promising.

However, the part concerning the microbiota is only one more piece of the puzzle that are ASDs, mainly because the etiology of such disorders remains elusive.

Acknowledgements

The following Mexican institutions supported this paper: National Autonomous University of Mexico, the National Institute of Perinatology and the National Polytechnic Institute.

Conflict of interest

The authors have declared that no competing interests exist.

Notes/thanks/other declarations

FJDG. Thanks to my beloved son, Manuel, a youngster with ASD who encourages me to understand how the world is seen through their eyes.

Author details

Francisco Javier Díaz-García[1]*, Saúl Flores-Medina[2,3] and
Diana Mercedes Soriano-Becerril[2]

1 Department of Biology, Faculty of Chemistry, National Autonomous University of Mexico, Mexico City, Mexico

2 Department of Infectology, National Institute of Perinatology "Isidro Espinosa de los Reyes", Health Ministry of Mexico, México City, México

3 Center of Scientific and Technological Studies No. 15 "DAE". National Polytechnic Institute, México City, México

*Address all correspondence to: jdiazgr@hotmail.com

References

[1] Carabotti M, Scirocco A, Maselli MA, Severi C. The gut-brain axis: Interactions between enteric microbiota, central and enteric nervous systems. Annals of Gastroenterology. 2015;**28**:203-209

[2] Cani PD, Knauf C. How gut microbes talk to organs: The role of endocrine and nervous routes. Molecular Metabolism. 2016;**5**(9):743-752. DOI: 10.1016/j.molmet.2016.05.011

[3] Martin CR, Osadchiy V, Kalani A, Mayer EA. The brain-gut-microbiome axis. Cellular and Molecular Gastroenterology and Hepatology. 2018;**6**(2):133-148. DOI: 10.1016/j.jcmgh.2018.04.003

[4] Dinan TG, Cryan JF. Brain-gut-microbiota axis and mental health. Psychosomatic Medicine. 2017;**79**:920-926. DOI: 10.1097/PSY.0000000000000519

[5] Cryan JF, Dinan TG. Mind-altering microorganisms: The impact of the gut microbiota on brain and behavior. Nature Reviews. Neuroscience. 2012;**13**:701-712. DOI: 10.1038/nrn3346

[6] Groen RN, de Clercq NC, Nieuwdorp M, Hoenders HJR, Groen AK. Gut microbiota, metabolism and psychopathology: A critical review and novel perspectives. Critical Reviews in Clinical Laboratory Sciences. 2018;**55**(4):283-293. DOI: 10.1080/10408363.2018.1463507

[7] Mannion A, Leader G. Gastrointestinal symptoms in autism spectrum disorder: A literature review. Review Journal of Autism and Developmental Disorders. 2014;**1**:11-17. DOI: 10.1007/s40489-013-0007-0

[8] Hsiao EY. Gastrointestinal issues in autism spectrum disorder. Harvard Review of Psychiatry. 2014;**22**(2):104-111. DOI: 10.1097/HRP.0000000000000029

[9] Hahler EM, Elsabbagh M. Autism: A global perspective. Current Developmental Disorders Reports. 2015;**2**:58-64. DOI: 10.1007/s40474-014-0033-3

[10] Rodakis J. An n=1 case report of a child with autism improving on antibiotics and a father's quest to understand what it may mean. Microbial Ecology in Health and Disease. 2015;**26**:26382. DOI: 10.3402/mehd.v26.26382

[11] Sandler RH, Finegold SM, Bolte ER, Buchanan CP, Maxwell AP, Väisänen ML, et al. Short-term benefit from oral vancomycin treatment of regressive-onset autism. Journal of Child Neurology. 2000;**15**(7):429-435. DOI: 10.1177/088307380001500701

[12] Frye RE, Slattery J, MacFabe DF, Allen-Vercoe E, Parker W, Rodakis J, et al. Approaches to studying and manipulating the enteric microbiome to improve autism symptoms. Microbial Ecology in Health and Disease. 2015;**26**:26878. DOI: 10.3402/mehd.v26.26878

[13] Sommer F, Bäckhed F. The gut microbiota--masters of host development and physiology. Nature Reviews. Microbiology. 2013;**11**(4):227-238. DOI: 10.1038/nrmicro2974

[14] Wang HX, Wang YP. Gut microbiota-brain Axis. Chinese Medical Journal. 2016;**129**:2373-2380. DOI: 10.4103/0366-6999.190667

[15] Icaza-Chávez ME. Gut microbiota in health and disease. Revista de Gastroenterología de México. 2013;**78**(4):240-248. DOI: 10.1016/j.rgmx.2013.04.004

[16] Mayer EA, Tillisch K, Gupta A. Gut/brain axis and the microbiota. The Journal of Clinical Investigation. 2015;**125**(3):926-938. DOI: 10.1172/JCI76304

[17] Rosenfeld CS. Microbiome disturbances and autism Spectrum disorders. Drug Metabolism and Disposition. 2015;**43**:1557-1571. DOI: 10.1124/dmd.115.063826

[18] Sender R, Fuchs S, Milo R. Revised estimates for the number of human and bacteria cells in the body. PLoS Biology. 2016;**14**(8):e1002533. DOI: 10.1371/journal.pbio.1002533

[19] Morgan XC, Huttenhower C. Chapter 12: Human microbiome analysis. PLoS Computational Biology. 2012;**8**(12):e1002808. DOI: 10.1371/journal.pcbi.1002808

[20] Le Chatelier E, Nielsen T, Qin J, Prifti E, Hildebrand F, Falony G, et al. Richness of human gut microbiome correlates with metabolic markers. Nature. 2013;**500**(7464):541-546. DOI: 10.1038/nature12506

[21] Cotillard A, Kennedy SP, Kong LC, Prifti E, Pons N, Le Chatelier E, et al. Dietary intervention impact on gut microbial gene richness. Nature. 2013;**500**(7464):585-588. DOI: 10.1038/nature12480

[22] Arumugam M, Raes J, Pelletier E, Le Paslier D, Yamada T, Mende DR, et al. Enterotypes of the human gut microbiome. Nature. 2011;**473**(7346):174-180. DOI: 10.1038/nature09944

[23] Human Microbiome Project Consortium. Structure, function and diversity of the healthy human microbiome. Nature. 2012;**486**(7402):207-214. DOI: 10.1038/nature11234

[24] Integrative HMP (iHMP) Research network consortium. The integrative human microbiome project. Nature. 2019;**569**(7758):641-648. DOI: 10.1038/s41586-019-1238-8

[25] Felice VD, O'Mahony SM. The microbiome and disorders of the central nervous system. Pharmacology, Biochemistry, and Behavior. 2017;**160**:1-13. DOI: 10.1016/j.pbb.2017.06.016

[26] Farhadi A, Banan A, Fields J, Keshavarzian A. Intestinal barrier: An interface between health and disease. Journal of Gastroenterology and Hepatology. 2003;**18**:479-497

[27] Groschwitz KR, Hogan SP. Intestinal barrier function: Molecular regulation and disease pathogenesis. The Journal of Allergy and Clinical Immunology. 2009;**124**:3-20. DOI: 10.1016/j.jaci.2009.05.038

[28] Johansson ME, Larsson JM, Hansson GC. The two mucus layers of colon are organized by the MUC2 mucin, whereas the outer layer is a legislator of host-microbial interactions. Proceedings of the National Academy of Sciences of the United States of America. 2011;**108**(Suppl 1):4659-4665. DOI: 10.1073/pnas.1006451107

[29] Durack J, Lynch SV. The gut microbiome: Relationships with disease and opportunities for therapy. The Journal of Experimental Medicine. 2018;**216**(1):20-40. DOI: 10.1084/jem.20180448

[30] Kelly CJ, Zheng L, Campbell EL, Saeedi B, Scholz CC, Bayless AJ, et al. Crosstalk between microbiota-derived short-chain fatty acids and intestinal epithelial HIF augments tissue barrier function. Cell Host & Microbe. 2015, 2015;**17**:662-671. DOI: 10.1016/j.chom.2015.03.005

[31] Zheng L, Kelly CJ, Battista KD, Schaefer R, Lanis JM, Alexeev EE, et al. Microbial-derived butyrate promotes epithelial barrier function through

IL-10 receptor-dependent repression of Claudin-2. Journal of Immunology. 2017;**199**:2976-2984. DOI: 10.4049/jimmunol.1700105

[32] Chang PV, Hao L, Offermanns S, Medzhitov R. The microbial metabolite butyrate regulates intestinal macrophage function via histone deacetylase inhibition. Proceedings of the National Academy of Sciences of the United States of America. 2014;**111**:2247-2252. DOI: 10.1073/pnas.1322269111

[33] Furusawa Y, Obata Y, Fukuda S, Endo TA, Nakato G, Takahashi D, et al. Commensal microbe-derived butyrate induces the differentiation of colonic regulatory T cells. Nature. 2013;**504**:446-450. DOI: 10.1038/nature12721

[34] Smith PM, Howitt MR, Panikov N, Michaud M, Gallini CA, Bohlooly-Y M, et al. The microbial metabolites, short-chain fatty acids, regulate colonic Treg cell homeostasis. Science. 2013;**341**:569-573. DOI: 10.1126/science.1241165

[35] Kaisar MMM, Pelgrom LR, van der Ham AJ, Yazdanbakhsh M, Everts B. Butyrate conditions human dendritic cells to prime type 1 regulatory T cells via both histone deacetylase inhibition and G protein-coupled receptor 109A signaling. Frontiers in Immunology. 2017;**8**:1429. DOI: 10.3389/fimmu.2017.01429

[36] Singh N, Gurav A, Sivaprakasam S, Brady E, Padia R, Shi H, et al. Activation of Gpr109a, receptor for niacin and the commensal metabolite butyrate, suppresses colonic inflammation and carcinogenesis. Immunity. 2014;**40**:128-139. DOI: 10.1016/j.immuni.2013.12.007

[37] Simeoli R, Mattace-Raso G, Pirozzi C, Lama A, Santoro A, Russo R, et al. An orally administered butyrate-releasing derivative reduces neutrophil recruitment and inflammation in dextran sulphate sodium-induced murine colitis. British Journal of Pharmacology. 2017;**174**:1484-1496. DOI: 10.1111/bph.13637

[38] Khan S, Jena G. Sodium butyrate reduces insulin-resistance, fat accumulation and dyslipidemia in type-2 diabetic rat: A comparative study with metformin. Chemico-Biological Interactions. 2016;**254**:124-134. DOI: 10.1016/j.cbi.2016.06.007

[39] Tong LC, Wang Y, Wang ZB, Liu WY, Sun S, Li L, et al. Propionate ameliorates dextran sodium sulfate-induced colitis by improving intestinal barrier function and reducing inflammation and oxidative stress. Frontiers in Pharmacology. 2016;**7**:253. DOI: 10.3389/fphar.2016.00253

[40] Ciarlo E, Heinonen T, Herderschee J, Fenwick C, Mombelli M, Le Roy D, et al. Impact of the microbial derived short chain fatty acid propionate on host susceptibility to bacterial and fungal infections in vivo. Scientific Reports. 2016;**6**:37944. DOI: 10.1038/srep37944

[41] Trompette A, Gollwitzer ES, Yadava K, Sichelstiel AK, Sprenger N, Ngom-Bru C, et al. Gut microbiota metabolism of dietary fiber influences allergic airway disease and hematopoiesis. Nature Medicine. 2014;**20**:159-166. DOI: 10.1038/nm.3444

[42] den Besten G, Bleeker A, Gerding A, van Eunen K, Havinga R, van Dijk TH, et al. Short-chain fatty acids protect against high-fat diet–induced obesity via a PPARγ-dependent switch from lipogenesis to fat oxidation. Diabetes. 2015;**64**:2398-2408. DOI: 10.2337/db14-1213

[43] Zelante T, Iannitti RG, Cunha C, De Luca A, Giovannini G, Pieraccini G, et al. Tryptophan catabolites from microbiota engage aryl hydrocarbon receptor and balance mucosal reactivity via interleukin-22. Immunity.

2013;**39**:372-385. DOI: 10.1016/j.immuni.2013.08.003

[44] Bansal T, Alaniz RC, Wood TK, Jayaraman A. The bacterial signal indole increases epithelial-cell tight-junction resistance and attenuates indicators of inflammation. Proceedings of the National Academy of Sciences of the United States of America. 2010;**107**:228-233. DOI: 10.1073/pnas.0906112107

[45] Chimerel C, Emery E, Summers DK, Keyser U, Gribble FM, Reimann F. Bacterial metabolite indole modulates incretin secretion from intestinal enteroendocrine L cells. Cell Reports. 2014;**9**:1202-1208. DOI: 10.1016/j.celrep .2014.10.032

[46] Lamas B, Richard ML, Leducq V, Pham HP, Michel ML, Da Costa G, et al. CARD9 impacts colitis by altering gut microbiota metabolism of tryptophan into aryl hydrocarbon receptor ligands. Nature Medicine. 2016;**22**:598-605. DOI: 10.1038/nm.4102

[47] Hwang IK, Yoo KY, Li H, Park OK, Lee CH, Choi JH, et al. Indole-3-propionic acid attenuates neuronal damage and oxidative stress in the ischemic hippocampus. Journal of Neuroscience Research. 2009;**87**:2126-2137. DOI: 10.1002/jnr.22030

[48] Venkatesh M, Mukherjee S, Wang H, Li H, Sun K, Benechet AP, et al. Symbiotic bacterial metabolites regulate gastrointestinal barrier function via the xenobiotic sensor PXR and toll-like receptor 4. Immunity. 2014;**41**:296-310. DOI: 10.1016/j.immuni.2014.06.014

[49] Miyamoto J, Mizukure T, Park SB, Kishino S, Kimura I, Hirano K, et al. A gut microbial metabolite of linoleic acid, 10-hydroxy-cis-12-octadecenoic acid, ameliorates intestinal epithelial barrier impairment partially via GPR40-MEK-ERK pathway. The Journal of Biological Chemistry. 2015;**290**:2902-2918. DOI: 10.1074/jbc.M114.610733

[50] Kaikiri H, Miyamoto J, Kawakami T, Park SB, Kitamura N, Kishino S, et al. Supplemental feeding of a gut microbial metabolite of linoleic acid, 10-hydroxy-cis-12-octadecenoic acid, alleviates spontaneous atopic dermatitis and modulates intestinal microbiota in NC/nga mice. International Journal of Food Sciences and Nutrition. 2017;**68**:941-951. DOI: 10.1080/09637486.2017.1318116

[51] Viladomiu M, Hontecillas R, Bassaganya-Riera J. Modulation of inflammation and immunity by dietary conjugated linoleic acid. European Journal of Pharmacology. 2016;**785**:87-95. DOI: 10.1016/j.ejphar.2015.03.095

[52] SA1 S, MH1 V, C1 G, XD1 Z, Reynolds CM. Conjugated linoleic acid supplementation improves maternal high fat diet-induced programming of metabolic dysfunction in adult male rat offspring. Scientific Reports. 2017;**7**:6663. DOI: 10.1038/s41598-017-07108-9

[53] Garibay-Nieto N, Queipo-García G, Alvarez F, Bustos M, Villanueva E, Ramírez F, et al. Effects of conjugated linoleic acid and metformin on insulin sensitivity in obese children: Randomized clinical trial. The Journal of Clinical Endocrinology and Metabolism. 2017;**102**:132-140

[54] Suzuki T, Yoshida S, Hara H. Physiological concentrations of short-chain fatty acids immediately suppress colonic epithelial permeability. The British Journal of Nutrition. 2008;**100**(2):297-305. DOI: 10.1017/S0007114508888733

[55] Macfarlane S, Macfarlane GT. Regulation of short-chain fatty acid production. The Proceedings of the Nutrition Society. 2003;**62**:67-72. DOI: 10.1079/PNS2002207

[56] Le Galliard JF, Cote J, Fitze PS. Lifetime and intergenerational fitness

consequences of harmful male interactions for female lizards. Ecology. 2008;**89**:56-64. DOI: 10.1890/06-2076.1

[57] LeBlanc JG, Milani C, de Giori GS, Sesma F, van Sinderen D, Ventura M. Bacteria as vitamin suppliers to their host: A gut microbiota perspective. Current Opinion in Biotechnology. 2013;**24**:160-168. DOI: 10.1016/j.copbio.2012.08.005

[58] Tan J, McKenzie C, Potamitis M, Thorburn AN, Mackay CR, Macia L. The role of short-chain fatty acids in health and disease. Advances in Immunology. 2014;**121**:91-119. DOI: 10.1016/B978-0-12-800100-4.00003-9

[59] Hooper LV, Midtvedt T, Gordon JI. How host-microbial interactions shape the nutrient environment of the mammalian intestine. Annual Review of Nutrition. 2002;**22**:283-307. DOI: 10.1146/annurev.nutr.22.011602.092259

[60] Dai ZL, Li XL, Xi PB, Zhang J, Wu G, Zhu WY. Metabolism of select amino acids in bacteria from the pig small intestine. Amino Acids. 2012;**42**(5):1597-1608. DOI: 10.1007/s00726-011-0846-x

[61] Saulnier DM, Gibson GR, Kolida S. In vitro effects of selected synbiotics on the human faecal microbiota composition. FEMS Microbiology Ecology. 2008;**66**:516-527

[62] Agus A, Planchais J, Sokol H. Gut microbiota regulation of tryptophan. Cell Host & Microbe. 2018;**23**(6):716-724. DOI: 10.1016/j.chom.2018.05.003

[63] Wall R, Cryan JF, Ross RP, Fitzgerald GF, Dinan TG, Stanton C. Bacterial neuroactive compounds produced by psychobiotics. Advances in Experimental Medicine and Biology. 2014;**817**:221-239. DOI: 10.1007/978-1-4939-0897-4_10

[64] Lyte M. Microbial endocrinology in the microbiome-gut-brain axis: How bacterial production and utilization of neurochemicals influence behavior. PLoS Pathogens. 2013;**9**(11):e1003726. DOI: 10.1371/journal.ppat.1003726

[65] Dominguez-Bello MG, Costello EK, Contreras M, Magris M, Hidalgo G, Fierer N, et al. Delivery mode shapes the acquisition and structure of the initial microbiota across multiple body habitats in newborns. Proceedings of the National Academy of Sciences of the United States of America. 2010;**107**(26):11971-11975. DOI: 10.1073/pnas.1002601107

[66] O'Callaghan A, van Sinderen D. Bifidobacteria and their role as members of the human gut microbiota. Frontiers in Microbiology. 2016;**7**:925. DOI: 10.3389/fmicb.2016.00925

[67] Hughes HK, Rose D, Ashwood P. The gut microbiota and dysbiosis in autism spectrum disorders. Current Neurology and Neuroscience Reports. 2018;**18**(11):81. DOI: 10.1007/s11910-018-0887-6

[68] Galland L. The gut microbiome and the brain. Journal of Medicinal Food. 2014;**17**(12):1261-1272. DOI: 10.1089/jmf.2014.7000

[69] American Psychiatric Association. Diagnostic and Statistical Manual of Mental Disorders. 5th ed. Arlington, VA: American Psychiatric Association; 2013. 947 p. DOI: 10.1176/appi.books.9780890425596

[70] World Health Organisation. Autism spectrum disorders. [Internet]. 2018. Available from: https://www.who.int/en/news-room/fact-sheets/detail/autism-spectrum-disorders [Accessed: 03 August 2019]

[71] Centers for Disease Control and Prevention. Autism Spectrum Disorder

(ASD) [Internet]. 2019. Available from: https://www.cdc.gov/ncbddd/autism/data.html [Accessed: 04 August 2019]

[72] Kraneveld AD, Szklany K, de Theije CG, Garssen J. Gut-to-brain axis in autism spectrum disorders: Central role for the microbiome. International Review of Neurobiology. 2016;**131**:263-287. DOI: 10.1016/bs.irn.2016.09.001

[73] Finegold SM, Molitoris D, Song Y, Liu C, Vaisanen ML, Bolte E, et al. Gastrointestinal microflora studies in late-onset autism. Clinical Infectious Diseases. 2002;**35**(Suppl 1):s6-s16. DOI: 10.1086/341914

[74] Song Y, Liu C, Finegold SM. Real-time PCR quantitation of clostridia in feces of autistic children. Applied and Environmental Microbiology. 2004;**70**(11):6459-6465. DOI: 10.1128/AEM.70.11.6459-6465.2004

[75] Parracho HM, Bingham MO, Gibson GR, McCartney AL. Differences between the gut microflora of children with autistic spectrum disorders and that of healthy children. Journal of Medical Microbiology. 2005;**54**(Pt 10):987-991. DOI: 10.1099/jmm.0.46101-0

[76] Finegold SM, Dowd SE, Gontcharova V, Liu C, Henley KE, Wolcott RD, Youn E, Summanen PH, Granpeesheh D, Dixon D, Liu M, MolitorisDR,GreenJA3rd.Pyrosequencing study of fecal microflora of autistic and control children. Anaerobe 2010;**16**(4):444-453. DOI: 10.1016/j.anaerobe.2010.06.008

[77] Adams JB, Johansen LJ, Powell LD, Quig D, Rubin RA. Gastrointestinal flora and gastrointestinal status in children with autism–comparisons to typical children and correlation with autism severity. BMC Gastroenterology. 2011;**11**:22. DOI: 10.1186/1471-230X-11-22

[78] Martirosian G, Ekiel A, Aptekorz M, Wiechula B, Kazek B, Jankowska-Steifer E, et al. Fecal lactoferrin and *Clostridium* spp. in stools of autistic children. Anaerobe. 2011;**17**(1):43-45. DOI: 10.1016/j.anaerobe.2010.12.003

[79] Williams BL, Hornig M, Buie T, Bauman ML, Cho Paik M, Wick I, et al. Impaired carbohydrate digestion and transport and mucosal dysbiosis in the intestines of children with autism and gastrointestinal disturbances. PLoS One. 2011;**6**(9):e24585

[80] Williams BL, Hornig M, Parekh T, Lipkin WI. Application of novel PCR-based methods for detection, quantitation, and phylogenetic characterization of *Sutterella* species in intestinal biopsy samples from children with autism and gastrointestinal disturbances. MBio. 2012;**3**(1):e00261-11. DOI: 10.1128/mBio.00261-11

[81] Wang L, Christophersen CT, Sorich MJ, Gerber JP, Angley MT, Conlon MA. Low relative abundances of the mucolytic bacterium *Akkermansia muciniphila* and *Bifidobacterium* spp. in feces of children with autism. Applied and Environmental Microbiology. 2011;**77**(18):6718-6721. DOI: 10.1128/AEM.05212-11

[82] Wang L, Christophersen CT, Sorich MJ, Gerber JP, Angley MT, Conlon MA. Elevated fecal short chain fatty acid and ammonia concentrations in children with autism spectrum disorder. Digestive Diseases and Sciences. 2012;**57**(8):2096-2102. DOI: 10.1007/s10620-012-2167-7

[83] Wang L, Christophersen CT, Sorich MJ, Gerber JP, Angley MT, Conlon MA. Increased abundance of *Sutterella* spp. and *Ruminococcus torques* in feces of children with autism spectrum disorder.

Molecular Autism. 2013;**4**:42. DOI: 10.1186/2040-2392-4-42

[84] Gondalia SV, Palombo EA, Knowles SR, Cox SB, Meyer D, Austin DW. Molecular characterisation of gastrointestinal microbiota of children with autism (with and without gastrointestinal dysfunction) and their neurotypical siblings. Autism Research. 2012;**5**(6):419-427. DOI: 10.1002/aur.1253

[85] Kang DW, Park JG, Ilhan ZE, Wallstrom G, Labaer J, Adams JB, et al. Reduced incidence of Prevotella and other fermenters in intestinal microflora of autistic children. PLoS One. 2013;**8**(7):e68322. DOI: 10.1371/journal.pone.0068322

[86] De Angelis M, Piccolo M, Vannini L, Siragusa S, De Giacomo A, Serrazzanetti DI, et al. Fecal microbiota and metabolome of children with autism and pervasive developmental disorder not otherwise specified. PLoS One. 2013;**8**(10):e76993. DOI: 10.1371/journal.pone.0076993

[87] Son JS, Zheng LJ, Rowehl LM, Tian X, Zhang Y, Zhu W, et al. Comparison of fecal microbiota in children with autism Spectrum disorders and Neurotypical siblings in the Simons simplex collection. PLoS One. 2015;**10**(10):e0137725. DOI: 10.1371/journal.pone.0137725

[88] Tomova A, Husarova V, Lakatosova S, Bakos J, Vlkova B, Babinska K, et al. Gastrointestinal microbiota in children with autism in Slovakia. Physiology & Behavior. 2015;**138**:179-187. DOI: 10.1016/j.physbeh.2014.10.033

[89] Luna RA, Oezguen N, Balderas M, Venkatachalam A, Runge JK, Versalovic J, et al. Distinct microbiome-neuroimmune signatures correlate with functional abdominal pain in children with autism spectrum disorder. Cellular and Molecular Gastroenterology and Hepatology. 2016;**3**(2):218-230. DOI: 10.1016/j.jcmgh.2016.11.008

[90] Kang DW, Ilhan ZE, Isern NG, Hoyt DW, Howsmon DP, Shaffer M, et al. Differences in fecal microbial metabolites and microbiota of children with autism spectrum disorders. Anaerobe. 2018;**49**:121-131. DOI: 10.1016/j.anaerobe.2017.12.007

[91] de Theije CG, Wu J, da Silva SL, Kamphuis PJ, Garssen J, Korte SM, et al. Pathways underlying the gut-to-brain connection in autism spectrum disorders as future targets for disease management. European Journal of Pharmacology. 2011;**668**(Suppl 1):S70-S80. DOI: 10.1016/j.ejphar.2011.07.013

[92] de Magistris L, Familiari V, Pascotto A, Sapone A, Frolli A, Iardino P, et al. Alterations of the intestinal barrier in patients with autism spectrum disorders and in their first-degree relatives. Journal of Pediatric Gastroenterology and Nutrition. 2010;**51**(4):418-424. DOI: 10.1097/MPG.0b013e3181dcc4a5

[93] Audet MC, Jacobson-Pick S, Wann BP, Anisman H. Social defeat promotes specific cytokine variations within the prefrontal cortex upon subsequent aggressive or endotoxin challenges. Brain, Behavior, and Immunity. 2011;**25**(6):1197-1205. DOI: 10.1016/j.bbi.2011.03.010

[94] Haba R, Shintani N, Onaka Y, Wang H, Takenaga R, Hayata A, et al. Lipopolysaccharide affects exploratory behaviors toward novel objects by impairing cognition and/or motivation in mice: Possible role of activation of the central amygdala. Behavioural Brain Research. 2012;**228**(2):423-431. DOI: 10.1016/j.bbr.2011.12.027

[95] van Heesch F, Prins J, Konsman JP, Westphal KG, Olivier B, Kraneveld AD, et al. Lipopolysaccharide-induced

anhedonia is abolished in male serotonin transporter knockout rats: An intracranial self-stimulation study. Brain, Behavior, and Immunity. 2013;**29**:98-103. DOI: 10.1016/j.bbi.2012.12.013

[96] Emanuele E, Orsi P, Boso M, Broglia D, Brondino N, Barale F, et al. Low-grade endotoxemia in patients with severe autism. Neuroscience Letters. 2010;**471**(3):162-165. DOI: 10.1016/j.neulet.2010.01.033

[97] Anderson GM, Freedman DX, Cohen DJ, Volkmar FR, Hoder EL, McPhedran P, et al. Whole blood serotonin in autistic and normal subjects. Journal of Child Psychology and Psychiatry. 1987;**28**(6):885-900

[98] Hanley HG, Stahl SM, Freedman DX. Hyperserotonemia and amine metabolites in autistic and retarded children. Archives of General Psychiatry. 1977;**34**(5):521-531. DOI: 10.1001/archpsyc.1977.01770170031002

[99] Schain RJ, Freedman DX. Studies on 5-hydroxyindole metabolism in autistic and other mentally retarded children. The Journal of Pediatrics. 1961;**58**:315-320. DOI: 10.1016/s0022-3476(61)80261-8

[100] Chugani DC1, Muzik O, Behen M, Rothermel R, Janisse JJ, Lee J, Chugani HT. Developmental changes in brain serotonin synthesis capacity in autistic and nonautistic children. Annals of Neurology. 1999;**45**(3):287-295

[101] Marler S, Ferguson BJ, Lee EB, Peters B, Williams KC, McDonnell E, et al. Brief report: Whole blood serotonin levels and gastrointestinal symptoms in autism spectrum disorder. Journal of Autism and Developmental Disorders. 2016;**46**(3):1124-1130. DOI: 10.1007/s10803-015-2646-8

[102] McDougle CJ, Naylor ST, Cohen DJ, Aghajanian GK, Heninger GR, Price LH. Effects of tryptophan depletion in drug-free adults with autistic disorder. Archives of General Psychiatry. 1996;**53**(11):993-1000. DOI: 10.1001/archpsyc.1996.01830110029004

[103] Foley KA, MacFabe DF, Vaz A, Ossenkopp KP, Kavaliers M. Sexually dimorphic effects of prenatal exposure to propionic acid and lipopolysaccharide on social behavior in neonatal, adolescent, and adult rats: Implications for autism spectrum disorders. International Journal of Developmental Neuroscience. 2014;**39**:68-78. DOI: 10.1016/j.ijdevneu.2014.04.001

[104] Foley KA, MacFabe DF, Kavaliers M, Ossenkopp KP. Sexually dimorphic effects of prenatal exposure to lipopolysaccharide, and prenatal and postnatal exposure to propionic acid, on acoustic startle response and prepulse inhibition in adolescent rats: Relevance to autism spectrum disorders. Behavioural Brain Research. 2015;**278**:244-256. DOI: 10.1016/j.bbr.2014.09.032

[105] Wang L, Conlon MA, Christophersen CT, Sorich MJ, Angley MT. Gastrointestinal microbiota and metabolite biomarkers in children with autism spectrum disorders. Biomarkers in Medicine. 2014;**8**(3):331-344. DOI: 10.2217/bmm.14.12

[106] Yang Y, Tian J, Yang B. Targeting gut microbiome: A novel and potential therapy for autism. Life Sciences. 2018;**194**:111-119. DOI: 10.1016/j.lfs.2017.12.027

Section 4

Industrial Uses

Chapter 12

Bioconversion of Weedy Waste into Sugary Wealth

Prajakta Prakash Kamble, Suresh Shivaji Suryawanshi, Maheshkumar Vishnu Kore, Nahid Irani, Jyoti Prafulla Jadhav and Yasmin Chand Attar

Abstract

Efforts put in overriding the inulin abundant invader nastiest category I weeds are infeasible that lead into its impermanent confiscation. Hence, their heedful exploitation is obligatory. These invasive weeds have ample amount of inulin, which serves as a renewable, cheap raw substrate for inulinase production. Therefore, they have enticed intention of many researchers toward exploring more idiosyncratic inulinase producing microbial strains that utilize invasive inulin-rich weeds as substrate for fructose liberation. Plenteous industrial applications of inulinases have marked it distinctly crucial in recent biotechnological epoch. This review thus elaborates the literature on infused footprints embedded by the substituted low calorie healthy sweetener in new advancing fields.

Keywords: invasive, weed, inulinases, low calorie, healthy sweetener, fructose

1. Introduction

Weeds are plants that grow luxuriantly in unsolicited places with no special assistance of human. These plants spread rapidly by shading abundant seedlings, making land unfit for agriculture, forestry, and livestock. Their characterized adaptability to extensive range of soils and weathers has proved a boon for their survival at any piece of earth. They also have capabilities of tumbling inherent plant ecosystems and fluctuating natural biota in injurious ways. Due to the competency of piercing and interchanging indigenous flora, majority of them are recognized as environmental weeds or exotic or noxious aggressive invaders [1].

Apart from this negative side, countless constructive purposes of weeds as a part of their control strategies have still remained unnoticed. Therefore, this research theme was needed to be explored and expanded. Consequently, when seen from a different standpoint, such weeds have high inulin in them. Thus, this work explores inulin-rich weeds as a veritable and bioconvertible resource for sugary wealth creation, using efficient inulinase producing microbes.

2. Inulin

Inulin is an allocated polysaccharide mixture composed of α-D-glucopyranosyl-[β-(2,1)-D-fructofuranosyl]-D-fructofuranosides linked by β-(2,1)-D-fructosyl-fructose

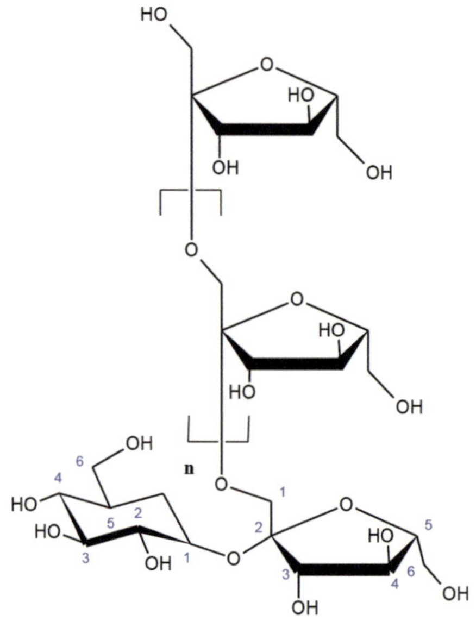

Figure 1.
Inulin structure.

bonds, and each of this chain is terminated by fructose moiety. The linking and bonding in inulin moiety are designated in **Figure 1**. Inulin is a reservoir of nondigestible carbohydrate known as fructans. It constitutes the bulk of glycosidic bonds joining fructosyl-fructose. The inulin-type fructans stored in Dicotyledonous species are connected with linear β(2→1) fructofuranosyl units, whereas monocots encompass branched complex-type fructans [2].

3. Plant sources of fructans

Inulin is abundant in structures such as bulbs, tubers, and tuberous roots of grasses and flowering plants belonging to Liliaceae (3500 species) and Compositae (25,000 species) families. Such plants, for example, asparagus, wheat, rye, and dahlias, mostly lack starch and thus synthesize inulin as energy store house. A wide array of inulin-rich plants with their inulin content is symbolized graphically (**Figure 2**) [3].

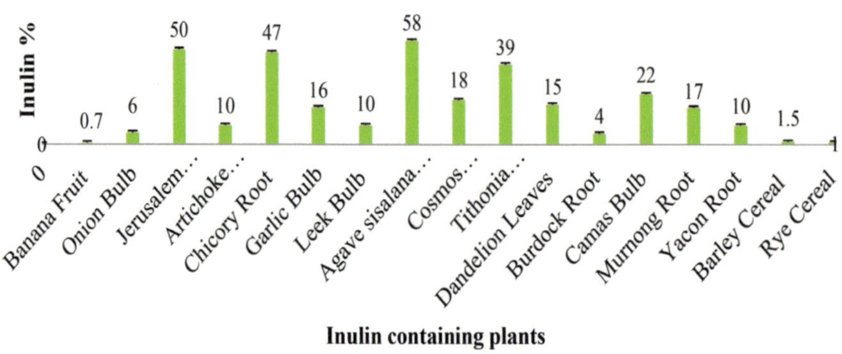

Figure 2.
Schematic depiction of inulin content in variety of inulin consisting plants.

Jerusalem artichoke (*Helianthus tuberosus*) and chicory (*Cichorium intybus*) are common commercialized inulin source available in market. The fleshy tap root of chicory serves as warehoused of inulin (70–80%) [4]. Depending upon the growth stage of chicory, either inulin or oligofructose can be obtained captivatingly. After full root development and inflorescence axis arrival, endoinulinase hydrolyzes inulin into oligofructose, and exoinulinase further converts it into fructose. Most European countries have officially recognized inulin, oligofructose, and fructose as natural food ingredients, thereby having vast fascinating functional features that are beneficial to satisfy the needs of industries for imminent healthy food formulations. The present work currently focuses on two invasive home-grown (*Tithonia rotundifolia* and *Cosmos bipinnatus*) and one universally studied (*Agave sisalana*) inulin-rich weed species (**Figure 3**).

3.1 Agave

Agave is the most taxonomically diverse members of family Agavaceae. They are been surviving in extreme conditions by adapting themselves morphologically and physiologically. To escape transpirational water loss, they conduct crassulacean acid metabolism, thus liberating fructans as the chief photosynthetic product. *A. sisalana* was the common species found throughout Asia with rich inulin content, thus being used as substrate for alcohol and inulinase synthesis [5].

3.2 *Cosmos bipinnatus*

Cosmos bipinnatus of Asteraceae family is commonly famous as garden cosmos or Mexican aster, which is an inulin comprising weedy annual herb exotic for India. It has acclimatized on infertile, sandy soils along roadsides, exposed slopes, fence lines, hedgerows, or background areas as an ornamental plant getting transmuted into invasive weed [1].

3.3 *Tithonia rotundifolia* (Mill.) S. F. Blake

It belongs to family Asteraceae/Compositae and is commonly known as red sunflower, rooisonneblom, Japanese sunflower, shrub sunflower, and tree marigold. It is rich in inulin [6]. Thus, it serves as renewable raw material for fructose syrup (D-fructose) production. It is also grown as a green manure. But its high propagation frequency has forced to classify it as alien, invasive, competitive, allelopathic [7], noxious category 1 weed. There are reports on these weeds competing with crop plants and shading out native vegetation in the humid and subhumid tropics of South America, South East Asia, and tropical and subtropical Africa. Thus, the overall deleterious impressions put forth by this weed need to be rectified by an ecofriendly way.

Figure 3.
Inulin-rich weeds under present investigation: (a) Agave sisalana, *(b)* Cosmos bipinnatus, *and (c)* Tithonia rotundifolia.

4. Weed management strategies

Control majors like manual irradiation of these inulin-rich weeds are a tough job since they induce allergic effects [8]. Chemical practice can be used, but there are reports reviling the incidences of herbicide resistance weed expansion. Additionally, a striking raise in expenditure (>30$ ha^{-1}) of such weed remedy is too observed [9]. Accumulation of chemical scums in groundwater was another problem that emerges by the application of herbicides [10].

Thus, the pressure was to lessen herbicide usage and to reevaluate its environmental safety, development of alternative weed-control options was cheered. The best proposed avenue is to use microbial weed treating strategy, where actively propagating microorganism is subjected on target weed to achieve rapid control by its enzymatic hydrolysis into cost-effective product. Thus, exploiting inulinase producing soil microbes has been crucial tool in our efforts to renovate these weeds into fructose: a profitable calorie condensed sweeteners [11]. Microbial bioconversion finally is the best defense evident against this invasive attack.

5. Inulinase

Inulinases are fructofuranosyl hydrolases that cleave inulin into fructose moieties. Fructo-sugars, fructooligosaccharides (FOSs), or simply oligofructoses are the fructose oligomers formed after the action of inulinase on inulin [12]. Inulinase is an industrially crucial class of enzyme incorporated into glycoside hydrolase families 32 and 91. Based on their mode of action (**Figure 4**) on inulin, inulinases are alienated into dualistic types: (1) exoinulinase (β-D-fructanfructohydrolase, E.C. 3.2.1.80) and (2) endoinulinase (2, 1-β-D-fructanfructanohydrolase, E.C. 3.2.1.7) [13].

5.1 Microbial sources of inulinase

Phylogentically diverse microorganisms comprising bacteria, filamentous fungi, yeasts, and actinomycetes were testified to synthesize inulinase enzyme [14]. Due to

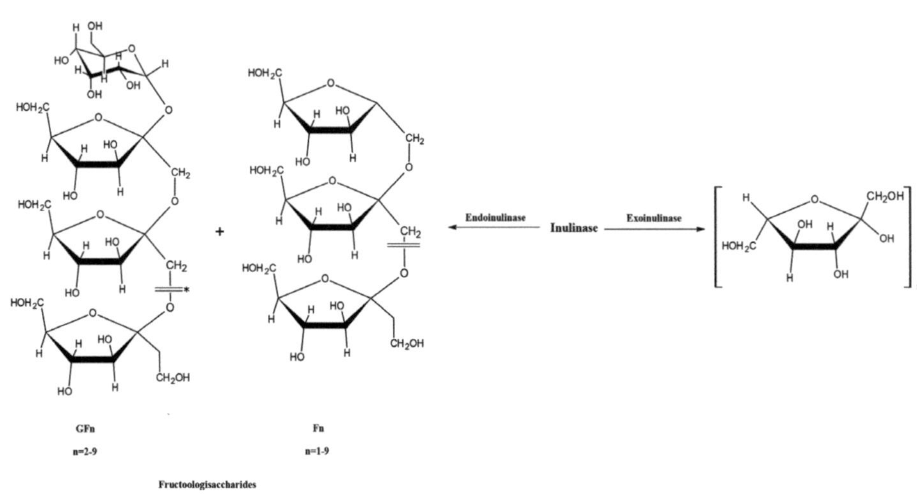

Figure 4.
Enzymatic hydrolysis of inulin-rich weed where (*) signifies site of inulinase activity on repeating β-(1-2)-D-fructosyl units of inulin.

the easy cultivation and higher enzyme yield, bacterial spp. are being commercially exploited to produce inulin hydrolyzing enzymes. The literature published recently [9, 13, 15–21] regarding the inulinase producers yielding maximum enzyme is embodied in **Figure 5**.

5.2 Substrates for inulinase production

Media complexity and culture conditions influence the enzyme production critically. The morphogenesis and metabolic pathway involved in enzyme induction can be noticeably affected by altering the media components and the growth parameters. Therefore, this substitution may accelerate biocatalysis of substrate into desirable products.

Inulin, starch, sucrose, and inulin-rich plant extracts are been widely utilized as exclusive, cheap, and best carbon source for biosynthesis of inulinase by several microbes. This polyfructan along with naturally occurring inulin-rich material and mixed substrates contributes as potent inducers for inulinase production. This plant-derived abundant storage polysaccharide is also present in roots and tubers of Compositae and Gramineae plants and numerous invader weeds. The review mentions a wide substrate used for inulinase production mutant [9, 19]. Dahlia (*Dahlia pinnata*), rhizosphere of Jerusalem artichoke (*H. tuberosus*), chicory (*C. intybus*) roots, kuth (*Saussurea lappa*) roots, *Allium sativum,* and *Allium cepa* have broadly been exploited for this perseverance. Mature *C. intybus* root was found to be the best substrate for receiving maximum extracellular inulinase from *Fusarium oxysporum* [22].

Figure 5.
Glance on varied inulinase producing microorganisms [55–78].

5.3 Inulinase production

Enzyme production is critically influenced by media complexity and culture conditions. Alterations in these two factors noticeably affect the morphogenesis and metabolic pathway involved in enzyme induction. It may accelerate biocatalysis of substrate into desirable products.

Inulinase enzymes are commercially produced consuming synthetic inulin and agroindustrial residues by submerged fermentation as well as by solid-state fermentation (SSF). Microorganisms, substrate, and cultivation method for inulinase production in certain studies reported in the literature [23, 24] are described later. The records show a resilient inclination to substitute high value synthetic inulin by agroindustrial substrates so as to make this enzyme production process cost effective. *Kluyveromyces* genus is reported to be the excellent inulinase producers [25]. Researchers explained that under optimum condition, the *Kluyveromyces marxianus* NRRL Y-7571 extracellular enzyme concentration extended to 391.9 U/g of dry fermented bagasse. Thus, due to the high availability and low rate sugarcane and corn industries, deposits (sugarcane bagasse, molasses, and corn steep liquor) can be economically attractive [26].

5.4 Factorial design

The escalated microbial growth and enzyme yield throughout the fermentation need to be keenly monitored. This is well accomplished by optimizing the fermentation conditions. The single-dimensional traditional simple frequently employed optimization method encompasses fluctuation of one independent variable at given level and maintaining others constant. Since it lacks the possible interactions among factors, it is least preferred. Thus, an effectual experimental scheme like response surface method is adopted to operate optimal conditions for multivariable systems. It aids in appreciating interaction of parameters and recognizing optimal range for higher yield. It also includes variety of statistical techniques used for experimental design and model erection that measures and scrutinizes the optimum conditions. Effective optimization of fungal, bacterial, and yeast inulinase production consuming diverse substrates such as Jerusalem artichoke, sugarcane bagasse, and molasses in submerged or solid-state cultivation was stated in the literature [27]. Diagrammatic depiction of microorganisms and optimized experimental variables is accessible in **Figure 6** [13, 16, 28–33].

5.5 Purification and properties of inulinases

The nature, interaction, and additional specific properties can be well understood in case of pure enzymes than the crude ones. Enzyme purification thus serves as a crucial footstep. The efficacious purification is reliant on complexity, charge distribution, and physicochemical properties of enzyme. Size, polarity, ligand interactions, and solubility are few of the strategic factors that define the choice of purification techniques to be applied for purifying inulinase. Some common purification techniques hired are salt or solvent precipitation, ion exchange, affinity, hydrophobic interaction, gel exclusion chromatography, and ultrafiltration [34].

Implication of ammonium sulfate precipitation method followed with column chromatography, boosted *X. oryzae* endoinulinase recovery by 2.9-folds [35]. Thermostable endoinulinase from *Bacillus smithii* was purified by ammonium sulfate precipitation and ion exchange chromatography. The exoinulinase synthesized by *Arthrobacter* spp., *Arthrobacter globiformis*, *Bacillus stearothermophilus, Pseudomonas mucidolens,* and *Thermotoga maritima* was recovered and purified for further studies. Salt precipitation functioned better in bacterial inulinase

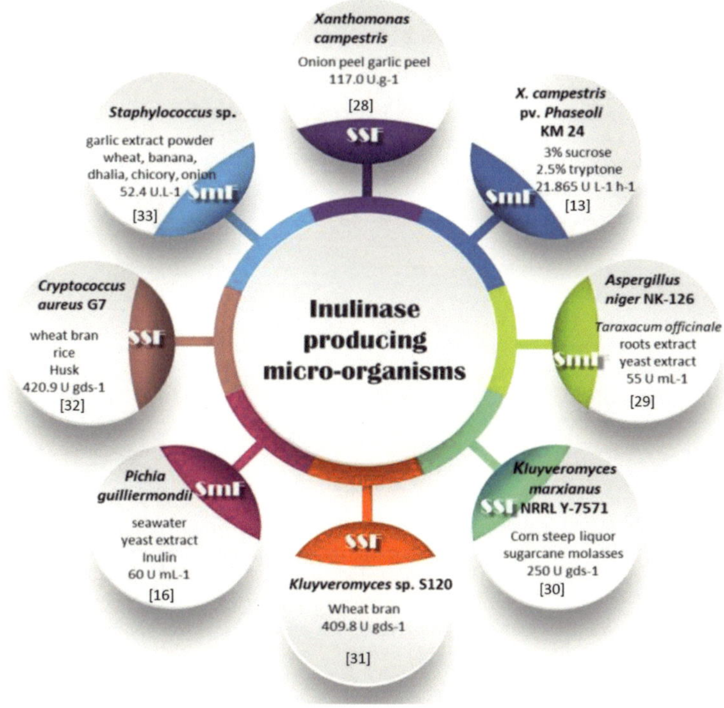

Figure 6.
Highlight on inulinase production by various microbes under specific fermentation conditions.

purification, whereas organic solvent precipitation was preeminent for fungal inulinases. The extraordinary solubility of ammonium sulfate in water makes it more preferential for salt precipitation. This ammonium sulfate after cleavage gets converted into two ionic forms, thus sustaining its top most position in Hofmeister series. Structural integrity of protein is least exaggerated by this salt during the salting out progression. The increased probability of protein repression in organic solvent existence reduces its utility in enzyme purification. Maximum reports on use of ion exchange and gel exclusion chromatography followed by high selective affinity chromatography are noticed for biomolecule purification. The chemical structure and function of bacterial and fungal inulinase decide which purification techniques are to be employed for its purification. These techniques are reliable in convalescing interested protein in short time. The requisite factors like widely oscillating temperature and pH stability of inulinase, along with other vital characters, before being exploited for industrial applications need to be thoroughly inspected.

Physical elements such as molecular weight (M_r), Michaelis-Menten constant (K_m), and maximal velocity (V_{max}) are significantly imperative to characterize an enzyme. Heteromeric structure and any conformational variations are well enlightened by molecular weight studies of an enzyme. K_m and V_{max} values illuminate the enzyme kinetics and also emphasize on the specificity and affinity of inulinase for varied substrates. This affinity is designated by K_m. K_m is the substrate concentration that engages half of enzyme's active site. Lower K_m illustrates higher affinity of enzyme toward specific substrate and vice versa.

5.6 Structural peculiarities of purified inulinases

The molecular masses of bacterial and fungal inulinases oscillate in the range from 28 to 450 kDa as denoted in **Figure 7** [36]. Most of the fungal inulinases have

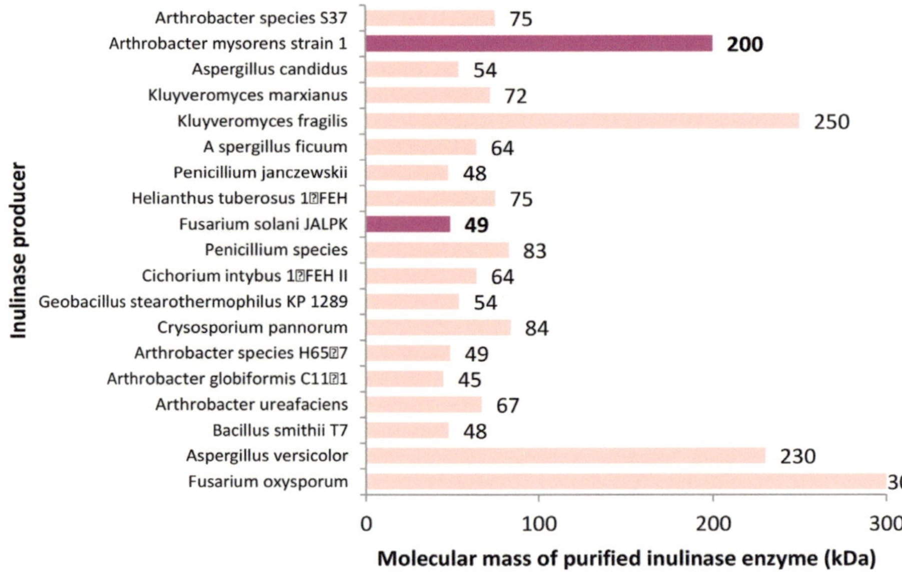

Figure 7.
Comparison of molecular masses of inulinase from numerous microbial sources obtained after SDS-PAGE electrophoresis.

molecular weight exceeding 50.0 kDa. Three inulinases with molecular masses 42, 65, and 57 kDa were isolated and purified from *Kluyveromyces* species Y 85.

Characterization of fungal and bacterial endoinulinases is also investigated after its purification. The purified endoinulinase harvested from *Penicillium* sp. TN-88 has molecular mass of 68.0 kDa [37]. *Arthrobacter* sp. S37 also produced extracellular endoinulinase, which was purified and found to have approximately 75 kDa [38].

5.7 Profitable approach of inulinase efficacy

Owing to the scenarios in food, pharmaceutical and nutraceutical industries, microbial hydrolysis and bioconversion of inulin have established a new source of revenue to several workers [39].

Inulinase offers exciting perceptions in view of the budding need for the Ultrahigh-Fructose Syrup (UHFS) production from inulin. Approximate 95% pure fructose can be obtained by enzymatic hydrolysis of inulin in the presence of inulinase. Thereby, inulinase-producing microbes are been extensively exploited by numerous industries so as to get value-added UHFS from inulin-rich weeds.

Inulinase and inulinase producers along with superfluous microorganism amalgamation are prominently affianced for simultaneous saccharification and fermentation (SSF) of diverse substrates in ethanol production methods [39–41]. Ethanol is the greatest hired liquid biofuel either as a fuel or as a gasoline complement [42]. Agave, chicory, dahlia, Jerusalem artichoke tuber, and many other inulin-rich weeds aid as the finest raw resources for fuel ethanol production. Certain wild-type microbes were mutated to offer maximum yield. Various experimentations were performed on sugar-beet molasses and numerous plant extracts so as to be used as feedstock to gain ethanol.

Inulinases are furthermore broadly subjugated in commercialization of inulo [43], gluconic acid, sorbitol, pullulan, acetone-butanol [44], and other key products.

6. Product formed after inulinolytic hydrolysis

The hydrolysis of inulin feedstock by inulinase yields astonishing amount of fructose in fermented broth. Carbohydrate, particularly fructose, is an indispensible chunk of the human diet. It owes exceptional properties and is nearly 1.5 times sweeter than sucrose, thus enhancing the palate and pleasure of several foodstuffs. It is recovered by passing through carpet bag filters containing activated charcoal and is further crystalized using chilled solvents, ethanol specifically.

Beyond 30 proceeding years, pure crystalline fructose has stood at the heights in the market as a health supplement in food and beverage. Purity is the pivotal feature that draws a distinguishing sharp line between crystalline fructose and high fructose corn syrup (HFCS). Crystalline fructose products are characteristically 100% pure fructose, while HFCS comprehends nearly equivalent shares of fructose and glucose-like sucrose (table sugar). As pure crystalline fructose is bounteously sweeter than sugar, its minor amount is also adequate to accomplish the same level of sweetness. Thus, lower-sugar and trifling calorie foods typically contain pure crystalline fructose. Food genii company also favors pure crystalline fructose as it owns supplementary properties beyond sweetness, which marks it very lucrative in drinks and candy, cakes, and other food industries [45].

6.1 Purification of fructose

The separation of FOS and fructose is frequently accomplished by reckonable chromatographic techniques. In dietetic products, optimal FOS separation is done by implementing glass-packed precoated silica gel with sodium acetate. Liquid chromatography (LC) with acetonitrile as a mobile phase is executed to purify nonstructural carbohydrates such as sugars and FOS with 3–19 degrees of polymerization. Auxiliary cost-effective methods exploiting activated charcoal fixed bed column with 80% degree of purification and 97.8% recovery of Fructose are superfluously proficient [43]. Purified fructose is assessed by diverse techniques such as NMR, MALDI-MS, MALDITOF, GC-MS, and ESI-MS [46]. The prebiotic fructose metabolism in microorganisms can be premeditated through microarrays [43].

6.2 Commercial applications of fructose

Pure fructose along with FOSs is finely specified to exist in voluminous natural foods. Gigantic companies are manufacturing these extensively applicable healthy and calorie-free products via hydrolyzing inulin weeds by exploiting microbial inulinases. Few lucrative applications of fructose emphasized in the review [47].

6.2.1 In food industries

Fructose serves as one of the key ingredients in food products such as energy and sports drinks, flavor boosted water, carbonated sodas and drinks, beverages, low-calorie food options, cereals, oatmeal, and yogurts and baked goods [3].

6.2.2 Fortification of nominated fruit juice beverages

Investigation reveals that sucrose employed as fruit juice sweetener, with no considerably quality loss can be replaced with FOS and fructose.

6.2.3 Fructose in medicine

Fructose is very frequently found as sweetener in cough suppressants, decongestant drops, rubs, and liquids for children and adults. Many pharmaceutical tablets, syrups, and solutions commonly have fructose as an excipient [48].

6.2.4 Proficient sweetener for diabetics

Inulinase from *Aspergillus oryzae* carries out hydrolysis liberating fructose and FOSs. Existence of mono to pentasaccharides without toxic microbial metabolites in the hydrolyzed product was assessed with NMR spectroscopy and LC-MS, thus excavating its application as a food ingredient [49].

6.2.5 Supplementing oral electrolyte solutions as diarrhea control remedy

The retrieval of overall bacterial counts amplified by ingestion of OES and fructose to pigs with acute diarrhea induced by cholera toxin was the most attention grabbing finding [50].

6.2.6 Dietary intonation of the human colonic microbiota

A trifling prebiotic effect with no gastrointestinal distress in pediatric patients with cancer was found to be induced by FOS, especially fructose [51].

6.2.7 Immunomodulatory effect

Clinical trials direct that fructose and FOS supplementation can reduce the influx of clinical inflammation, abridged level in cytokine interleukin (IL)-1α, and necrosis factor-α in ulcerative colitis by *Bacillus longum*. A shoot-up in IL10 positive mucosal dendritic due to inulin, fructose, and FOS intake was displayed in patients with Crohn's illness [52].

6.2.8 Cancer treatment

Incidence of cancer has been rapidly decreasing due to the use of FOS and fructose. Tumor growth, cell differentiation, and upregulate apoptosis were vetoed by the Butyrate manufactured by FOS and fructose [53].

6.2.9 Antibiotic therapy

Damage of normal protective intestinal microflora was a common observation found to be accompanied with acute diarrhea after been treatment with penicillin, cephalosporin, and clindamycin antibiotics. Double-blind randomized controlled trials were set, which efficaciously explain that in the course of antibiotic treatment reoccurrence of diarrhea was shortened in patients ingesting fructose.

6.2.10 Antioxidant properties

Mesa and his coworkers studied protein glycation and cross linking along with the effect of elevated temperature and proteolysis on antioxidant properties of the Maillard reaction mixtures of soy protein isolates, FOS, and fructose with appropriate controls [54].

6.2.11 Enhancing Salmonella vaccine efficacy

An elevation in specific blood immunoglobulin G specific to *Salmonella* and fecal immunoglobulin A was recorded in mice fed on the fructose and inulin encompassing diet as compared with control mice when infected with LD100 of virulent *S. typhimurium* for tolerable time interlude [49].

7. Concluding remark

The current review discloses the elucidations of many global researchers specifically highlighting on the isolation of novel inulinase-producing rhizospheric microbial flora to hydrolyze high inulin content in weeds, thereby serving as a potential, abundant, and profitable avenue of fructose production with vast industrial applications. The food and pharmaceutical preparations with fructose have extended at the top in the market demand list of health conscious modern era. Thereby, the enzyme production expenses linger to be the logjam in understanding its commercial application.

Thus, this review explores the exploitation of inulin containing weeds such as *Tithonia* and *Cosmos* as low-value and efficacious replacement of synthetic inulin as substrates for inulinase production. The research embarked on the health implications of dietary and pharmaceutical fructoses was underlined in the review. Finally, electrifying new uses of fructan polysaccharides such as drug stabilizers, scrupulous release drug delivery systems, and vaccine adjuvants proclaims evolution in pharmaceutical applications of this extremely multipurpose plant-derived sugar.

Acknowledgements

The author thanks the following contemporaries for their service in data assembly: Nishigandha D. Patil, Rohit. R. Joshi, Pramod R. Kamble, Deepali D. Jadhav, Arun D. Jadhav, Samita S. Bhosale, and other Rajaramians. Special thanks to Shivaji University, Kolhapur, for granting Golden Jubilee Research Fellowship to Prajakta P. Kamble, as a financial support to execute the present work.

Conflict of interest

The authors declare that they have no conflicts of interest.

Author details

Prajakta Prakash Kamble[1], Suresh Shivaji Suryawanshi[2], Maheshkumar Vishnu Kore[1], Nahid Irani[3], Jyoti Prafulla Jadhav[4] and Yasmin Chand Attar[5]*

1 Department of Microbiology, Shivaji University Kolhapur, India

2 Department of Biochemistry, Shivaji University Kolhapur, India

3 Department of Biological Science, University of Illinois, Chicago, IL, USA

4 Department of Biotechnology, Shivaji University Kolhapur, India

5 Department of Microbiology, Rajaram College, Kolhapur, India

*Address all correspondence to: ycamicro@gmail.com

References

[1] Henderson L. Invasive, naturalized and casual alien plants in southern Africa: A summary based on the Southern African Plant Invaders Atlas (SAPIA). Bothalia. 2007;**37**:8

[2] Rosa M et al. Fungal inulinases as potential enzymes for application in the food industry. Advance Journal of Food Science and Technology. 2013;**5**:1031-1042

[3] Barclaya T et al. The chemistry and sources of fructose and their effect on its utility and health implications. Journal of Excipients and Food Chemicals. 2012;**3**:67-82

[4] Apolinário AC et al. Inulin-type fructans: A review on different aspects of biochemical and pharmaceutical technology. Carbohydrate Polymers. 2014;**101**:368-378

[5] Sánchez-Marroquin A, Hope PH. Agave juice: Fermentation and chemical composition studies of some species. Journal of Agricultural and Food Chemistry. 1953;**1**:246-249

[6] Kamble PP et al. Comparison of optimization conditions for elevated bacterial and fungal inulinase. International Journal of Researches in Biosciences, Agriculture and Technology. 2017;**V**(3):225-228

[7] Oyeniyi TA et al. Allelopathic effects of *Tithonia diversifolia* extracts on biochemical parameters and growth of *Vigna unguiculata*. International Journal of Biology. 2016;**8**:45

[8] Bailey KL. Microbial weed control: An off-beat application of plant pathology. Canadian Journal of Plant Pathology. 2004;**26**:239-244

[9] Cruz-Guerrero AE et al. Inulinase-hyperproducing strains of Kluyveromyces sp. isolated from aguamiel (Agave sap) and pulque. World Journal of Microbiology & Biotechnology. 2006;**22**:115-117

[10] Mason PG. Biological control in Ontario 1952-2012: A summary. Journal of the Entomological Society Of Ontario. 2013;**144**:1952-2012

[11] Kennedy A. Soil microorganisms for weed management. Journal of Crop Production. 1999;**2**:123-138

[12] Chi ZM et al. Biotechnological potential of inulin for bioprocesses. Bioresource Technology. 2011;**102**: 4295-4303

[13] Chi Z et al. Inulinase-expressing microorganisms and applications of inulinases. Applied Microbiology and Biotechnology. 2009;**82**:211-220

[14] Kango N, Jains SC. Production and properties of microbial inulinases: Recent advances. Food Biotechnology. 2011;**25**:1532-4249

[15] Zhou J et al. Cold-active and NaCl-tolerant exo-inulinase from a cold-adapted Arthrobacter sp. MN8 and its potential for use in the production of fructose at low temperatures. Journal of Bioscience and Bioengineering. 2015;**119**:267-274

[16] Li Y et al. Overexpression of the endo-inulinase gene from Arthrobacter sp. S37 in *Yarrowia lipolytica* and characterization of the recombinant endo-inulinase. Journal of Molecular Catalysis B: Enzymatic. 2012;**74**:109-115

[17] Gao L et al. Inulinase-producing marine yeasts: Evaluation of their diversity and inulin hydrolysis by their crude enzymes. Microbial Ecology. 2007;**54**:722-729

[18] Rawat HK et al. Biotechnological potential of microbial inulinases:

Recent perspective. Critical Reviews in Food Science and Nutrition. 2017;**57**:3818-3829

[19] Zhang L et al. Inhibition of glucose on an exoinulinase from *Kluyveromyces marxianus* expressed in Pichia pastoris. Process Biochemistry. 2005;**40**:1541-1545

[20] Kim CH, Rhee SK. Fructose production from Jerusalem artichoke by inulinase immobilized on chitin. Biotechnology Letters. 1989;**11**:201-206

[21] Gill PK et al. Comparative analysis of thermostability of extracellular inulinase activity from aspergillus fumigatus with commercially available (Novozyme) inulinase. Bioresource Technology. 2006;**97**:355-358

[22] Gupta AK et al. Fructose and inulinase production from waste *Cichorium intybus* roots. Biological Wastes. 1989;**29**:73-77

[23] Treichel H et al. A review on the production and partial characterization of microbial inulinases. Global Journal of Biochemistry. 2011. Available from: http://scholar.google.com/scholar?hl=en&btnG=Search&q=intitle:A+review+on+the+production+and+partial+characterization+of+microbial+inulinases#0

[24] Singh P, Gill PK. Production of inulinases: Recent advances. Food Technology and Biotechnology. 2006;**44**:151-162

[25] Lane MM, Morrissey JP. *Kluyveromyces marxianus*: A yeast emerging from its sister's shadow. Fungal Biology Reviews. 2010;**24**:17-26

[26] Couto SR, Sanromán MÁ. Application of solid-state fermentation to food industry—A review. Journal of Food Engineering. 2006;**76**:291-302

[27] Kamble PP et al. Statistical optimization of process parameters for inulinase production from Tithonia weed by Arthrobacter mysorens strain no.1. Journal of Microbiological Methods. 2018;**149**:55-66

[28] Ayyachamy M et al. Production of inulinase by Xanthomonas campestris pv phaseoli using onion (*Allium cepa*) and garlic (*Allium sativum*) peels in solid state cultivation. Letters in Applied Microbiology. 2007;**45**:439-444

[29] Kango N. Production of inulinase using tap roots of dandelion (*Taraxacum officinale*) by aspergillus Niger. Journal of Food Engineering. 2008;**85**:473-478

[30] Sheng J et al. Use of response surface methodology for optimizing process parameters for high inulinase production by the marine yeast *Cryptococcus aureus* G7a in solid-state fermentation and hydrolysis of inulin. Bioprocess and Biosystems Engineering. 2009;**32**:333-339

[31] Xiong C et al. Optimization of solid-state medium for the production of inulinase by Kluyveromyces S120 using response surface methodology. Biochemical Engineering Journal. 2007;**34**:179-184

[32] Chen QH et al. Optimization of ultrasonic-assisted extraction (UAE) of betulin from white birch bark using response surface methodology. Ultrasonics Sonochemistry. 2009;**16**:599-604

[33] Sharma AD et al. Inulinase production using garlic (*Allium sativum*) powder as a potential substrate in Streptomyces sp. Journal of Food Engineering. 2006;**77**:486-491

[34] Singh RS et al. A panorama of bacterial inulinases: Production, purification, characterization and industrial applications. International Journal of Biological Macromolecules. 2017;**96**:312-322

[35] Cho Y, Yun J. Purification and characterization of an endoinulinase from *Xanthomonas oryzae* No. 5. Process Biochemistry. 2002;**37**:1325-1331

[36] Artyukhov VG et al. Structural and functional properties of inulinases. Ways to regulate their activity. Biophysics (Oxf). 2013;**58**:493-501

[37] Nakamura T et al. Production, purification and properties of an endoinulinase of Penicillium sp. TN-88 that liberates inulotriose. Journal of Fermentation and Bioengineering. 1997;**84**:313-318

[38] Kang SI et al. Purification and properties of an endo-inulinase from an Arthrobacter sp. Biotechnology Letters. 1998;**20**:983-986

[39] Pandey A et al. Recent developments in microbial inulinases its production, properties, and industrial applications. Applied Biochemistry and Biotechnology, Part A, Enzyme Engineering and Biotechnology. 1999;**81**:35-52

[40] Bourgi J et al. Isolation of a *Kluyveromyces fragilis* derepressed mutant Hyperproducer of inulinase for ethanol production from Jerusalem artichoke. Journal of Fermentation Technology. 1986;**64**:239-243

[41] Onsoy T et al. Ethanol production from Jurusalem artichoke by *Zymomonas mobilis* in batch fermentation. Kmitl. 2007;7:55-60

[42] Bakker GJ et al. Sarcoidosis of the liver: To treat or not to treat? The Netherlands Journal of Medicine. 2012;**70**:349-356

[43] Kuhn RC, Filho FM. Purification of fructooligosaccharides in an activated charcoal fixed bed column. New Biotechnology. 2010;**27**:862-869

[44] Oiwa H et al. Acetone-butanol production from dahlia inulin by Clostridium pasteurianum var. I-53. Agricultural and Biological Chemistry. 1987;**51**:2819-2820

[45] Willdenow L. European Pharmacopoeia. 5th ed. Strasbourg: Council of Europe; 2013

[46] Arrizon J et al. Comparison of the water-soluble carbohydrate composition and fructan structures of *Agave tequilana* plants of different ages. Food Chemistry. 2010;**122**:123-130

[47] Bali V et al. Fructo-oligosaccharides: Production, purification and potential applications. Critical Reviews in Food Science and Nutrition. 2015;**55**:1475-1490

[48] Evans RK et al. Development of stable liquid formulations for adenovirus-based vaccines. Journal of Pharmaceutical Sciences. 2004;**93**:2458-2475

[49] Mabel MJ et al. Physicochemical characterization of fructooligosaccharides and evaluation of their suitability as a potential sweetener for diabetics. Carbohydrate Research. 2008;**343**:56-66

[50] Oli MW et al. Evaluation of fructooligosaccharide supplementation of oral electrolyte solutions for treatment of diarrhea. Recovery of the intestinal bacteria. Digestive Diseases and Sciences. 1998;**43**:138-147

[51] Zheng S et al. Nutritional support of pediatric patients with cancer consuming an enteral formula with fructooligosaccharides. Nutrition Research. 2006;**26**:154-162

[52] Seifert S, Watzl B. Inulin and oligofructose: Review of experimental data on immune modulation. The Journal of Nutrition. 2007;**137**:2563-2567

[53] Pierre F et al. Short-chain fructo-oligosaccharides reduce the occurrence

of colon tumors and develop gut-associated lymphoid tissue in min mice. Cancer Research. 1997;**57**:225-228

[54] Mesa MD et al. Antioxidant properties of soy protein-fructooligosaccharide glycation systems and its hydrolyzates. Food Research International. 2008;**41**:606-615

[55] Elyachioui M et al. General properties of extracellular bacterial inulinase. The Journal of Applied Bacteriology. 1992;**73**:514-519

[56] Baron M et al. Difructose anhydride-forming bacterial inulinase II and fructogenic fungal inulinase I: Free and immobilized forms. Applied Biochemistry and Biotechnology, Part A, Enzyme Engineering and Biotechnology. 1996;**57-58**:605-615

[57] Kim DH et al. Production of inulo-oligosaccharides using endo-inulinase from a pseudomonas sp. Biotechnology Letters. 1997;**19**:369-371

[58] Drent WIMJ et al. A thermophilic anaerobic bacterium isolated from various habitats. Mycologia. 1991;**57**:455-462

[59] Drent WJ, Gottschal JC. Fermentation of inulin by a new strain of clostridium thermoautotrophicum isolated from dahlia tubers. FEMS Microbiology Letters. 1991;**78**:285-292

[60] Selvakumar P, Pandey A. Solid state fermentation for the synthesis of inulinase from staphylococcus sp. and *Kluyveromyces marxianus*. Process Biochemistry. 1999;**34**:851-855

[61] Efstathiou I et al. A study of inulinase activity in the clostridium acetobutylicum strain ABKn8. Applied Microbiology and Biotechnology. 1986;**25**:143-149

[62] Allais J-J et al. Isolation and characterization of bacterial strains with inulinase activity. Applied and Environmental Microbiology. 1986;**52**:1086-1090

[63] Of P et al. Biotechnology letters. Biotechnology Letters. 1979;**1**:102-102

[64] Rouwenhorst RJ et al. Production, distribution, and kinetic properties of Inulinase in continuous cultures of *Kluyveromyces marxianus* CBS 6556. Applied and Environmental Microbiology. 1988;**54**:1131-1137

[65] Kochhar A et al. Purification and immobilisation of inulinase from aspergillus candidus for producing fructose. Journal of the Science of Food and Agriculture. 1999;**79**:549-554

[66] Guiraud JP et al. Isolation of a respiratory-deficient Kluyveromyces fragilis mutant for the production of ethanol from Jerusalem artichoke. Biotechnology and Bioengineering. 1987;**29**:850-858

[67] Xiao R et al. Purification and some properties of endoinulinase from *Chrysosporium pannorum*. Journal of Fermentation and Bioengineering. 1989;**67**:244-248

[68] Azhari R et al. Purification and characterization of endo- and exo-lnulinase. Biotechnology and Applied Biochemistry. 1989;**11**:105-117

[69] Tsang EWT, GrootWassink JWD. Stability of exo-inulase production on lactose in batch and continuous culture of a *Kluyveromyces fragilis* hyperproducing mutant. Enzyme and Microbial Technology. 1988;**10**:297-301

[70] Rouwenhorst RJ et al. Production and localization of β-fructosidase in asynchronous and synchronous chemostat cultures of yeasts. Applied and Environmental Microbiology. 1991;**57**:557-562

[71] Passador-Gurgel GC et al. Application of a microtitre reader system to the screening of inulinase nulinase-producing yeasts. Applied Microbiology and Biotechnology. 1996;**45**:158-161

[72] Cruz-guerrero A et al. Marxianus CDBB-L-278: A wild inulinase hyperproducing. Strain. 1995;**80**:159-163

[73] Paper O. DaGwL k | lzIfs k | lzIf0f tflnd ; xof] uL k ' l: tsf. 1994. DOI: 10.1007/BF00173914

[74] Fructanohydrolase F. Production of inulinase (/? - by KLUYVEROMYCES. Chemical Engineer. 1986;**50**:1085-1087

[75] Anil K, Gill A. Revised 20th may. 1994;**16**:733-734

[76] Nakamura T et al. Continuous production of fructose syrups from inulin by immobilized inulinase from *Aspergillus niger* mutant 817. Journal of Fermentation and Bioengineering. 1995;**80**:164-169

[77] Xiao R et al. Purification and characteristics of two exoinulinases from *Chrysosporium pannorum*. Journal of Fermentation and Bioengineering. 1989;**67**:331-334

[78] Rawat HK et al. Production and properties of inulinase from Penicillium sp. NFCC 2768 grown on inulin-rich vegetal infusions. Biocatalysis & Biotransformation. 2015;**33**:61-68

Chapter 13

Plastics Polymers Degradation by Fungi

José Maria Rodrigues da Luz,
Marliane de Cássia Soares da Silva,
Leonardo Ferreira dos Santos
and Maria Catarina Megumi Kasuya

Abstract

The studies on plastic degradation are very important for the development of biodegradable plastics, and for reduction of pollution, since plastic waste can remain in the environment for decades or centuries. We have showed the degradation of oxo-biodegradable plastic bags and green polyethylene by *Pleurotus ostreatus*. This fungus can also produce mushrooms using these plastics. The plastic degradation was possibly by three reasons: (a) presence of pro-oxidant ions or plant polymer, (b) low specificity of the lignocellulolytic enzymes, and (c) the presence of endomycotic nitrogen-fixing microorganisms. In this chapter, the plastic bags' degradation by abiotic and microbial process using the exposure to sunlight and the use of a white-rot fungus will described. The physical, chemical, and biological alterations of plastic were analyzed after each process of degradation. The degradation of plastic bags was more effective when the abiotic and biotic degradations were combined.

Keywords: oxo-biodegradable, green polyethylene, *Pleurotus* sp., plastic bag, sunlight, landfills

1. Introduction

About 800 million metric tons (Mt) of plastics were produced worldwide in the last 67 years, and 79% of this production is accumulated in the environment [1]. According to these authors, in 2050 is estimated an accumulation of about 12,000 Mt in landfills or in the natural environment that represent an annual accumulation of ~339 Mt. Therefore, the development of efficient degradation process is very important to avoid this annual accumulation.

The human population on the Earth in 2050 will be about 10 billion people [2]. Thus, adequate disposal of plastics wastes is important for the maintenance of the natural resources to supply this population. Furthermore, the development of degradable plastics is necessary to prevent the accumulation of plastic waste in the ocean [3].

In Brazil, the National Solid Waste Policy [4] establishes the selective collection, separation of solid waste, recycling and the shared responsibility for the appropriate management of these wastes among manufacturers, distributors, consumers and the government. However, in these 7 years of law it has observed satisfactory results only in selective collection. In 2014, with the deadline for replacing dumps to landfills, new deadlines for 2021 are in discussion in the National Congress [5].

Our results of biotic and fungal degradation of oxo-biodegradable plastics and green polyethylene could contribute for development of a process of degradation of these residues using white rot fungi. These microorganisms can grow under adverse temperature, nutrient and moisture conditions that facilitate composting and fermentation processes.

The plastics polymers degradation is analyzed by alterations in mechanical, optical or electrical characteristics, cracking, fission, corrosion, discoloration, phase separation, chemical transformations and formation of new functional groups after degradation process [6].

Unlike of the petroleum-derived synthetic polymers, the biodegradable plastics polymers, when discarded in the environment, can be degraded by non-biological and biological processes [7]. Exposure to ultraviolet light, thermal heating, and treatment with acidic or basic substances function as term initiators or photo-oxidation of polyethylene [6]. After this oxidation fragments of polyethylene are degraded by action of microbial enzymes [7].

Oxo-biodegradable or d2W plastics are polymers that contain a pro-oxidant additive to accelerate photo or thermo-oxidation [8, 9]. So, these polymers when exposed to ultraviolet light or at high temperatures are cleaved in low molecular mass compounds that are assimilated by microorganisms [8]. Several studies have shown the biodegradable plastics degradation, after exposed to ultraviolet light or heat, by bacteria and fungi [10–16].

The plastic bags of green polyethylene are produced using low-density polyethylene (LDPE) and green polymers obtain of sugarcane [17]. We have showed the green polyethylene degradation by *Pleurotus ostreatus* PLO6 [15]. However, little information regarding to the biodegradation of these bags is available.

The microbial enzymes, such as depolymerase, esterase and lignolytic ones, that cleave the polymers in small chain compounds, may be involved in the plastics degradation [6, 13, 18, 19]. Thus, white rot fungi have a great potential, because they are enzymes producers and have shown their ability for treatment of industrial waste [20–22].

The white rot, *P. ostreatus*, is a potent degrader of lignin, cellulose and hemicellulose, which lives as saprophyte in wood. This fungus has also been used in the bioconversion of agricultural residues, in biodegradation of organic pollutants, xenobiotics, and industrial effluents, in the cellulose bleaching and production of food and enzymes [23–25]. We showed that *P. ostreatus* PLO6 are capable to degrade oxo-biodegradable plastics and green polyethylene [13–15]. Furthermore, this fungus form edible mushroom that is source of proteins, fibers, minerals and carbohydrates.

Thus, in this chapter described the plastics bags degradation, by abiotic and microbial process, using the exposure to sunlight and *P. ostreatus*.

2. Methods and Results

The degradation of two plastic polymers used in the production of supermarket plastic bags was evaluated (**Figure 1**, I). The oxo-biodegradable and green polyethylene polymers were submitted the abiotic and biotic degradation (**Figure 1**, II). The oxo-biodegradable bags contain titanium oxide as pro-oxidant additive and low-density polyethylene [14].

The abiotic degradation of the plastic bags was the exposure to sunlight up to 120 days (**Figure 1**, III). This exposure was in the summer time in a green house. In this season, the sunlight is from 6:00 am to 5:00 pm.

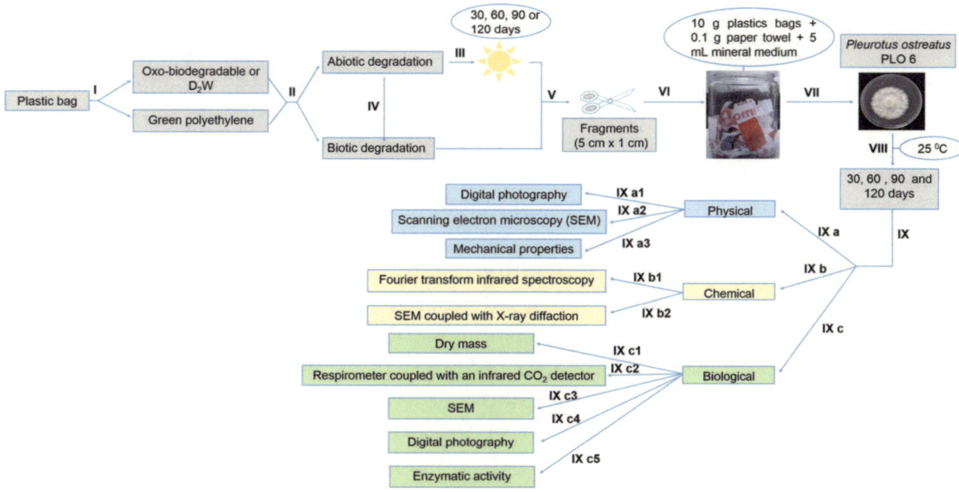

Figure 1.
Steps and technics used for monitoring the abiotic and biotic (fungal) degradation of oxo-biodegradable e green polyethylene plastics bags.

For the biotic degradation (**Figure 1**, IV at VIII) the plastic polymers without (**Figure 1**, II) or with the exposure to sunlight (**Figure 1**, IV) was used *P. ostreatus* PLO6 (GenBank accession number KC782771). These polymers were cut in fragments of 5 cm^2 (**Figure 1**, V) and placed in a glass flask (100 mL) containing paper towel fragments (5–10 cm^2) and mineral medium (**Figure 1**, VI). The proportion of plastics and paper towel was of 99:1.

In each glass flask fours discs of agar (6–8 mm) containing the mycelium of *P. ostreatus* PLO6 were inoculated (**Figure 1**, VII). This fungus was cultivated in 20 mL of potato dextrose lignin (0.1%) agar (PDLA) for 15 days. The initial inoculum was obtained from the collection of the Department of Microbiology of Universidade Federal de Viçosa. The stock culture is maintained on PDLA at 4°C.

After inoculation the glass flask were incubated at 25°C for 30, 60, 90 and 120 days (**Figure 1**, VIII).

The alterations in plastic polymers (**Figure 1**, IX) after each time of incubation were performed (**Figure 1**, III, IV at VIII). These alterations were compared with analysis done before of the exposure to sunlight.

Physical alterations (**Figure 1**, IX a), such as wrinkles on the surface, formation of holes and cracks, crumbling, discoloration, were performed by digital photograph and scanning electron microscopy (SEM) with a magnification of 50,000 (**Figure 1**, IX a2). Mechanical properties, such as, energy at break and load at tensile strength were made in universal testing equipment (Instron model 3367) (**Figure 1**, IX a3).

Chemical changes (**Figure 1**, IX b) by Fourier transform infrared spectroscopy (FTIR) (**Figure 1**, IX b1) and SEM coupled with X-ray diffraction (**Figure 1**, IX b2) were determined. These alterations were the disappearance or formation of new functional groups in spectrum of FTIR with scanning of 500 at 4000 cm^{-1} wavenumbers and the decrease in-oxidant additive concentration by spectrum of X-ray diffraction.

The mycelial growth (**Figure 1**, IX c), the main agent of the biological alterations, was evaluated by dry mass (**Figure 1**, IX c1), respiratory activity (**Figure 1**, IX c2) determined without interruption for 120 days of incubation, electronic micrograph (**Figure 1**, IX c3), digital photography (**Figure 1**, IX c3) and lignocellulolytic enzymes activity (**Figure 1**, IX c5).

The capacity of *P. ostreatus* to produce mushrooms in plastics waste may be evaluated under the same growing conditions, steps and procedures shown in **Figure 1**. However, for mushrooms formation it is need, after the mycelial growth (about 20 days), a thermal shock that can be performed by reducing the incubation temperature to 4°C for 24 h and returning to 25°C. During the mushrooms growth, the flasks should be kept in a place at 18 ± 2°C and a relative humidity of 80%.

A total of 240 days were the time applied to degradation of the plastic bags, being 120 days of exposure to sunlight and 120 days of fungal incubation. According to the manufacturer, depending on environmental conditions, for example, the exposure to oxygen and outdoor element, oxo-biodegradable plastic bags decompose within a maximum period of 18 months after disposal [26, 27]. They also add that in only 121 days the biodegradability index of d2W plastics was 88.86% [28]. Our time of abiotic degradation is the same those used for calculating the biodegradability index and corresponds to ¼ of the required time for the decomposition of these bags. However, after 4 months of exposure to sunlight we did not observe any fragmentation of the plastic bags, only the appearance of small cracks and the bleaching of the film were observed (**Figure 2**). Da Luz et al. [14, 15] also showed changes in mechanical properties of oxo-biodegradable and green polyethylene after 120 days of exposure to sunlight. According to them, this time of exposure is insufficient for other physical or chemical changes, concluding that the mechanical properties alterations, such as the reduction of breaking energy and elasticity facilitated the fungal colonization of plastic waste. The chemical and physical changes in the low-density polyethylene (LDPE) was observed after pretreated of the LDPE sheets with low discharge plasma (O_2, 3.0×10^{-2} mbar, 600 V) for 6 minutes [29, 30]. According to authors, this pretreated was important by plastics biodeterioration by *P. ostreatus*.

In a new experiment, we observed a fragmentation of oxo-biodegradable plastics bags after 21 months of exposure to sunlight (**Figure 3**). The control samples were cut with a scissors (**Figure 3A**), but after that exposure to sunlight, it was no longer possible to cut the bags. These plastics were easily fragmented using the hand, resulting in a powder (**Figure 3B, C**). This result shows that there are needed more than 18 months of exposure to sunlight to completely degradation of the plastics. However, this result is promising, shows the ability of abiotic degradation of these bags and enables new testing using these bags with exposure to sunlight in a period equal to or greater than 18 months and inoculation of microorganisms to complete degradation of the remaining polymers. Degradation analysis with *P. ostreatus* has not been performed in this experiment.

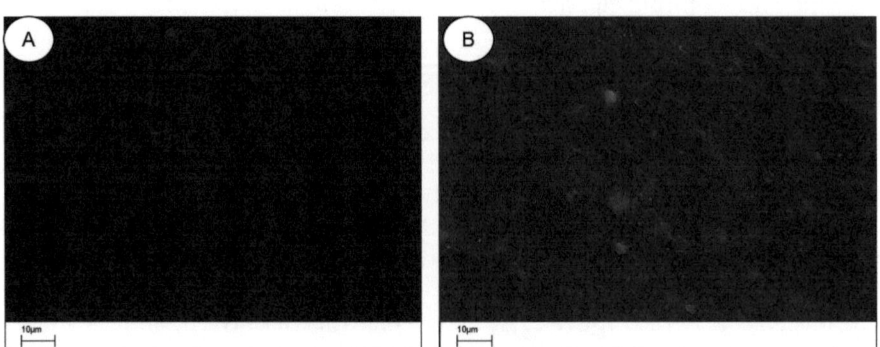

Figure 2.
Scanning electron micrograph of oxy-biodegradable plastics before (A) and after 120 days of exposure to sunlight (B).

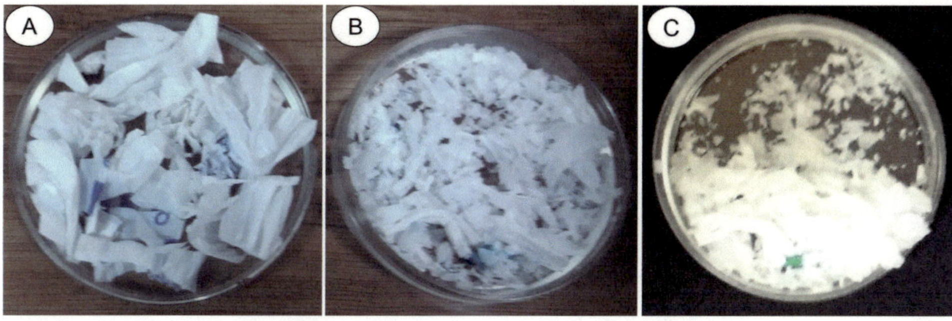

Figure 3.
Oxo-biodegradable plastic bags before (A) and after 21 months of exposure to sunlight (B and C). The letter C shows the fragmented plastic.

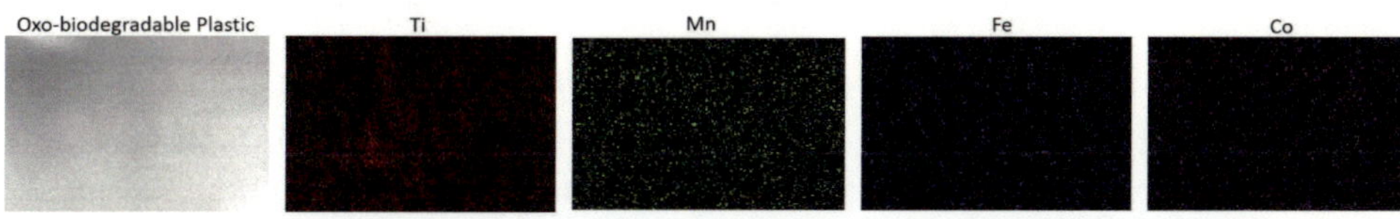

Element	Intensity (c/s)	Error	Concentration (% wt)
Si	12.65	2.461	9.402
Ti	76.30	2.946	57.847
Mn	1.45	1.344	1.646
Fe	0.84	1.260	1.027
Co	1.07	1.158	1.494
Cu	7.46	1.253	14.974
Zn	4.78	1.083	11.556
Cd	1.62	2.074	2.055

Figure 4.
Spatial distribution and relative concentration of the main elements found in oxo-biodegradable bags. Analysis carried out by a scanning electron microscope coupled to the X-ray diffraction detector.

The oxo-biodegradable polyethylene degradation, assessed by carbonyl index, was observed through exposure to sunlight, up to 90 days, in soil with of moisture and pH control [28]. However, these authors concluded that the polyethylene films without pro-oxidant additive had greater structural and superficial modifications, than the films with the additive. Thus, action of the pro-oxidants by the effect of sunlight depends on e conditions and time of exposure to sunlight.

In the plastic bags the presence of titanium was identified, a component of the pro-oxidant additive (**Figure 4**). This element presents a higher relative concentration than the other elements analyzed and it is uniformly distributed on the surface of the bags. This homogeneous distribution was also observed to manganese, iron and cobalt (**Figure 4**). Furthermore, with the exception of titanium and cadmium, the other elements analyzed are important for fungal metabolism (**Figure 4**). These micronutrients may be elicitors or enzyme cofactors. Thus, the presence of these elements may also have contributed to the *P. ostreatus* growth on the surface of plastic bags.

Mycelial growth of *P. ostreatus* was observed on the surface of the paper towel (**Figure 5A**) and the plastic waste (**Figure 5B**). This figure shows an example of mycelial growth in oxo-biodegradable plastics after 30 (**Figure 5A**) and 90 days

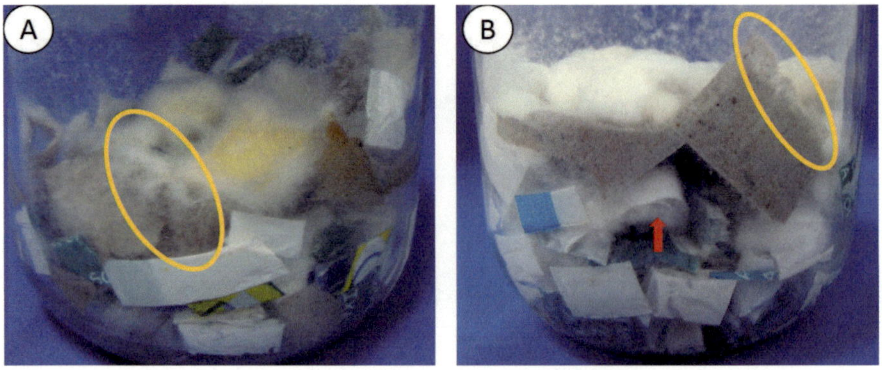

Figure 5.
Mycelial growth of Pleurotus ostreatus *after 30 days of incubation in paper towel and oxo-biodegradable plastic bags after 30 (A) and 90 (B) days of exposure to sunlight. The arrow and circle show mycelial growth.*

(**Figure 5B**) of exposure to sunlight and 30 days incubation with the fungus. The yellow circle and the red arrow show, respectively, the mycelial growth on the paper towel and plastic. This paper was added in the culture medium to retain moisture and to be an inducer of fungal growth and stimulates the synthesis of lignocellulolytic enzymes that degrades the paper itself and the plastic (**Figure 5**).

The respiratory activity of *P. ostreatus* was influenced by time exposure to sunlight. This result confirms that the physical changes caused by sunlight contributed to the fungal growth in plastic bags. Furthermore, we did not observe reduction of this activity until 90 days of incubation showing a cellular activity for a long period. The activity of lignocellulolytic enzymes, like laccase, cellulase and xylanase, during 45 days of incubation of *P. ostreatus* in oxo-biodegradable plastics was observed [13].

The *P. ostreatus* growth on the surface of oxo-biodegradable plastic was also observed by SEM (**Figure 6**). In this micrograph, the red arrows show hyphae in plastic waste after 60 days of exposure to sunlight and 30 days of incubation. Da Luz et al. [13–15] showed the formation of mycelium on the surface of d2w plastic and green polyethylene with different time of exposure to sunlight and of fungal incubation. They also reported the morphological characteristics of the mycelia of *P. ostreatus* PLO6.

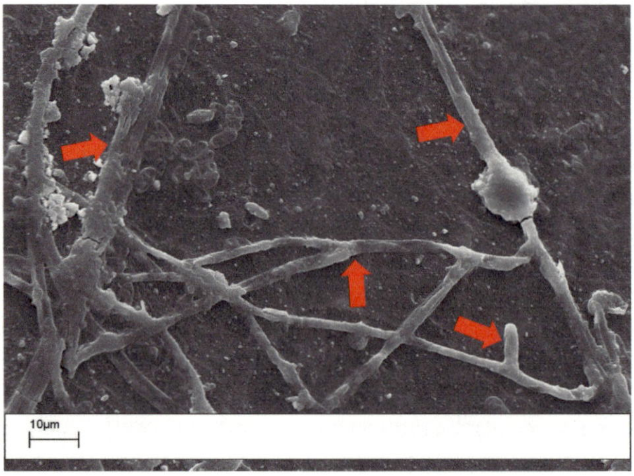

Figure 6.
Scanning electron micrograph of oxo-biodegradable plastics after 60 days of exposure to sunlight and 30 days of incubation with Pleurotus ostreatus. *The arrows show the hyphae.*

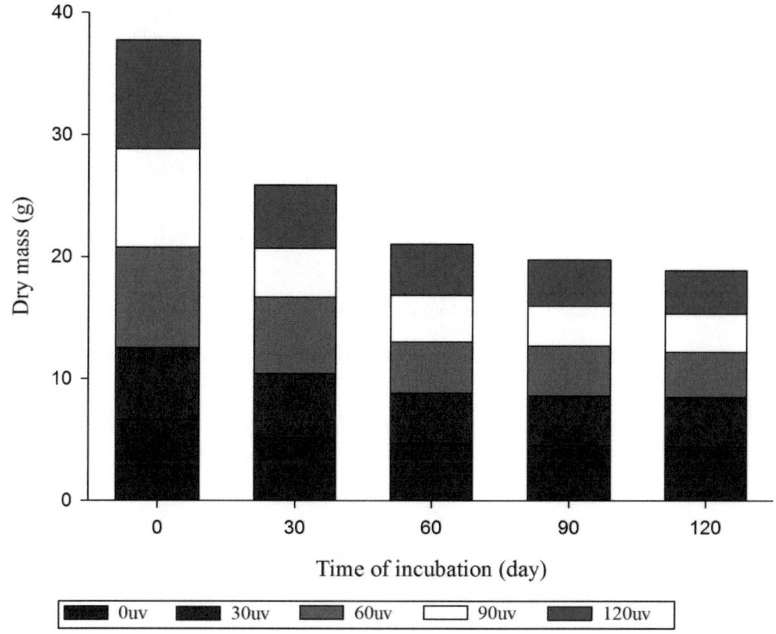

Figure 7.
Loss of dry mass plastic after inoculated with Pleurotus ostreatus *in oxo-biodegradable plastic bags before (0uv) or after 30, 60, 90 and 120 days of exposure to sunlight.*

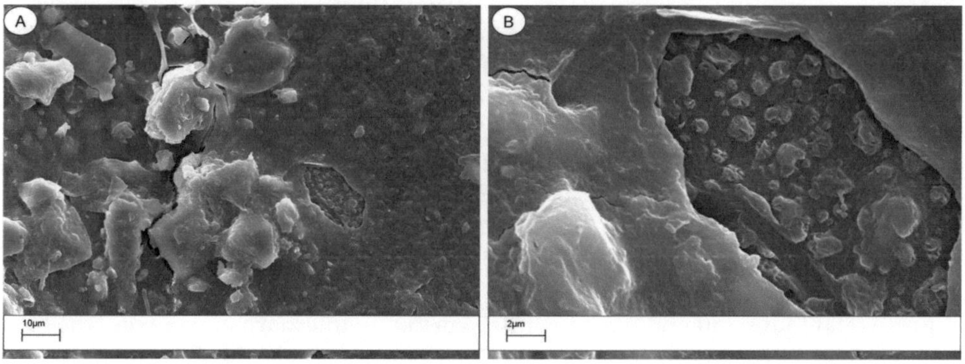

Figure 8.
Scanning electron micrograph of oxo-biodegradable plastics bags after 30 days of exposure to sunlight and 30 days of incubation with Pleurotus ostreatus. *Micrograph without (A) and with a scale of 100 folds (B).*

The loss of plastic dry mass was influenced by the time of exposure to sunlight and fungal incubation (**Figure** 7). Fungal growth was lower in plastic polymers without exposure to sunlight than in others with different time of exposure to sunlight. This result shows that *P. ostreatus* can grow in plastic waste without or with exposure to sunlight. However, this exposure facilitates the fungal growth, as shown by Da Luz et al. [14, 15]. Thus, the combination of abiotic and biotic processes shows to be more efficient in the oxo-biodegradable plastic and green polyethylene degradation. In addition, the presence of other carbon sources from marine sediments and lack of abiotic degradation as the initiator were the main factors of the lack of biodegradation of polyethylene and biodegradable plastic bags after 100 days of incubation with benthic microbes [3].

In this study, we observed the formation of cracks and holes in oxo-biodegradable plastics and green polyethylene after fungal growth (**Figures 8** and **9**).

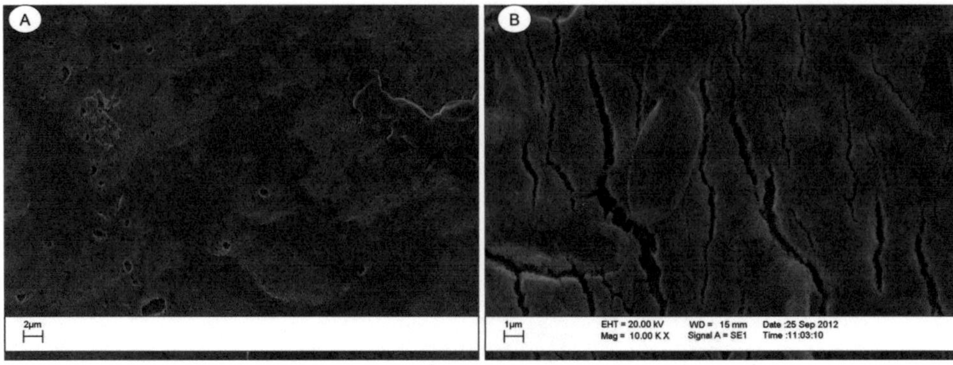

Figure 9.
Scanning electron micrograph of green polyethylene plastics bags after 30 days of exposure to sunlight and 30 days of incubation with Pleurotus ostreatus. *Micrograph without (A) and with scaling of 100 fold (B).*

Comparing the **Figure 2B** and **8** it is observed that these changes in plastic polymers were caused by *P. ostreatus* growth. The fungi, *Penicillium oxalicum*, *Penicillium chrysogenum*, *Myceliophthora* sp., *Phanerochaete chrysosporium*, and *Trametes versicolor*, also exhibit the ability to degrade polyethylene [16, 31, 32].

In a simulation according to ASTM G160–03 of polyethylene films degradation with and without pro-oxidant additive through the exposure to sunlight on the soil, different genera or microbial groups, *Geothrichum* spp., *Mucor* spp., *Rhizopus* spp., *Thichoderma* spp., *Penicillium* spp., *Aspergillus* spp. and Zygomycota, were identified [28]. These results indicate that (1) the plastic films did not alter or inhibit the development of the microbial community of the soil, since these microorganisms are part of the natural microbial community of the soil or (2) the growth of these microorganisms was due to the use of the films as source of carbon and energy. According to the authors, the biodegradation of polyethylene without or with pro-oxidant additive can be shown by the adhesion and surface erosion of the films, microbial colonization and presence of fruiting bodies and hyphae on the plastic surface.

The **Figures 8** and **9** show the plastic degradation with 30 days of exposure to sunlight and 30 days of incubation with different scale enlargements. According to Da Luz et al. [14], the low specificity of the lignocellulolytic enzymes and presence of pro-oxidant ions and endomycotic nitrogen-fixing microorganisms were the main reasons for the biotic degradation of oxo-biodegradable plastics. Gómez-Méndez et al., [29] observed activities of laccase, manganese peroxidase (MnP) and lignin peroxidase during *P. ostreatus* growth in plasma pretreated Low-density polyethylene (LDPE) sheets. These authors showed that LDPE biodeterioration was due to activities of these fungal enzymes. Furthermore, the LDPE after mycelia fungal may be used by biochar production [30].

The laccase produced by the fungus *Cochliobolus* sp. isolated from plastic dumped soils showed capacity for polyvinyl chloride degradation [33]. This enzyme produced by *Myceliophthora* sp. was also able to degrade polyethylene [30] and polyurethane [34]. Manganese peroxidase from white rot fungi, *Phanerochaete chrysosporium*, is involved in the degradation of nylon and polyethylene [35]. The laccase and manganese peroxidase activity of *Penicillium* sp. are responsible by degradation of polyethylene [31, 36] and natural rubber [37]. These studies confirm the low specificity of these enzymes to the substrate.

After 120 days exposure to sunlight, no changes in the FTIR spectrum of oxo-biodegradable plastics was observed. This result shows that pro-oxidant oxidation by sunlight was not sufficient for cleavage of the polymer chain or it there is no oxidation thereof. However, in a previous study, a reduction of the relative concentration of titanium on the surface of oxo-biodegradable plastics wastes after exposure to sunlight was observed [14]. According to these authors, the oxidation of the pro-oxidant may have occurred initially by sunlight and then by co-metabolism with the extracellular fungal enzymes. The authors concluded that the presence of this pro-oxidant proved to be important to cause the breakage of this chain in fragments that were used as a source of carbon and energy by fungus.

In polyethylene green, which contain none pro-oxidant additive, no changes in the FTIR spectrum after exposure to sunlight was observed.

The formation of bands of the bonds oxygen-hydrogen and carbon-hydrogen at 3500–3000 cm^{-1} and carbon–oxygen and ether or peroxide at 1500–1000 cm^{-1} were the main changes in the FTIR spectra observed in plastic waste after *P. ostreatus* growth. The carbon-hydrogen bond band may be evidence of the fragmentation of the polyethylene chain. The other bands observed indicate that an oxidation has occurred, which may have contributed to the fungal colonization in the plastic polymers (13–15).

In studies on the plastics degradation for *P. ostreatus*, the authors also observed chemical and physical changes similar to the observed in our study [29, 30].

The intensity of the degradation was higher in the green polyethylene than in the oxo-biodegradable polymers (**Figures 8** and **9**). The green polyethylene degradation by fungus was possible due to the presence of sugarcane polymers in the composition of the bags, low specificity of the lignocellulolytic enzymes and presence of endomycotic nitrogen-fixing microorganisms. In addition, Da Luz et al. [15] was observed mineralization in green polyethylene with longer times of exposure to sunlight and fungal incubation.

Similar to Da Luz et al. [13], during the time of incubation we also observed the mushrooms formation in the plastic (**Figure 10**). The conversion of plastic waste into fungal biomass and mushrooms would be a very important biotechnological innovation for plastic waste degradation that has been increased by millions of tons in recent years [1, 3, 16] and for environmental sustainability. However, the presence of toxic compounds and heavy metals, and also due to the low productivity and high costs are the main limitations to mushrooms production. Productivity in mushrooms can be increased by altering the composition of substrate, as for example, adding different proportions of agroindustrial residue and plastic.

Figure 10.
Mycelial growth and Pleurotus ostreatus *mushrooms (arrows) formation in substrate containing oxo-biodegradable plastics and paper towels (99: 1 m/m).*

3. Conclusions

The exposure to sunlight up to 120 days is insufficient to initiate degradation of oxo-biodegradable and green polyethylene plastic bags. However, this exposure is important for *P. ostreatus* growth. Therefore, these plastics degradation occurs more efficiently with the combination of abiotic and biotic process. These plastics degradation may be due to the activities of lignocellulolytic enzymes that are produced during fungal growth on the plastics sheets.

Acknowledgements

The authors are grateful to the Coordenação de Aperfeiçoamento de Pessoal de Nível Superior (Capes), Conselho Nacional de Desenvolvimento Científico e Tecnológico (CNPq), Fundação de Amparo à Pesquisa do Estado de Minas Gerais (FAPEMIG), Fundação Arthur Bernardes (FUNARBE) and Núcleo de Microscópia e Microanálise da UFV.

Conflict of interest

The authors declare no conflict of interest.

Author details

José Maria Rodrigues da Luz[1*], Marliane de Cássia Soares da Silva[2], Leonardo Ferreira dos Santos[3] and Maria Catarina Megumi Kasuya[2]

1 Programa de Pós-Graduação Multicêntrico na Área de Bioquímica e Biologia Molecular, Instituto de Ciências Farmacêuticas, Universidade Federal de Alagoas, Maceió, AL, Brazil

2 Department of Microbiology, Universidade Federal de Viçosa, Viçosa, Minas Gerais, Brazil

3 Centro Universitário FIPMOC, Brazil

*Address all correspondence to: josemarodrigues@yahoo.com.br

References

[1] Geyer R, Jambeck JR, Law KL. Production, use, and fate of all plastics ever made. Science Advances. 2017;**3**:e1700782. DOI: 10.1126/sciadv.1700782

[2] United Nations, Department of Economic and Social Affairs, Population Division. New York, USA: World Population Ageing ST/ESA/SER.A/348; 2013. Available from: https://www.un.org/en/development/desa/population/publications/pdf/ageing/WorldPopulationAgeing2013.pdf

[3] Nauendorf A, Krause S, Bigalk N, Gorb EV, Gorb SN, Haeckel M, et al. Microbial colonization and degradation of polyethylene and biodegradable plastic bags in temperate fine-grained organic-rich marine sediments. Marine Pollution Bulletin. 2016;**103**:168-178

[4] Brazil. Law number 12.30, Brasília, Brazil; 2010. Available from: http://www.mma.gov.br/port/conama/legiabre.cfm?codlegi=636 [Accessed: 12 June 2019]

[5] Brazil. Law project 2289/15. Brasília, Brazil; 2015. Available from: http://www.camara.gov.br/proposicoesWeb/fichadetramitacao?idProposicao=1555331 [Accessed: 12 June 2019]

[6] Shah AA, Hasan F, Hameed A, Ahmed S. Biological degradation of plastics: A comprehensive review. Biotechnology Advances. 2008;**26**:246-265. DOI: 10.1016/j.biotechadv.2007.12.005

[7] Gross RA, Kalra B. Biodegradable polymers for the environment. Science. 2002;**297**:803-807. DOI: 10.1126/science.297.5582.803

[8] Koutny M, Sancelme M, Dabin C, Pichon N, Delort AM, Lemaire J. Acquired biodegradability of polyethylenes containing pro-oxidant additives. Polymer Degradation and Stability. 2006;**91**:1495-1503. DOI: 10.1016/j.polymdegradstab.2005.10.007

[9] Scott G. Oxobiodegradable plastic. Bioplastics Magazine. 2009;**4**:28-30. Available from: http://www.bioplasticsmagazine.com/bioplasticsmagazine-wAssets/docs/article/0905_p30_bioplasticsMAGAZINE.pdf

[10] Yamada-Onodera K, Mukumoto H, Katsuyaya Y, Saiganji A, Tani Y. Degradation of polyethylene by a fungus, *Penicillium simplicissimum* YK. Polymer Degradation and Stability. 2001;**72**:323-327. DOI: 10.1016/S0141-3910(01)00027-1

[11] Bonhomme S, Cuer A, Delort AM, Lemaire J, Sancelme M, Scott G. Environmental biodegradation of polyethylene. Polymer Degradation and Stability. 2003;**81**:441-452

[12] Ojeda TFM, Dalmolin E, Forte MMC, Jacques RJS, Bento FM, Camargo FAO. Abiotic and biotic degradation of oxo-biodegradable polyethylenes. Polymer Degradation and Stability. 2009;**94**:965-970

[13] Da Luz JMR, Paes SA, Nunes MD, da Silva MCS, Kasuya MCM. Degradation of oxo-biodegradable plastic by *Pleurotus ostreatus*. PLoS One. 2013;**8**:e69386

[14] Da Luz JMR, Paes SA, Bazzolli DMS, Tótola MR, Demuner AJ, Kasuya MCM. Abiotic and biotic degradation of oxo-biodegradable plastic bags by *Pleurotus ostreatus*. PLoS One. 2014;**9**(11):e107438

[15] Da Luz JMR, Paes SA, Ribeiro KVG, Mendes IR, Kasuya MCM. Degradation of green polyethylene by *Pleurotus ostreatus*. PLoS One. 2015;**10**(6):e0126047

[16] Ojha N, Pradhan N, Singh S, Barla A, Shrivastava A, Khatua P, et al. Evaluation of HDPE and LDPE degradation by fungus, implemented by statistical optimization. Nature Scientific Reports. 2017;**7**:39515

[17] Brasken. Green Polyethylene Biopolymer, Innovation Transforming Plastic into Sustainability. 2014. Available from: http://www.braskem.com.br/Portal/Principal/Arquivos/ModuloHTML/Documentos/846/AF_Catalogo_PE%20Verde_2014_ING_site.pdf [Accessed: 15 August 2017]

[18] Gu JD. Microbiological deterioration and degradation of synthetic polymeric materials: Recent research advances. International Biodeterioration & Biodegradation. 2003;**52**:69-91

[19] Sato S, Saika A, Shinozaki Y, Watanabe T, Suzuki K, Sameshima-Yamashita Y, et al. Degradation profiles of biodegradable plastic films by biodegradable plastic-degrading enzymes from the yeast *Pseudozyma antarctica* and the fungus *Paraphoma* sp. B47-9. Polymer Degradation and Stability. 2017;**141**:26-32

[20] Purnomo AS, Mori T, Kamei I, Nishii T, Kondo R. Application of mushroom waste medium from *Pleurotus ostreatus* for bioremediation of DDT-contaminated soil. International Biodeterioration & Biodegradation. 2010;**64**:397-402

[21] Nunes MD, Da Luz JMR, Paes AS, Torres DP, Kasuya MCM. Jatropha seed cake supplementation for improved fungal growth and later use as animal feed. African Journal of Microbiology Research. 2014;**8**:3457

[22] Oliveira SF, Da Luz JMR, Kasuya MCM, Ladeira, LO, Correa AJ. Enzymatic extract containing lignin peroxidase immobilized on carbon nanotubes: Potential biocatalyst in dye decolourization. Saudi Journal of Biological Sciences. 2018;**25**:651-659

[23] Da L, Nunes JMR d, Paes MD, Torres SA, Silva DP, Kasuya M d CS d, et al. Lignocellulolytic enzyme production of *Pleurotus ostreatus* growth in agroindustrial wastes. Brazilian Journal of Microbiology. 2012;**43**:1508-1515

[24] Ergun SO, Urek RO. Production of ligninolytic enzymes by solid state fermentation using *Pleurotus ostreatus*. Annals of Agrarian Science. 2017;**15**:273-277

[25] Purnomo AS, Nawfa R, Martak F, Shimizu K, Kamei I. Biodegradation of Aldrin and Dieldrin by the white-rot fungus *Pleurotus ostreatus*. Current Microbiology. 2017;**74**:320-324

[26] Omniplast. Oxo-Biodegradable. Merchandise Bags... It's All About Green. 2017. Available from: http://www.omniplast.ca/assets/chooseoxobio.pdf [Accessed: 15 July 2019]

[27] Resbrasil. RES d2w™ Biodegradável. 2015. Available from: http://www.resbrasil.com.br/embalagens-plasticas-inteligentes/res-d2w-biodegradavel/ [Accessed: 15 July 2019]

[28] Gomes LB, Klein JM, Brandalise RN, Zeni M, Zoppas BC, Grisa AMC. Study of oxo-biodegradable polyethylene degradation in simulated soil. Materials Research. 2014;**17**(1):121-126

[29] Gómez-Méndez LD, Moreno-Bayona DA, Poutou-Piñales RA, Salcedo-Reyes JC, Pedroza-Rodríguez AM, Vargas A, et al. Biodeterioration of plasma pretreated LDPE sheets by *Pleurotus ostreatus*. PLoS One. 2018;**13**(9):e0203786. DOI: 10.1371/journal. pone.0203786

[30] Moreno-Bayona DA, Gómez-Méndez LD, Blanco-Vargas A, Castillo-Toro A, Herrera Carlosama L,

Poutou-Piñales RA, et al. Simultaneous bioconversion of lignocellulosic residues and oxodegradable polyethylene by *Pleurotus ostreatus* for biochar production, enriched with phosphate solubilizing bacteria for agricultural use. PLoS One. 2019;**14**(5):e0217100. DOI: 10.1371/journal.pone.0217100

[31] Iiyoshi Y, Tsutsumi Y, Nishida T. Polyethylene degradation by lignin-degrading fungi and manganese peroxidase. Journal of Wood Science. 1998;**44**:222-229

[32] Khalil MI, Ramadan NA, Albarhawi RK. Biodegradation of polymers by fungi isolated from plastic garbage and the optimum condition assessment of growth. Journal of Environmental Management. 2013;**1**:33-43

[33] Sumathi T, Viswanath B, Lakshmi AS, DVR SG. Production of laccase by Cochliobolus sp. isolated from plastic dumped soils and their ability to degrade low molecular weight PVC. Hindawi Publishing Corporation, Biochemistry Research International. 2016;**9519527**:1-10

[34] Loredo-Treviño A, García G, Velasco-Téllez A, Rodríguez-Herrera R, Aguilar CN. Polyurethane foam as substrate for fungal strains. Advances in Bioscience and Biotechnology. 2011;**2**:52-58

[35] Shimao M. Biodegradation of plastics. Current Opinion in Biotechnology. 2001;**12**:242-247

[36] Santo M, Weitsman R, Sivan A. The role of the copper-binding enzyme-laccase-in the biodegradation of polyethylene by the actinomycete *Rhodococcus ruber*. International Biodeterioration and Biodegradation. 2013;**84**:204-210

[37] Nayanashree G, Tyhippeswam B. Natural rubber degradation by laccase and manganese peroxidase enzymes of *Penicillium chrysogenum*. International Journal of Environmental Science and Technology. 2015;**12**:2665-2672. DOI: 10.1007/s13762-014-0636-6

Chapter 14

Microorganisms as Biocatalysts and Enzyme Sources

Arturo Cano-Flores, Javier Gómez, Iker S. Escalona-Torres and Benjamín Velasco-Bejarano

Abstract

Microbial-catalyzed biotransformations have considerable potential for the generation of an enormous variety of structurally diversified organic compounds, especially natural products with complex structures like triterpenoids, flavonoids, steroids, steroidal saponins, and sesquiterpenoids. They offer efficient and economical ways to produce semisynthetic analogues and novel lead molecules. Microorganisms such as bacteria and fungi could catalyze *chemo*-, *regio*-, and *stereo*specific hydroxylations of diverse substrates that are extremely difficult to produce by chemical routes. During recent years, considerable research has been performed on the microbial transformation of bioactive compounds, in order to obtain biologically active molecules with diverse structural features. In green chemistry, biotransformations are an important chemical methodology toward more sustainable industrial processes.

Keywords: microorganisms, fungi, bacteria, microbial transformation, natural products, enzymes

1. Introduction

Microbial transformation is regarded as an enzymatic reaction by using the metabolic activities of microorganisms to modify the chemical structures of bioactive substrates for finding the new chemical derivatives with the potent bioactivities and physical-chemical characteristics. It has a number of advantages over chemical synthesis such as higher *stereo*- and *regio*selectivity but is also enantiospecific, allowing the production of chiral products from racemic mixtures. The conditions for biotransformations are mild, and in the majority of cases, they do not require the protection of pre-existing functional groups. Furthermore, some reactions that do not occur when using chemical approaches are easily carried out by microbial transformation. Microbial factories show advantages, for instance, growing rapidly and ease of large-scale production [1–3].

The use of microorganisms may be a highly efficient method of production of these compounds. The reactions involved in biotransformation of organic compounds by whole cells of various microorganisms include oxidation, reduction, hydroxylation, esterification, methylation, demethylation, isomerization, hydrolysis, glycosylation, and hydrogenation [4, 5].

Biotransformation may be carried out with isolated enzyme systems or with intact organism. Although isolated enzyme systems may be more specific and efficient for certain biotransformation, these reactions may involve isolating the enzyme system, and, for some classes of enzyme-catalyzed reaction, a recycling sequence may be required to regenerate the enzyme [6].

Fungi are playing a prominent role in the catalysis of organic compounds and in the production of commercially and industrially important compounds, because of their ability to catalyze novel reactions [7]. Fungi are commonly used in the industry for production of fermented beverages, foods, physiologically active substances, solvents, organic acids, polysaccharides, antibiotics, etc. Of the zygomycota, *Mucor* and *Rhizopus* are commonly used in the industry. *Rhizopus* strains are important in citric acid production. *Mucor* strains make a significant number of important lipases and catalyze the hydroxylation of a wide range of chemical compounds [2–4].

The use of the microbial model offers a number of advantages over the use of animals in metabolism studies, mainly: (1) simple, easy, and can be prepared at low cost; (2) screening for a large number of strains is a simple repetitive process; (3) the large number of metabolites formed allows easier detection, isolation, and structural identification; (4) newer metabolites can be isolated; (5) utilized for synthetic reactions involving many steps; (6) useful in cases where *regio*- and *stereo*specificity is required; (7) maintenance of stock cultures of microorganisms is simpler and cheaper than the maintenance of cell or tissue cultures or laboratory animals; (8) ease of setup and manipulation; and (9) more reliable and reproducible [8, 9].

The objective of this review is to highlight the importance of microorganisms or enzymes isolated from them in the biotransformation process of natural products or xenobiotic compounds, according to green chemistry or white biotechnology.

2. Microbiological transformations of some selected natural products with different microorganisms

2.1 Sesquiterpene lactone

Artemisinin (**1**), a sesquiterpene lactone endoperoxide and an antimalarial drug, is effective against chloroquine-resistant parasites; but its toxicities and low solubility in water hamper its therapeutical use. Studies on modification of **1** through biological and chemical methodologies have been reported to yield more effective and water-soluble derivatives. A wide array of microbial transformations of **1** involve oxidation, reduction, and degradation reactions by different microorganisms, such as *Aspergillus niger*, *A. flavus*, *A. adametzi* (ATCC 10407), *Cunninghamella echinulata, Caenorhabditis elegans, Mucor polymorphous, M. rammanianus, Streptomyces griseus, Penicillium simplicissimum, P. chrysogenum, P. purpuresceus, Pestalotiopsis guepini* (P-8), *Eurotium amstelodami, Trichoderma viride* (T-58), *Saccharomyces cerevisiae, and Pichia pastoris.* Biotransformation of **1** usually includes the processes such as hydroxylation of methyl, methine and methylene groups, deoxidation reactions, hydration and acetylation reactions, epimerization, and breakdown of heterocyclic rings (**Table 1**).

2.2 Triterpene

Ursolic acid (3β-hydroxy-urs-12-en-28-oic acid, UA, **2**), a natural pentacyclic triterpene, is broadly used in food, cosmetics, and biomedical industries. As a ubiquitous constituent in the plant kingdom and the major component of many traditional medicine herbs, ursolic acid remarkably exhibits a lot of biological

Microorganism	Products	Action	Reference
A. niger	3β-hydroxy-4,12-epoxy-1-deoxyartemisinin Artemisinin G 3,13-epoxyartemisinin 4α-hydroxy-1-deoxyartemisinin	Epoxidation, hydroxylation C-3β site Endoperoxide function reduction Breakdown of heterocyclic rings Epoxidation C-3 and C-13 Hydroxylation C-4, endoperoxide function reduction	[10]
A. flavus (MTCC 9167)	14-hydroxyartemisinin Artemisinin G 4α-hydroxydeoxyartemisinin Deoxyartemisinin	Hydroxylation of C-14 site Breakdown of heterocyclic rings Hydroxylation of C-4α site Endoperoxide function reduction	[11, 12]
C. elegans (ATCC 9245)	7β-hydroxy-9α-artemisinin 4α-hydroxy-1-deoxoartemisinin 7β-hydroxyartemisinin 6β-hydroxyartemisinin 7α-hydroxyartemisinin 6β,7α-dihydroxyartemisinin	Hydroxylation C-7β site Epimerization C-9 Hydroxylation C-4α site Hydroxylation C-7β site Hydroxylation C-6β site Hydroxylation C-7α site Hydroxylation C-6β and C-7α sites	[13, 14]
P. simplicissimum	9β-acetoxyartimisinin 9α-hydroxyartemisinin	Acetylation of C-9β site Hydroxylation of C-9α site	[15]
R. stolonifer	Deoxyartemisinin 1α-hydroxyartemisinin 10β-hydroxyartemisinin	Endoperoxide function reduction Hydroxylation of C-1 site	[16]
S. griseus (ATCC 13273)	9-artemisitone 9α-hydroxyartemisinin 9β-hydroxyartemisinin 3α-hydroxyartemisinin	Oxidation of C-9 site Hydroxylation of C-9α site Hydroxylation of C-9β site Hydroxylation of C-3α site	[17]
N. corallina	Deoxyartemisinin	Endoperoxide function reduction	[11]

Table 1.
Products obtained from the biotransformation of artemisinin (1) by different microorganisms.

activities, such as antibacterial, anti-allergic, antioxidative, anti-inflammation, and antitumor activities [18].

Microbial transformation of ursolic acid (**2**) by *Bacillus megaterium* CGMCC 1.1741 yielded five metabolites identified as 3-oxo-urs-12-en-28-oic acid (**3**, 6.2%); 1β,11α-dihydroxy-3-oxo-urs-12-en-28-oic acid (**4**, 13.5%); 1β-hydroxy-3-oxo-urs-12-en-28,13-lactone (**5**, 5.0%); 1β,3β,11α-trihydroxy-urs-12-en-28-oic acid (**6**, 26.9%); and 1β,11α-dihydroxy-3-oxo-urs-12-en-28-O-β-D-glucopyranoside (**7**, 8.6%) [19]. The biotransformation studies of **2** by *Alternaria longipes* AS 3.2875 have led to the isolation of six products of hydroxylation or glycosylation. Their structures were identified as 3-carbonyl-ursolic acid-28-O-β-D-glucopyranosyl ester (**8**), ursolic acid-3-O-β-D-glucopyranoside (**9**), ursolic acid-28-O-β-D-glucopyranosyl ester (**10**), 2α,3β-dihydroxy-ursolic acid-28-O-β-D-glucopyranosyl ester (**11**), 3β,21β-dihydroxy-ursolic acid-28-O-β-D-glucopyranosyl ester (**12**), and 3-O-(β-D-glucopyranosyl)-ursolic acid-28-O-(β-D-glucopyranosyl) ester (**13**). Glycosylation reaction on pentacyclic triterpenoid fulfilled with difficulty in the process of chemical synthesis is facile by microbial transformation [20]. Biotransformation of **2** by *A. alternata* eight metabolites were found to be 2α,3β-dihydroxyurs-12-en-28-oic acid (corosolic acid, **14**), urs-12-en-2α,3β,28-triol (**15**), 3β,23-dihydroxyurs-12-en-28-oic acid (**16**), 2α,3β,23-trihydroxyurs-12-en-28-oic acid (**17**), 2α,3β,23,24-tetrahydroxyurs-12-en-28-oic acid (**18**), 3β,28-dihydroxy-12-ursene (**19**), urs-12-en-3β-ol (**20**), and urs-12-en-2α,3β-diol (**21**). The reduction of the C-28 carboxyl group and hydroxylation at C-2, 23, and 24 are steps in the metabolic pathway of **2** [21].

Biotransformation of UA by *S. racemosum* (3.2500) yielded five metabolites 3β,7β,21β-trihydroxy-urs-12-en-28-oic acid (**22**); 3β,21β-dihydroxy-urs-11-en-28-oic acid-13-lactone (**23**); 1β,3β,21β-trihydroxy-urs-12-en-28-oic acid (**24**); 3β,7β,21β-trihydroxy-urs-1-en-28-oic acid-13-lactone (**25**); and 1β,3β-dihydroxy-urs-12-en-21-oxo-28-oic acid (**26**) which were afforded [22]. Additionally, of the biotransformation of **2** with by *S. racemosum* compounds **27–30** and 11,26-epoxy-3β-21β-dihydroxy-urs-12-en-28-oic acid were obtained (**31**), (**Figure 1**) [23].

The endophytic fungi *Pestalotiopsis microspora* isolated from medical plant *Huperzia serrata* can transform **1** to afforded 3-oxo-15β,30-dihydroxy-urs-12-en-28-oic acid (**32**), 3β,15β-dihydroxy-urs-12-en-28-oic acid (**33**), 3β,15β,30-trihydroxy-urs-12-en-28-oic acid (**34**), and **30** [24].

Microbial transformation of ursolic acid by *Mucor spinosus* AS 3.3450 were isolated and their structures were identified as **9**, **22** and 3β,7β-dihydroxy-ursolic acid-28-ethanone (**35**) (**Figure 1**) [25].

The gum resin *Boswellia serrata* has been used for the treatment of inflammatory and arthritic diseases. Its major active constituents are ursane triterpenoids, which include 11-keto-β-boswellic acid (KBA, **36**), β-boswellic acid (BA), and acetyl-β-boswellic acid (ABA). Microbial transformation **36** by *Cunninghamella blakesleeana* (AS 3.970) yielded ten regioselective transformed products: 7β-hydroxy-11-keto-β-boswellic acid (**37**), 7β,15α-dihydroxy-11-keto-β-boswellic acid (**38**), 7β,16β-dihydroxy-11-keto-β-boswellic acid (**39**), 7β,16α-dihydroxy-11-keto-β-boswellic acid (**40**), 7β,22β-dihydroxy-11-keto-β-boswellic acid (**41**), 7β,21β-dihydroxy-11-keto-β-boswellic acid (**42**), 7β,20β-dihydroxy-11-keto-β-boswellic acid (**43**), 7β,30-dihydroxy-11-keto-β-boswellic acid (**44**), 3α,7β-dihydroxy-11-oxours-12-en,24,30-dioic acid (**45**), and 3α,7β-dihydroxy-30-(2-hydroxypropanoyloxy)-11-oxours-12-en, 24-oic acid (**46**). Bioconversion of **36** with *Bacillus megaterium* based on a recombinant cytochrome P450 system yielded *regio*- and *stereo*selective 15α-hydroxylation (**47**) of substrate (**Figure 2**) [26].

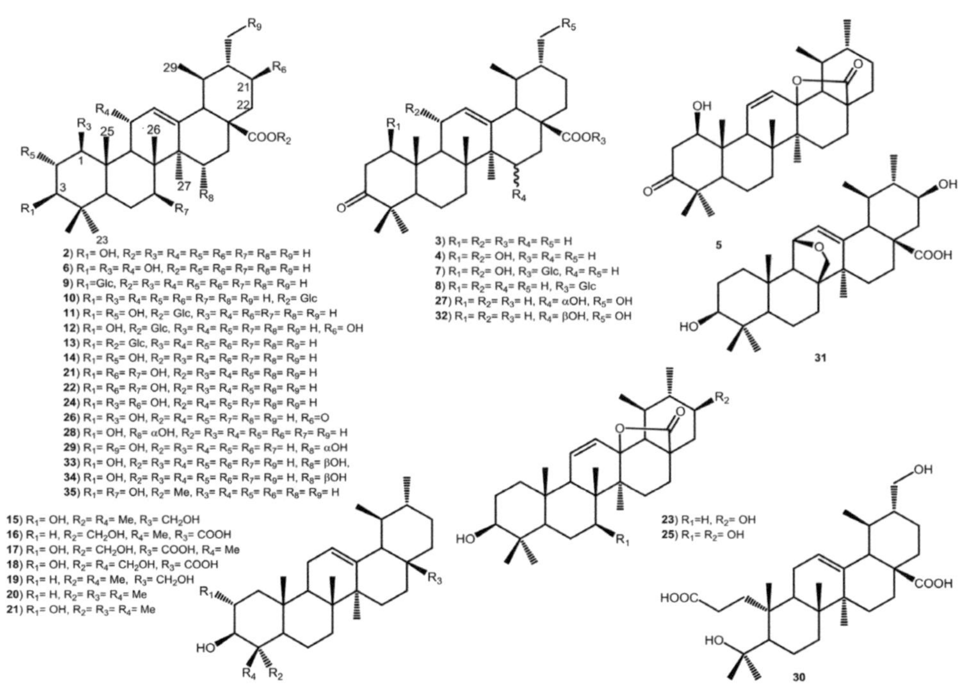

Figure 1.
Biotransformation products of ursolic acid (2).

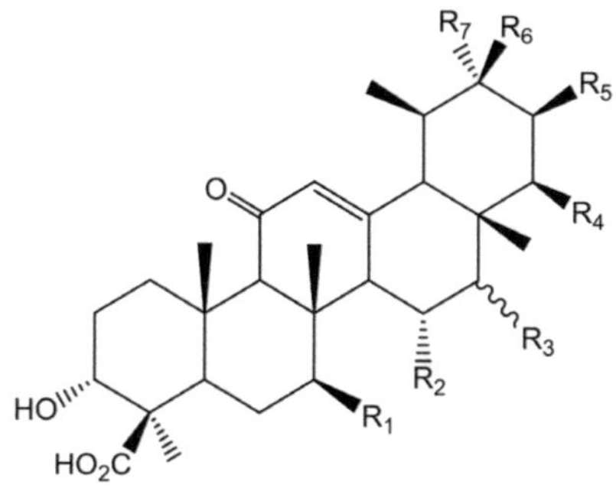

36) $R_1 = R_2 = R_3 = R_4 = R_5 = R_6 = H$, $R_7 = CH_3$
37) $R_1 = OH$, $R_2 = R_3 = R_4 = R_5 = R_6 = H$, $R_7 = CH_3$
38) $R_1 = R_2 = OH$, $R_3 = R_4 = R_5 = R_6 = H$, $R_7 = CH_3$
39) $R_1 = OH$, $R_3 = \beta OH$, $R_2 = R_4 = R_5 = R_6 = H$, $R_7 = CH_3$
40) $R_1 = OH$, $R_3 = \alpha OH$, $R_2 = R_4 = R_5 = R_6 = H$, $R_7 = CH_3$
41) $R_1 = R_4 = OH$, $R_2 = R_3 = R_5 = R_6 = H$, $R_7 = CH_3$
42) $R_1 = R_5 = OH$, $R_2 = R_3 = R_4 = R_6 = H$, $R_7 = CH_3$
43) $R_1 = R_6 = OH$, $R_2 = R_3 = R_4 = R_5 = H$, $R_7 = CH_3$
44) $R_1 = OH$, $R_2 = R_3 = R_4 = R_5 = R_6 = H$, $R_7 = CH_2OH$
45) $R_1 = OH$, $R_2 = R_3 = R_4 = R_5 = R_6 = H$, $R_7 = COOH$
46) $R_1 = OH$, $R_2 = R_3 = R_4 = R_5 = R_6 = H$, $R_7 = OCOCH(OH)CH_3$
47) $R_1 = R_3 = R_4 = R_5 = R_6 = H$, $R_2 = OH$, $R_7 = CH_3$

Figure 2.
*Biotransformation products of 11-keto-b-boswellic acid (**36**).*

18β-glycyrrhetinic acid (**48**) is the active form of glycyrrhizin which is the major pentacyclic triterpene found in licorice (*Glycyrrhiza glabra* L.). Glycyrrhetinic acid has been shown to possess several pharmacological activities, such as antiulcerative, anti-inflammatory, immunomodulating, antitumor, antiviral, antihepatitis effects, and anticancer. Biotransformation **48** with a fungus *C. blakesleeana* (AS 3.970) yielded 3-oxo-7β-hydroxyglycyrrhetinic acid (**49**) and 7β-hydroxyglycyrrhetinic acid (**50**) [27], while of **48** using *Absidia pseudocylindrospora* (ATCC 24169), *Gliocladium viride* (ATCC 10097) and *Cunninghamella echinulata* (ATCC 8688a) afforded seven derivatives: **51**, **52**, 7β,15α-dihydroxy-18β-glycyrrhetinic acid (**53**), 15α-hydroxy-18β-glycyrrhetinic acid (**54**), 1α-hydroxy-18β-glycyrrhetinic acid (**55**) and 13β-hydroxy-7α,27-oxy-12-dihydro-18β-glycyrrhetinic acid (**56**), and the epimer of compound **53** on C-17 (**Figure 3**) [28].

Ginsenoside Rb1 (**61**) is the most predominant protopanaxadiol-type ginsenoside in *Panax* species (ginseng). Several microbial transformations of this substrate (Ginsenoside Rb1) have been accomplished with an ample and varied group of microorganisms, all of these having β-glucosidase activities.

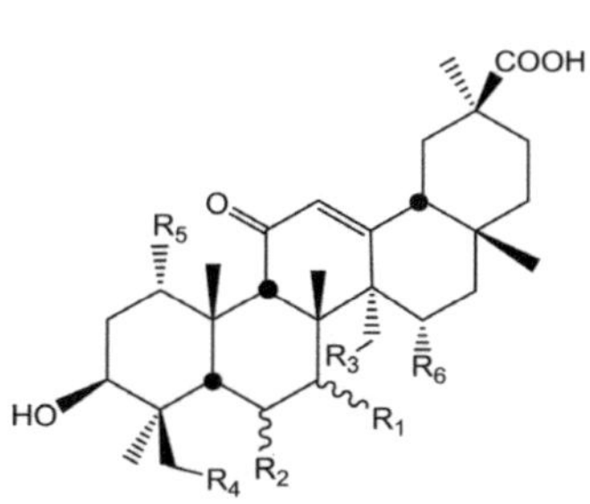

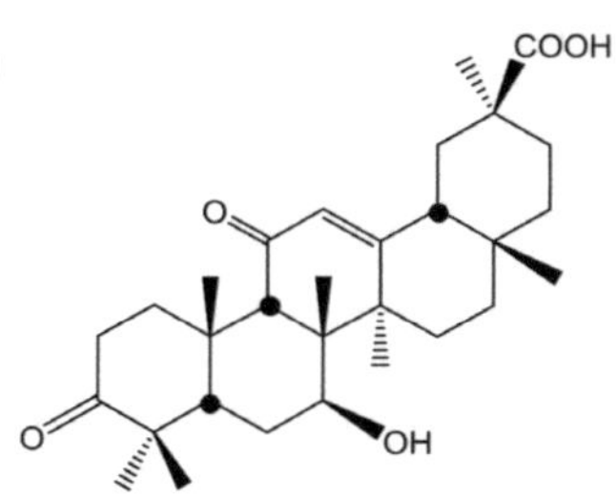

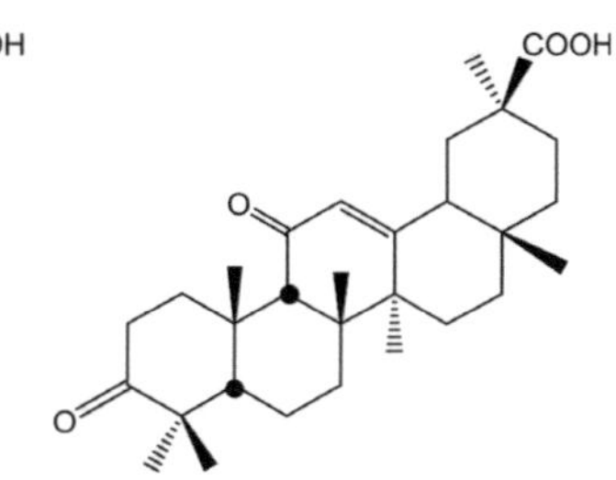

48) $R_1 = R_2 = R_3 = R_4 = R_5 = R_6 = H$
50) $R_1 = \beta OH$, $R_2 = R_3 = R_4 = R_5 = R_6 = H$
51) $R_1 = R_2 = \beta OH$, $R_3 = R_4 = R_5 = R_6 = H$
52) $R_1 = R_2 = R_4 = R_5 = R_6 = H$, $R_3 = OH$
53) $R_1 = R_2 = R_3 = R_5 = R_6 = H$, $R_4 = OH$
54) $R_1 = \alpha OH$, $R_2 = R_3 = R_4 = R_5 = R_6 = H$
57) $R_1 = R_2 = R_3 = R_4 = R_5 = H$, $R_6 = OH$
59) $R_1 = R_2 = R_3 = R_4 = R_6 = H$, $R_5 = OH$

49

56

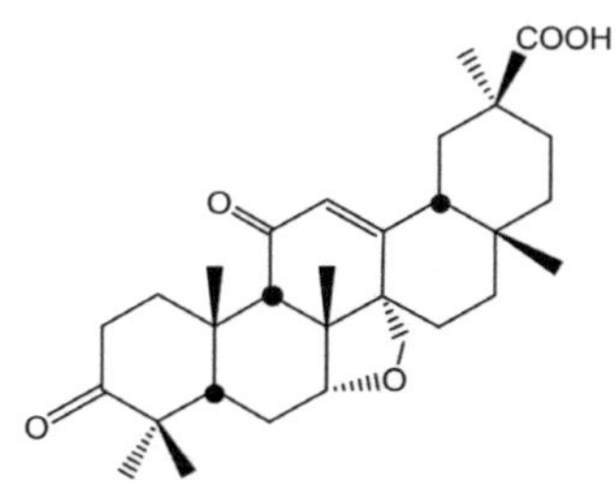

60

Figure 3.
*Biotransformation products of 18β-glycyrrhetinic acid (**48**).*

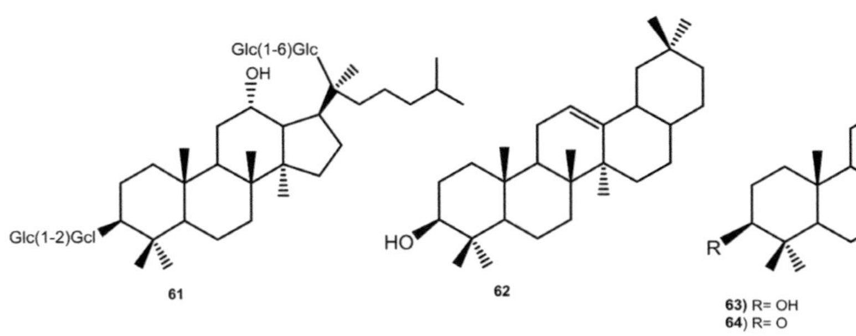

Figure 4.
Triterpenic acid: ginsenoside Rb1 (61), oleanolic acid (62), betulinic (63) and betulonic acid (64).

Deglycosylation appears to be the major transformation pathway, and the intermediate and the final hydrolysis products of **61** depended on the microorganisms used. The biotransformation of various triterpenes, such as **61–64**, has been described in the literature. For each triterpenoid, the transforming microorganism together with the type and site of the reaction catalyzed is given in **Table 2** (**Figure 4**) [29].

Biotransformation of oleanolic acid (**62**) with *Bacillus subtilis* (ATCC 6633) resulted in five more polar metabolites as 28-O-β-D-glucopyranosyl oleanic acid (**63**), 3β-O-β-D-glucopyranosyl oleanic acid (**64**), 3-O-(β-D-glucopyranosyl)-oleanic acid-28-O-β-D-glucopyranoside (**61**), 24-hydroxyl-oleanolic acid (**62**), and 3β-24-dihydroxy-olean-12-en-28-O-β-D-glucopyranosyl-oic acid (**63**), while echinocystic acid (**64**, 250 mg) was metabolized to three more polar metabolites as 28-O-β-D-glucopyranosyl echinocystic acid (**65**), 3-O-(β-D-glucopyranosyl)-echinocystic acid-28-O-β-D-glucopyranoside (**66**), and 24-hydroxyl-28-O-β-glucopyranosyl echinocytic acid (**67**), and then biotransformation of betulinic acid (**68**) contributed four metabolites as 28-O-β-D-glucopyranosyl betulinic acid (**69**), 3-O-(β-D-glucopyranosyl)-betulinic acid-28-O-β-D-glucopyranoside (**70**), 23-hydroxy-betulinic acid (**71**), and 23-hydroxy-28-O-D-β-glucopyranosyl betulinic acid (**72**). In this way there were two types of reactions in the biotransformation of triterpenic acids **58**, **64**, and **68**: hydroxylation and glycosylation [41]. Biotransformation of **58** by *C. muscae* yielded nine hydroxylated and glycosylated metabolites. The specific hydroxylation (7β, 15α, and 21β) was main reaction type. In addition, the selective glycosylation at C-28 was another main reaction type. It was also observed that the 3β-OH group was selectively dehydrogenated into carbonyl group [42].

A C-3 oxidized derivative of oleanolic acid **73** (3-oxoolean-12-en-28-oic acid) was transformed by the *Chaetomium longirostre* (RF-1095) into 4-hydroxy-3,4-seco-olean-12-ene-3,28-dioic acid (**74**) and the corresponding 21-hydroxylated derivative (**75**). Analogous ring-A cleavage oxidation reactions have been observed in the biotransformation of triterpenoid substrates with the fungi *Septomyxa affinis* ATCC 6737 and *Glomerella fusarioides* ATCC 9552. (**Figure 5**) [4, 43].

2.3 Steroidal saponins

Diosgenin [(25R)-spirost-5-en-3β-ol, **76**] is an important natural starting material in the pharmaceutical industry to produce steroid drugs and hormones since the last century. In recent years, a wide array of new biological activities of **76** has been disclosed. Diosgenin was subjected to several structural modification studies to secure new derivatives via microbial transformation. Several microorganisms have been found to be capable of degrading **76**, *Bacillus megaterium*, *Corynebacterium mediolanum*, *Mycobacterium fortuitum*, *M. phlei*, *Nocardia rhodochrous*, and *F. solani*.

Triterpenoid	Microorganism	Reaction	Reference
Ginsenoside Rb_1 (**61**)	*A. niger* (KTC 6909)	Deglycosylation at the C-3 and C-20 sites	[30]
	A. niger (AS 3.1858)	Deglycosylation at the C-3 and C-20 sites	[30]
	A. usamii (KTC 6956)	Deglycosylation at the C-3 and C-20 sites	[30]
	F. sacchari	Deglycosylation at the C-3 and C-20 sites	[31]
	P. oxalicum	Deglycosylation at the C-3 site	[32]
	C. lunata (AS 3.1109)	Deglycosylation at the C-20 site, hydration $\Delta^{24(25)}$ Formation of tertiary alcohol	[33]
	R. stolonifer (AS 3.822)	Deglycosylation at the C-3 and C-20 sites	[33]
Oleanolic acid (**62**)	*C. blakesleeana*	Diverse hydroxylation at the C-1β, C-7β, C-13β sites	[4]
	F. lini	Dehydrogenation C-13 and C-18. oleanderolide formation	[4]
	P. chrysogenum	Hydroxylation on C-21. Oxidation of the hydroxyl group in C-3	[4]
	C. phomoides	Hydroxylation in C-6β	[4]
	A. ochraceus (NG 1203)	Hydroxylation in C-11α	[4]
	Chaetomium longirostre	Oxidative ring A cleavage, hydroxylation at the C-21β sites	[34]
	Nocardia sp. (NRRL 5646)	Methyl esterification of the C-28 carboxyl group	[4]
	R. miehei (CECT 2749)	Hydroxylation of the C-7β, C-15α and C-30 sites Deshidrogenation $\Delta^{9(11)}$	[35]
Betulinic acid (**63**)	*B. megaterium* (ATCC 14581)	Dehydrogenation of the C-3 secondary alcohol group, hydroxylation at the C-6α and C-7β sites	[36]
	B. megaterium (ATCC 13368)	Dehydrogenation of the C-3 secondary alcohol group, hydroxylation at the C-7β and C-15α sites	[37]
	C. elegans (ATCC 9244)	Hydroxylation at the C-1β and C-7β sites	[4]
	Cunninghamella sp.	Introduction of a β-glucopyranosyl at the C-28 carboxylic acid group	[38]
Betulonic acid (**64**)	*B. megaterium* (ATCC 13368)	Ketone α-hydroxylation at the C-2 site	[37]
	Ch. longirostris	Oxidative ring A cleavage, hydroxylation, decarboxylation	[39]
	C. lunata (ATCC 13432)	Hydroxylations at the C-7β and (or) C-15β sites	[4, 40]

Table 2.
Examples of biotransformed triterpenes (61–64) with different microorganisms.

Three major products were accumulated, diosgenone (**77**), 1-dehydrodiosgenone (**78**), androst-4-en-3,17-dione (AD, **79**), and androsta-1,4-diene-3,17-dione (ADD, **80**) (**Table 3**) [44, 45]. In addition, two side-chain cleavage intermediates of **76** were produced by *C. elegans* and *Aspergillus nidulans*. Microbial transformation

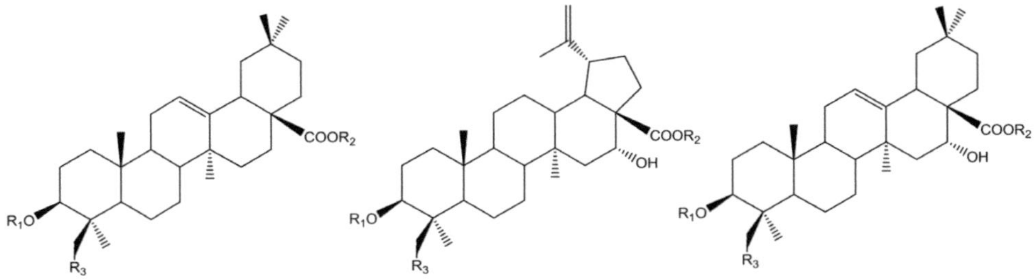

62) $R_1 = R_2 = R_3 = H$
63) $R_1 = R_3 = H$, $R_2 = Glc$
64) $R_1 = Glc$, $R_2 = R_3 = H$
65) $R_1 = R_2 = Glc$, $R_3 = H$
66) $R_1 = R_2 = H$, $R_3 = OH$
67) $R_1 = H$, $R_2 = Glc$, $R_3 = OH$

63) $R_1 = R_2 = R_3 = H$
72) $R_1 = R_3 = H$, $R_2 = Glc$
73) $R_1 = R_2 = Glc$, $R_3 = H$
74) $R_1 = R_2 = H$, $R_3 = OH$
75) $R_1 = H$, $R_2 = Glc$, $R_3 = OH$

68) $R_1 = R_2 = R_3 = H$
69) $R_1 = R_3 = H$, $R_2 = Glc$
70) $R_1 = Glc$, $R_2 = R_3 = H$
71) $R_1 = H$, $R_2 = Glc$, $R_3 = OH$

Figure 5.
*Biotransformation products of oleanolic acid (**62**), echnocystic acid (**68**) and betulinic acid (**63**).*

Fungi	Diosgenona (77)	AD (79)	ADD (80)	Progesterone (102)	16-AD
A. nidulans	++	++	++	++	
C elegans	++			++	
F. solani	++				++
Rhizopus sp.	++	++	++		

AD androst-4-ene-3,17-dione; ADD androsta1,4-diene-3,17-dione; 16-AD androst-4-ene-3,16-dione.

Table 3.
The ability of different fungi to transform diosgenin (76).

of **76** using white-rot fungus *Coriolus versicolor* afforded eight polyhydroxylated steroids, 7β-hydroxydiosgenin (**81**), (25R)-spirost-5-en-3β,7β,21-triol (**82**), (25R)-spirost-5-en-3β,7β,12β-triol (**83**), (25R)-spirost-5-en-3β,7α,15α,21-tetraol (**84**), (25R)-spirost-5-en-3β,7β,12β,21-tetraol (**85**), (25R)-spirost-5-en-3β,7α,12β,21-tetraol (**86**), and (25R)-spirost-5-en-3β,7β,11α,21-tetraol (**87**). The 3β-hydroxyl group and double bond in the B-ring of 76 were found to be important structural determinants for their activity [46].

Microbial transformation of **76** using *Cunninghamella blakesleeana* AS 3.970 afforded polyhydroxylated derivatives, such as (25R)-spirost-5-en-3β,7α,12β-triol (**88**), (25R)-spirost-5-en-3β,7α,12β,15α,21-pentaol (**89**), (25R)-spirost-5-en-3β,7α,12β,18-tetraol (**90**), (25R)-spirost-5-en-3β,7α,12β,15α-tetraol (**91**), (25R)-spirost-5-en-3β,7α,11α,21-tetraol (**92**), (25R)-spirost-5-en-3β,7β,15α,21-tetraol (**93**), and (25R)-spirost-5-en-3β,7β,12β,18-tetraol (**94**) [47], specifically, the hydroxylation, ketonization, and methoxylation by *Cunninghamella blakesleeana*, *C. elegans*, *Helicostylum piriforme*, and *Streptomyces virginiae*, at C-7, C-9, C-11, C-12, and C-25 positions of **76**. Biotransformation of **76** by *Syncephalastrum racemosum* afforded (25R)-spirost-5-en-3β,7α,9α-triol (**95**, 1%), (25R)-spirost-5-en-3β,9α,12α-triol-7-one (**96**, 2%), (25R)-spirost-5-en-3β,9α-diol-7,12-dione (**97**, 1.5%), (25R)-spirost-4-en-9α,12β,14α-triol-3-one (**98**, 0.66%), and (25S)-spirost-4-en-9α,14α,25β-triol-3-one (**99**, 0.66%) [48]. *C. echinulata* (CGMCC3.2716) metabolized **76** to afford **81** (0.9%), **83** (7.7%), (25R)-spirost-5-en-3β,7β-diol-11-one (**100**, 7, 1.5%), and (25R)-spirost-5-en-3β,7β,11α-triol (**101**, 6.2%) (**Figure 6**) [49].

2.4 Steroids

Microorganisms are able to hydroxylate steroids in different positions C-1 to C-21. These represent the most widespread type of steroid bioconversion carried

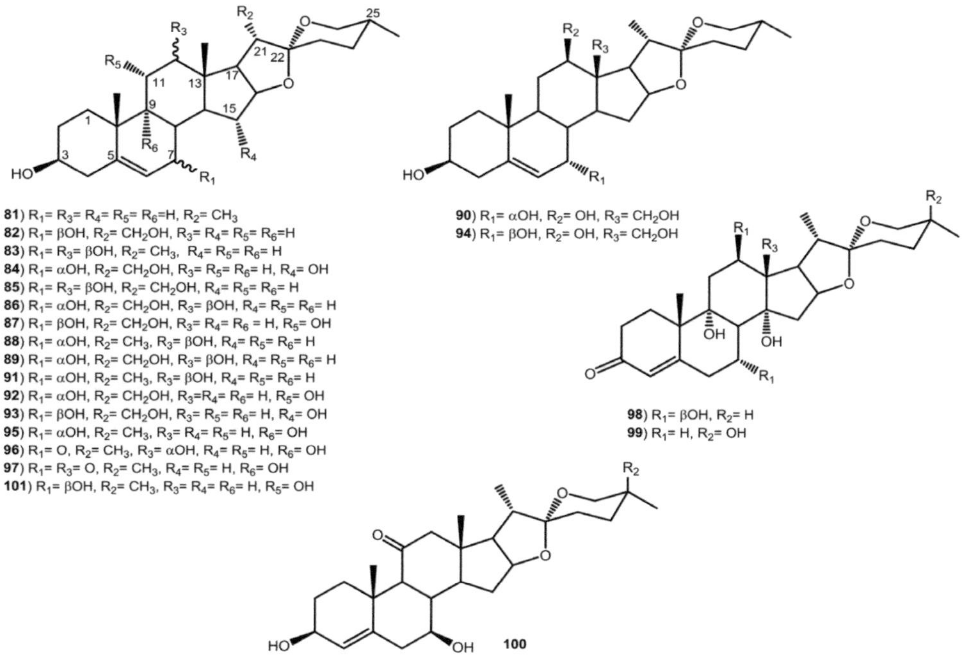

Figure 6.
Biotransformation products of diosgenin (76).

out by fungi. The commercialized microbial process in the steroid field was in the production of 11α-hydroxyprogesterone. This process was realized for the first time by Peterson and Murray (1952), which patented this process of 11α-hydroxylation of progesterone (**102**) by *Rhizopus* species [50]. Microbial hydroxylation of **102** by *A. griseola* produced two hydroxylated pregnane identified as 6β,14α-dihydroxy-progesterone (**103**) and 7α,14α-dihydroxyprogesterone (**104**). *R. pusillus* produced 6β,11α-dihydroxyprogesterone (**105**) with excellent yield (65.5%) and 7α,14α-dihydroxyprogesterone (**106**) (**Figure** 7) [51].

Industry, which is carried by different microorganisms, such as different species of *Curvularia* spp., *Cunninghamella* spp. and fungi *Trichoderma hamatum*, *Cochliobolus lunatus*. Structural transformation of steroidal compounds through microorganisms has emerged as an important application in the steroidal drug industry. Microbial conversions of steroids generally involve dehydrogenation, esterification, halogenation, isomerization, methoxylation, and side-chain modification of steroidal skeleton. Recently, *Mucor circinelloides lusitanicus* transformed 5-en-3β-ol steroids (**108** and **109**) into di- and trihydroxy products. The compound

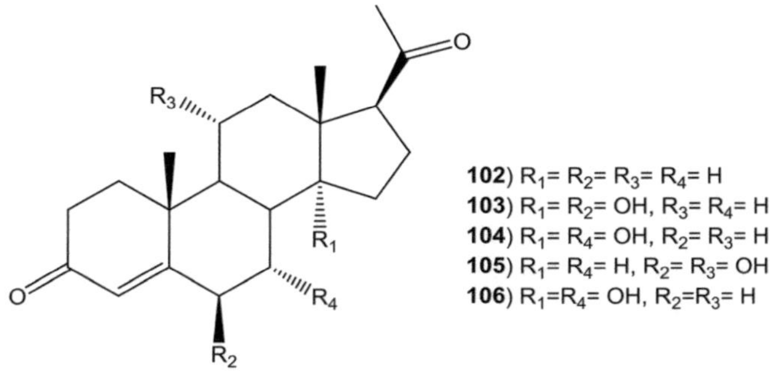

Figure 7.
Biotransformation products of progesterone (102).

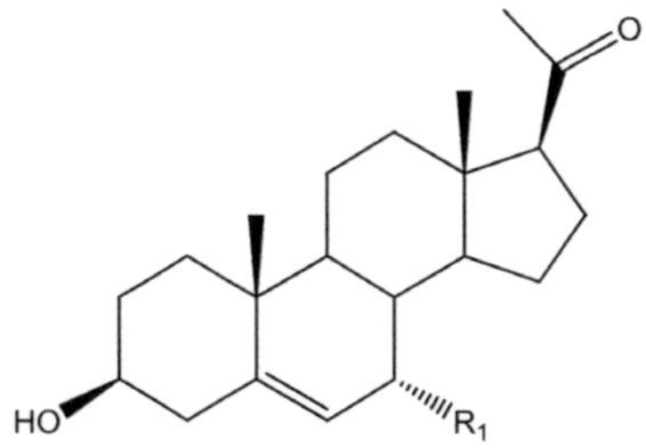

108) pregna-5-en-20-one, R_1= H
110) 3β,7α,11α-trihydroxypregna-5-en-20-one, R_1= OH

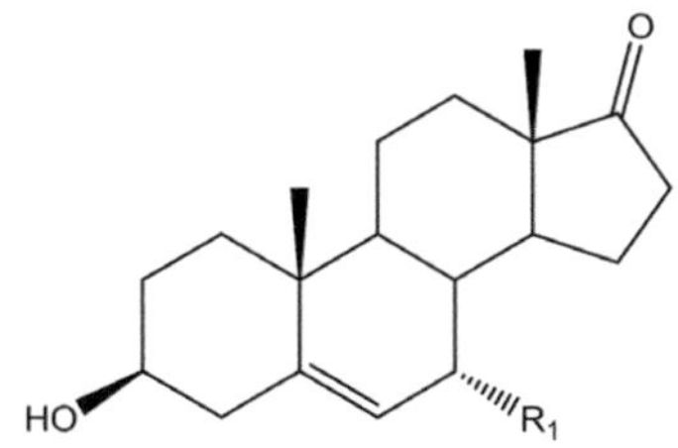

109) androst-5-en-17-one, R_1= H
111) 3β,7α-dihydroxi-androst-5-en-17-one, R_1= OH

Figure 8.
Biotransformation products of 5-en-3β-ol steroids.

108 yielded 3β,7α,11α-trihydroxypregna-5-en-20-one (**110**, 46.4%), and **109** afforded **111** (3β,7α-dihydroxyandrost-5-en-17-one, 43.6%) (**Figure 8**) [52].

Microbial transformation of (20S)-20-hydroxymethylpregna-1,4-dien-3-one (**112**) is by four filamentous fungi, *Cunninghamella elegans* (**113–119**), *Macrophomina phaseolina* (**115, 117, 120–122**), *Rhizopus stolonifer* (**113, 123**), and *Gibberella fujikuroi* (**115–117, 123**). These metabolites were obtained as a result of biohydroxylation of **112** at C-6β, 7β, 11α, 14α, 15β, 16β, and 17α positions (**Figure 9**) [53].

The 11α-, 11β-, 15α, and 16α-hydroxylations are currently established processes in the steroid industry mainly for the production of adrenal cortex hormones and their analogues. 11α-, 11β-, and 16α-hydroxylations are usually performed using *Rhizopus* spp. or *Aspergillus* spp., *Curvularia* spp. or *Cunninghamella* spp. and *Streptomyces* spp., respectively (**Figure 10**) (**Table 4**) [54].

Boldenone (**124**) is an important steroid hormone drug which is the derivative of testosterone. Biotransformation of **124** by *Arthrobacter simplex* and recombinant *Pichia pastoris* with 17β-hydroxysteroid dehydrogenase from *Saccharomyces cerevisiae* produces BD (**124**) from androst-4-ene-3,17-dione (**79**, AD) efficiently [65]. Many microorganisms such as *Mucor racemosus*, *Nostoc muscorum*, and *Arthrobacter oxydans* can utilize androst-1,4-diene-3,17-dione (**80**, ADD) as substrate to produce testosterone through 17β-carbonyl reduction reactions (**Table 5**). The ability of

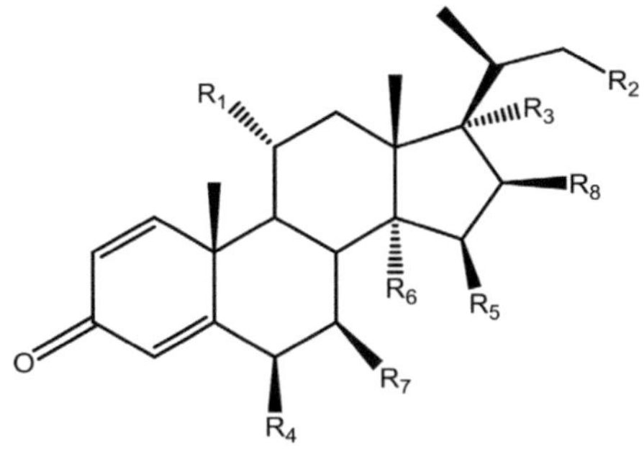

112) R_1= R_2= R_3= R_4= R_5= R_6= R_7= R_8= H
113) R_1= OH, R_2= OAc, R_3= R_4= R_5= R_6= R_7= R_8= H
114) R_1= R_4= R_5= R_6= R_7= R_8= H, R_2= R_3= OH
115) R_1= R_2= OH, R_3= R_4= R_5= R_6= R_7= R_8= H
116) R_1= R_2= R_4= OH, R_3= R_5= R_6= R_7= R_8= H
117) R_1= R_2= R_5= OH, R_3= R_4= R_6= R_7= R_8= H
118) R_1= R_2= R_3= OH, R_4= R_5= R_6= R_7= R_8= H
119) R_1= R_4= R_7= R_8= H, R2= R3= R_5= R_6= OH
120) R_1= R_3= R_4= R_5= R_6= R_8= H, R_2= R_7= OH
121) R_1= R_3= R_4= R_6= R_7= R_8= H, R_2= R_5= OH
122) R_1= R_2= R_3= R_6= R_8= H, R_2= R_5= R_7= OH
123) R_1= R_3= R_4= R_5= R_6= R_7= H, R_2= R_8= OH

Figure 9.
*Biotransformation products of (20S)-20-hydroxymethylpregna-1, 4-dien-3-one (**112**).*

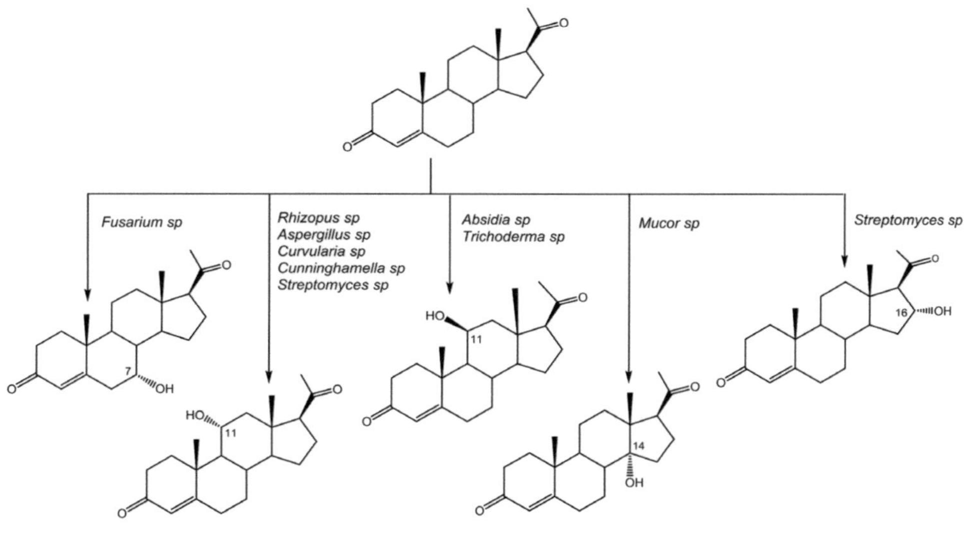

Figure 10.
The ability of different microorganism to transform progesterone (102).

Hydroxylation sites	Microorganisms	Applications	Reference
C-7α	*Fusarium* sp., *Gibberella* sp., *Nigrospora* sp., *Acremonium* sp., *Phycomyces* sp.	Production of bile acids and drugs for neuropsychiatry and immunology	[55, 56]
C-7β	*Mortierella* sp.	Obtaining drugs for prostate cancer	[56, 57]
11β	*Curvularia* sp., *Absidia* sp., *Cunninghamella* sp., *Trichoderma* sp., *Cochliobolus* sp.	Obtaining anti-inflammatory drugs, like hydrocortisone, prednisone acetate, dexamethasone	[56–60]
11α	*Aspergillus* sp., *Rhizopus* sp.	Obtaining of anti-inflammatory, immunosuppressive, anti-allergic drugs, and production of contraceptive drugs	[56, 61]
14α	*Mucor* sp.		[56]
15β	*Bacillus* sp.		[56, 62]
16α	*Streptomyces* sp.		[63, 64]

Table 4.
Some examples of steroid hydroxylation reactions promoted by microorganisms and their applications.

microorganisms to reduce 17-keto- to 17β-hydroxysteroids was evidenced for a wide variety of substrates and microorganisms of different taxonomy: bacteria, fungi, and yeast [54, 56, 57, 66].

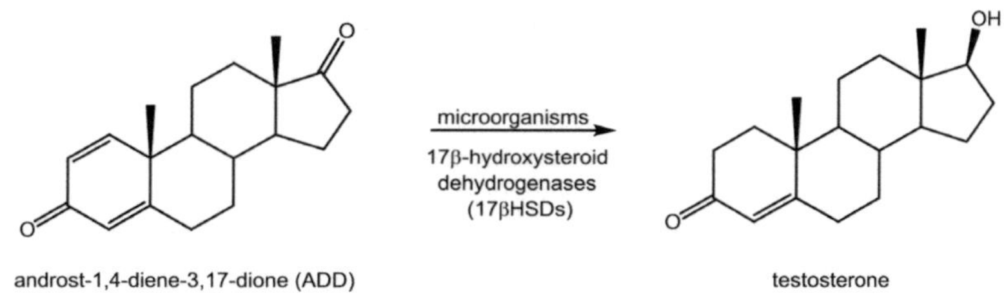

Fungi		Yeast	Bacteria
Actinomucor elegans	*Fusarium*	*Candida albicans*	*B. stearothermophilus*
Agaricus silvaticus	*culmorum*	*C. pelliculosa*	*Bacteroides fragilis*
A. pantherina	*F. oxysporum* var.	*C. pseudotropicalis*	*Brevibacterium sterolicum*
A. spissa	*cubense*	*C. robusta*	*Clostridium paraputrificum*
Armillaria mellea	*F. solani*	*C. tropicalis*	*Comamonas testosteroni*
Corticium centrifugum	*Mucor piriformis*	*C. utilis*	*Lactobacillus bulgaricus*
Fusarium spp.	*M. spinosus*	*Cryptococcus*	*Mycobacterium* spp.
Gibberella saubinetti	*P. chrysogenum*	*albidus*	*B. stearothermophilus*
Mucor spp.	*P. crustosum*	*C. laurentii*	*Bacteroides fragilis*
Penicillium spp.	*P. blakesleeanus*	*C. tsukubaensis*	*Brevibacterium sterolicum*
Aphanocladium album	*R. stolonifer*	*Debaryomyces*	*Clostridium paraputrificum*
Aspergillus chevalieri	*Septomyxa affinis*	*hansenii*	*Comamonas testosteroni (syn.*
A. flavus	*T. piriforme*	*D. kloeckeri*	*Pseudomonas testosteroni*)
A. oryzae	*Trichoderma*	*D. nicotianae*	*Lactobacillus bulgaricus*
A. tamarii	*viride*	*D. subglobosus*	*Pediococcus cerevisiae*
B. obtusa	*Zygodesmus* sp.	*D. vini*	*Sarcina lutea*
C. aphidicola		*Hansenula*	*Staphylococcus aureus*
Ceratocystis paradoxa		*anomala*	*Streptomyces globisporus*
C. lunatus		*H. califórnica*	*S. sphaeroides*
Colletotrichum musae		*H. schnegii*	*S. viridochromogenes*
C. radicicola		*H. suaveolens*	*S. hydrogenans*
Exophiala jeanselmei var.		*Kloeckera jensenii*	*S. lavendulae*
lecaniicorni		*Saccharomyces*	
		carlsbergensis	
		S. cerevisiae	
		S. fragilis	
		S. lactis	
		S. oviformis	
		S. turbidans	
		S. validus	
		Pichia farinosa	
		P.	
		membranaefaciens	
		Torulopsis spp.	
		Hortaea werneckii	
		Phaeotheca	
		triangularis	
		P. herbarum	
		P. ostreatus	
		Rhodotorula	
		aurantiaca	
		R. mucilaginosa	

Table 5.
Reduction of the C-17 carbonyl group of steroids by (17βHSDs) different microorganisms.

The oxidation of 17β-hydroxyl group was observed along with hydroxylation of steroids at C_5 (*Penicillium crustosum, P. chrysogenum*), C_6 (*Bacillus stearothermophilus, B. obtusa, P. blakesleeanus*), C_7 (α/β) (*A. coerulea*, *Botrytis cinerea*, *B. obtusa*, *P. blakesleeanus*, *Rhizopus stolonifer*), C_{10} (*Absidia glauca*), C11 (α/β) (*A. coerulea*, *B. obtusa*, *Cephalosporium aphidicola*, *R. stolonifer*), C_{12} (*A. glauca*, *B. obtusa*), C_{14} (*Bacillus* sp.), and C15 (*A. glauca*, *Aspergillus fumigatus*, *B. obtusa*) [54, 56, 57, 67–69]. The biotransformation of **79** with different microorganisms is shown. Compound **79** is an endogenous weak androgen steroid hormone and intermediate in the biosynthesis of estrone and of testosterone from dehydroepiandrosterone (DHEA) [70]. DHEA is an endogenous steroid hormone. It functions as a metabolic intermediate in the biosynthesis of the androgen and estrogen sex steroids. Various microorganisms have had the ability to biotransform steroidal compounds such as AD (**79**) [54], DHEA (**125**)

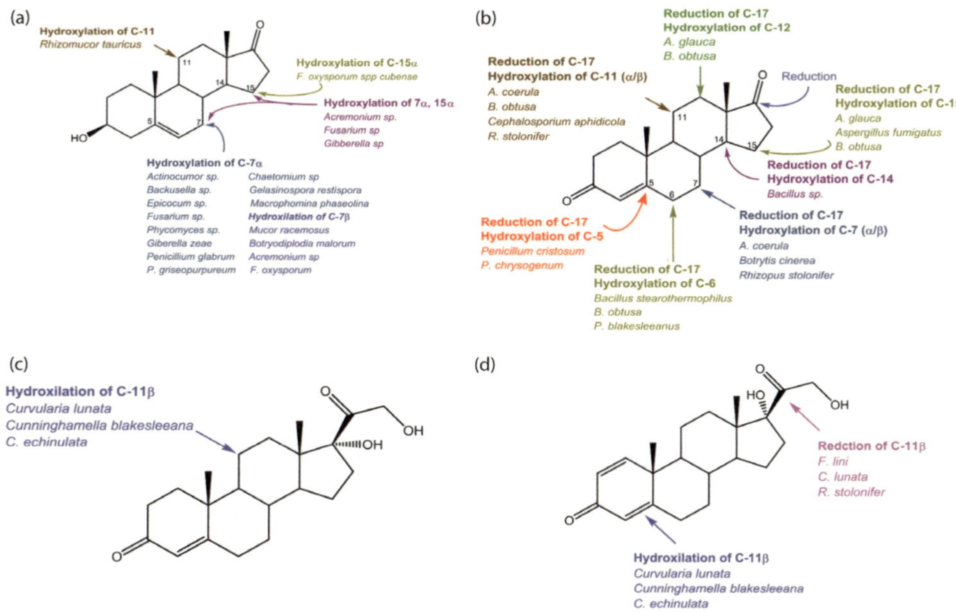

Figure 11.
The ability of different fungi to transform DHEA (125), testosterone, cortexolone (126) and prednisone (127). (a) Hydroxilation of 3β-hydroxy-5-androsten-17-one (DHEA) by various microorganisms. (b) Reduction of C-17 and hydroxilation of testosterone by various microorganisms. (c) Hydroxylation of cortexolone (123) by various microorganisms. (d) Reduction and hydroxylation of prednisone (126) microorganisms.

[54, 55, 70–73, 76, 77, 80, 81], testosterone [54, 55, 74–76, 81], cortexolone (**126**) [78, 79], and prednisone (**127**) (**Figure 11a–d**) [54, 55, 82].

2.5 Diterpene

Sclareolide (**128**) is a natural product isolated from several plant species which displays phytotoxic and cytotoxic activities against several human tumor cells lines. This compound has also been used as starting material for the synthesis of various bioactive products. Regarding the biotransformation of the **128** with different microorganisms, mono- (**130, 131, 135, 140–142**) and dihydroxylation (**132–134, 136, 139, 143, 146**), oxidation (**129, 144**), hydroxylation/oxidation (**145**), epimerization (**137**), and cyclization (**138**) products have been obtained [83]. The microbial transformation of **128** by *Curvularia lunata* yielded 3-ketoesclareolide (**129**), 1β-hydroxysclareolide (**130**), 3β-hydroxysclareolide (**131**), 1α,3β-dihydroxysclareolide (**133**), and 1β,3β-dihydroxysclareolide (**134**) [84]. The incubation of **128** with *Cunninghamella elegans* afforded **129, 131, 133,** and **135–137** [85]. *C blakesleeana* metabolized **128** to afford **129, 135, 134**, and **138–140**. Biotransformation of **128** with *C. echinulata* yielded 5-hydroxysclareolide (**141**) and 7β-hydroxysclareolide (**142**) [86]. Fermentation of **148** with *A. niger* using a nutrient-rich culture medium yielded **141** and **144–146** (**Figure 12**) [83].

2.6 Flavonoids

As most important phytochemicals in food, the dietary flavonoids exert a wide range of benefits for human health. Recent researches have explored diverse biological and pharmacological activities of natural flavonoids—antioxidant activity, anti-inflammatory activity, anti-Alzheimer's disease, antibacterial activity, antifungal activity, anti-HIV activity, anticoagulant activity, antileishmanial activity, and

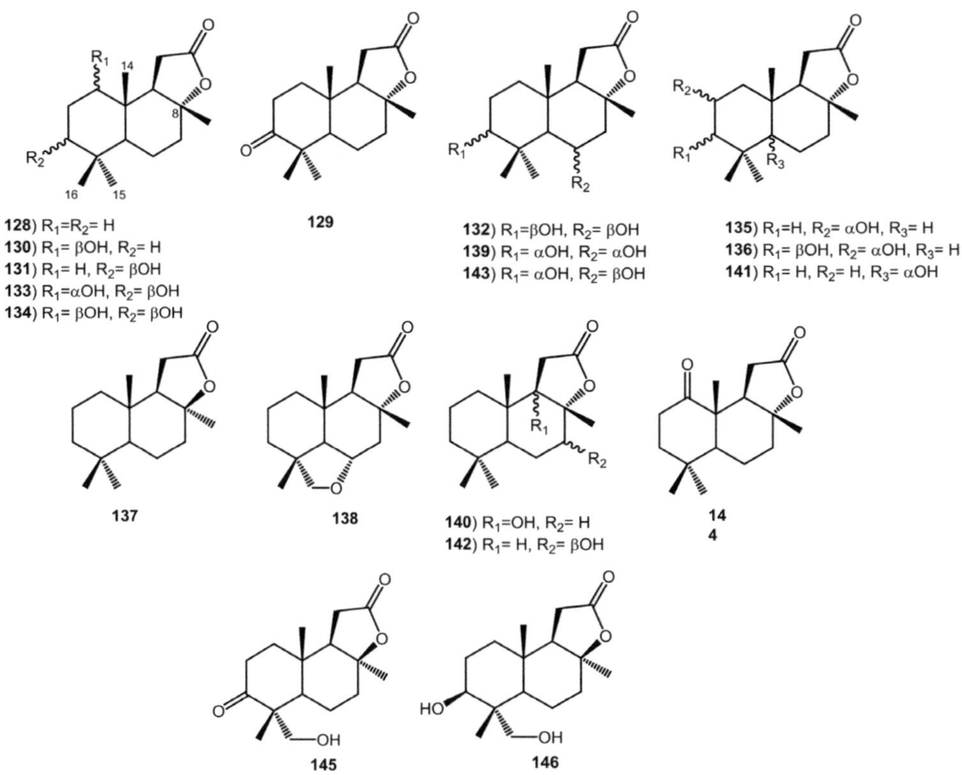

Figure 12.
*Biotransformation products of sclareolide (**128**).*

anti-obesity activity [87–91]. Microbial biotransformation strategies for production of flavonoids have attracted considerable interest because they allow yielding novel flavonoids, which do not exist in nature.

The main reactions during microbial biotransformation are hydroxylation, dehydroxylation, O-methylation, O-demethylation, glycosylation, deglycosylation, dehydrogenation, hydrogenation, C ring cleavage of the benzo-γ-pyrone system, cyclization, and carbonyl reduction. *Cunninghamella*, *Penicillium*, and *Aspergillus* strains are very popular to biotransform flavonoids, and they can perform almost all the reactions with excellent yields (**Figure 13**). Isoflavones are usually hydroxylated at the C-3′ position of the B ring by microorganisms. Chalcones **147-152** were regioselectively cyclized to flavanones (**Figure 14**). Hydrogenation of flavonoids was only reported on transformation of chalcones to dihydrochalcones (**Figure 14**) [92, 93].

Aspergillus niger is one of the most applied microorganisms in the flavonoids' biotransformation; for example, *A. niger* can transfer flavanone to flavan-4-ol, 2′-hydroxydihydrochalcone, flavone, 3-hydroxyflavone, 6-hydroxyflavanone, and 4′-hydroxyflavanone. The hydroxylation of flavones by microbes usually happens on the ortho position of the hydroxyl group on the A ring and C-4′ position of the B ring, and microbes commonly hydroxylate flavonols at the C-8 position. Natural flavonoids, such as naringenin (**166**), hesperetin (**167**), chrysin (**168**), apigenin (**169**), and luteolin (**170**) were subjected to microbiological transformations by *Rhodotorula glutinis* (KCh 735). Yeast was able to regioselectively C-8 hydroxylate **167, 168, 169**, and **170** to generate **171** (17%), **172** (31%), **173** (12.9%), and **174** (25%), respectively. Naringenin (**166**) was transformed to carthamidin (**175**) and isocarthamidin (**176**) in a ratio of 1:19, respectively (**Figure 15**) [94].

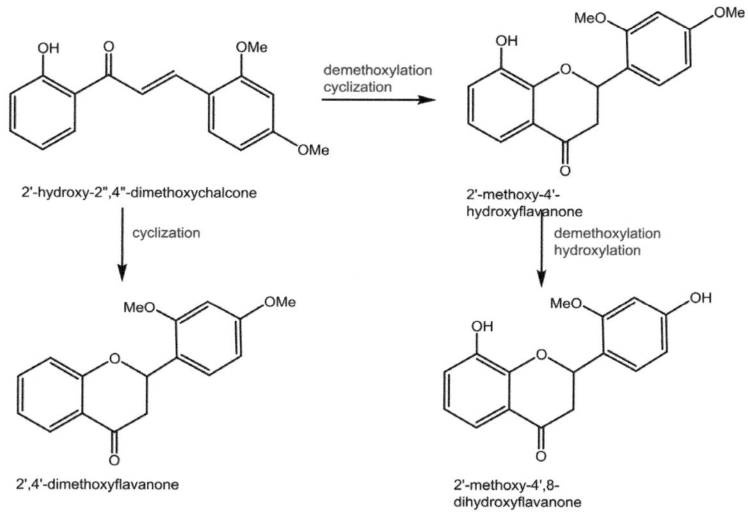

Figure 13.
The main reactions during biotransformation of chalcone whit microorganisms.

The microorganisms tend to hydroxylate flavanones at the C-5, 6, and 4′ positions; however, for prenylated flavanones, dihydroxylation often takes place on the $\Delta^{4(5)}$ double bond on the prenyl group (the side chain of A ring), although cyclization of the prenyl group to dihydrofurane derivatives is rather common biotransformation pathway of prenylated flavonoids. Prenylated flavanones are a unique class of naturally occurring flavonoids characterized by the presence of a prenylated side chain (prenyl, geranyl) in the flavonoid skeleton [95]. The prenyl chain generally refers to the 3,3-dimethylallyl substituent (3,3-DMA), geranyl and lavandulyl. It is proposed that the prenyl-moiety makes the backbone compound more lipophilic, which leads to its high affinity with cell membranes. The prenylation brings the flavonoids with enhancement of antibacterial, anti-inflammatory, antioxidant, cytotoxicity, larvicidal, as well as estrogenic activities. **Figure 16** demonstrated

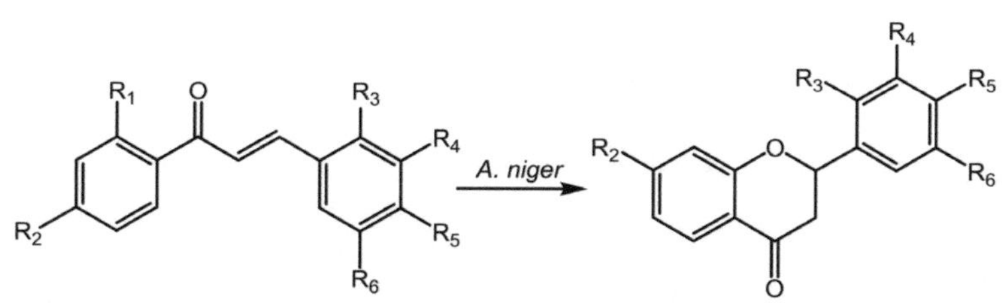

147) R_1= OH, R_2= H, R_3= OMe, R_4= H, R_5= OMe, R_6= H
148) R_1= OH, R_2= H, R_3= OMe, R_4= OMe, R_5= H, R_6= H
149) R_1= OH, R_2= H, R_3= H, R_4= OMe, R_5= OMe, R_6= H
150) R_1= OH, R_2= H, R_3= H, R_4= OMe, R_5= H, R_6= OMe
151) R_1= OH, R_2= OMe, R_3= OMe, R_4= OMe, R_5= H, R_6= H
152) R_1= OH, R_2= OMe, R_3= OMe, R_4= H, R_5= OMe, R_6= H

153) R_1=H, R_2= H, R_3= OMe, R_4= H, R_5= OMe, R_6= H
154) R_1= H, R_2= H, R_3= OMe R_4=H, R_5= OH, R_6=H
155) R_1=O, R_2= H, R_3= OMe, R_4= H, R_5= OH, R_6= H
156) R_1= H, R_2= H, R_3= OMe R_4= OMe, R_5= H, R_6= H
157) R_1= H, R_2= H, R_3= OMe, R_4= OH, R_5= H, R_6= H
158) R_1= H, R_2= H, R_3= H, R_4= OMe, R_5= OMe, R_6= H
159) R_1= H, R_2=H, R_3=H, R_4= OH, R_5= OH, R_6= H
160) R_1= H, R_2= H, R_3= H, R_4= OMe, R_5= H, R_6= OMe
161) R_1=H, R_2= H, R_3= H, R_4= OH, R_5= H, R_6= OMe
162) R_1= H, R_2= OMe, R_3= OMe, R_4= OMe, R_5= H, R_6= H
163) R_1=H, R_2= OMe, R_3= OMe, R_4= OH, R_5= H, R_6= H
164) R_1=H, R_2= OMe, R_3= OMe, R_4= H, R_5= OMe, R_6= H
165) R_1=H, R_2= OMe, R_3= OH, R_4= H, R_5= OMe, R_6= H

Figure 14.
Biotransformation products obtained from biotransformation of chalcones **147-152** *with A. niger.*

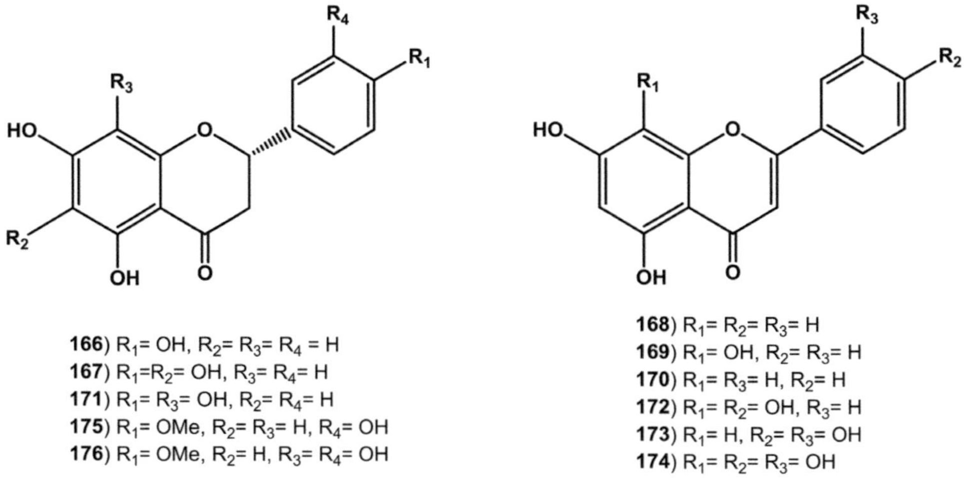

Figure 15.
Biotransformation products of flavanone (166, 167) and flavone (168–170).

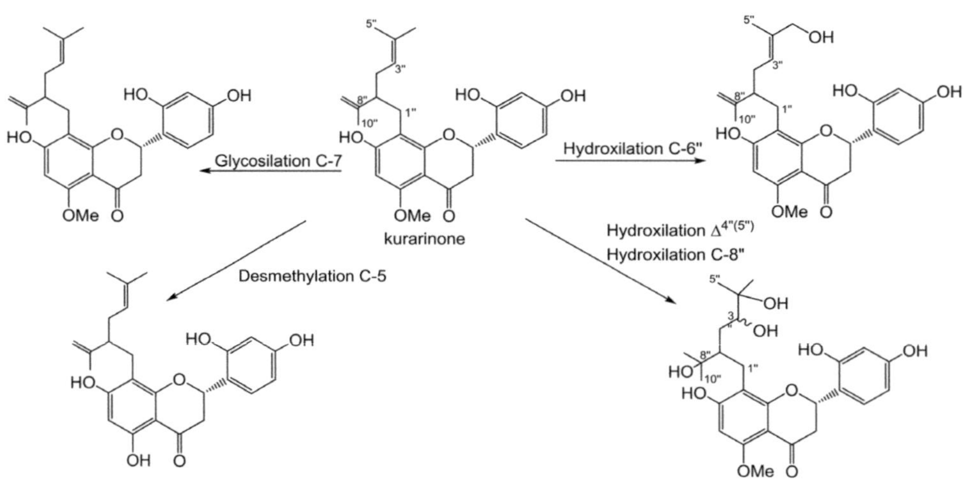

Figure 16.
Microbial biotransformation of kurarinone (177) using C. echinulate and C. militaris.

the microbial biotransformation of kurarinone (**177**) using *C. echinulata* and *C. militaris* [96, 97].

Incubation of *Absidia coerulea* (AM93) with prenylnaringenin (**178**) led to metabolite **179** (8-prenylnaringenin 7-O-β-D-glucopyranoside, 49.3%), while *B. bassiana* transformed **178** into **180** (8-prenylnaringenin 7-O-β-D-4″-O-methyl-glucopyranoside, 32.9%); the metabolites **179** and **180** originated in Sabouraud medium. In the absence of glucose in the culture of *A. coerulea*, the sulfation of substrate **178** (8-prenylnaringenin-7-sulfate, **181**, 31.1%) occurs, while *B. bassiana* into the same product (**180**). The capacity of some fungi—*Cunninghamella elegans*, *Streptomyces fulvissimus*, *Mucor ramannianus*, and *B. bassiana*—in the sulfation of certain phenolic compounds has been reported (**Figure 17**) [98].

Regioselective glycosylation of biologically active flavonoid aglycones catalyzed by microorganisms is an interesting and desired reaction, which significantly increases the water solubility of the compound and, therefore, may improve bio-availability of flavonoids. *Absidia glauca* AM177, *A. coerulea* AM93, *Rhizopus nigricans* UPF701, *Beauveria bassiana* AM278, and *B. bassiana* AM446 are able to conjugate

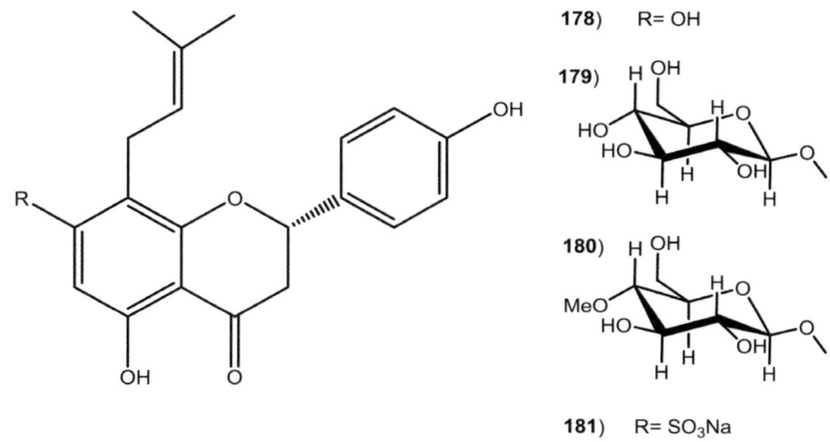

Figure 17.
Biotransformation products of prenylnaringenin (178).

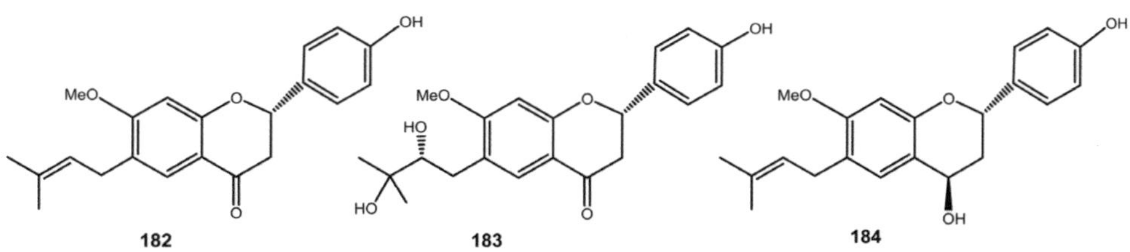

Figure 18.
Biotransformation products of bavachinina (182).

sugar moiety to chalcones, flavanones, and isoflavanones with high regioselectivity. Therefore, it is possible to use *Beauveria* and *Absidia* for the microbial transformation of simple or prenylated flavonoids by glycosidation reactions [97, 99].

Bavachinin (**182**) is one kind compound of flavanones and isolated from the aerial parts and dried fruits of *Psoralea corylifolia*, and bavachinin displays a broad range of biological activities, such as antioxidant, antibacterial, antifungal, anti-inflammatory, antitumor, anti-pyretic, and analgesic properties [100, 101]. Bavachinin (**182**) was subject to biotransformation by cultured cells of *A. flavus* (ATCC 30899); *C. elegans* (CICC 40250) afforded the same product **183** [(S)-6-((R)-2,3-dihydroxy-3-methylbutyl)-2-(4-hydroxyphenyl)-7-methoxychromen-4-one]. On the other hand, one major product **184** [(2S,4R)-2-(4-hydroxyphenyl)-7-methoxy-6-(3-methylbut-2-en-1-yl)-chromen-4-ol] was obtained by *P. raistrickii* (ATCC 10490) by the reduction at the position of ketone group of the C-ring (**Figure 18**) [102].

The biotransformation of xanthohumol (**185**), a prenylated chalcone isolated from hops by selected fungi, *Absidia coerulea* (AM93), *Rhizopus nigricans* (UPF701), *Mortierella mutabilis* (AM404), and *Beauveria bassiana* (AM446), was investigated. The incubation of *A. coerulea* with **185** resulted in the isolation of xanthohumol 4′-O-β-D-glucopyranoside (**186**, 29%). This metabolite was also produced by *R. nigricans* (**186**, 14.2%). Biotransformation of **185** with *B. bassiana* and *M. mutabilis* yielded xanthohumol 7-O-β-D(4‴-O-methyl)-glucopyranoside (**187**, 23%) and isoxanthohumol 7-O-β-glucopyranoside (**188**, 49%), respectively (**Figure 19**) [103]. The compounds **188** (9.3%) and **186** (12%) were also observed as products of **185** transformation by *Cunninghamella elegans* [104]. Another way to obtain **188** is by the transformation of isoxanthohumol (**189**, 61.6%) with *Absidia glauca*; although

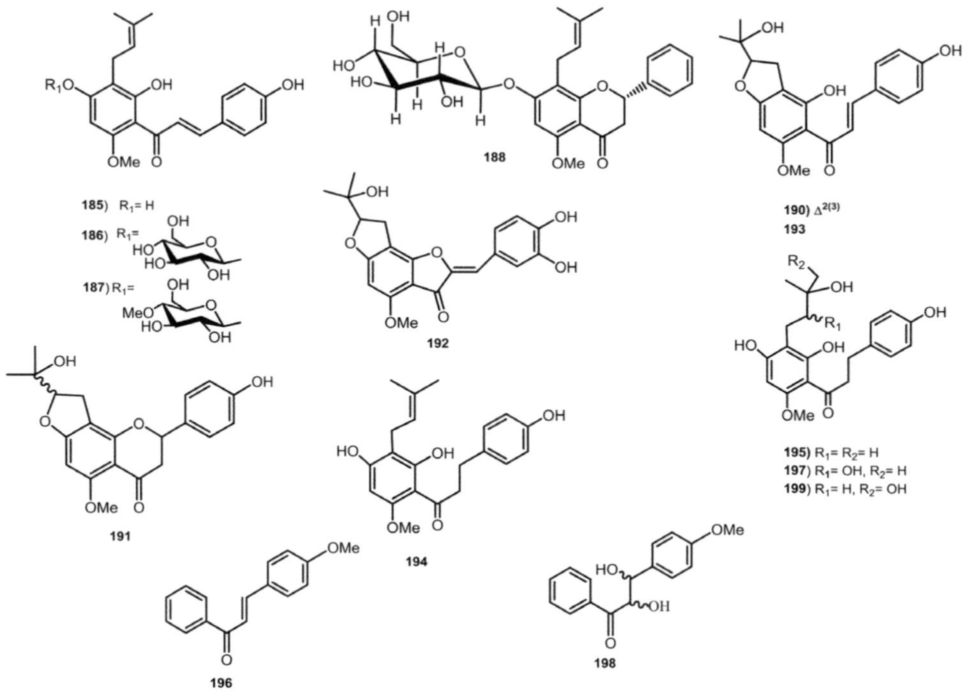

Figure 19.
Biotransformation products of xanthohumol (185).

the efficiency of this process was high (61.6% yield), it required the chemical isomerization of **185** to **189**, prior to biotransformation [105].

2″-(2″-hydroxyisopropyl)-dihydrofurano-[4″,5″:3′,4′]-4,2′-dihydroxy-6′-methoxychalcone (**190**), mixture of diastereoisomers of (2S, 2″S) and (2S, 2″R) 2″-(2″-hydroxyisopropyl)-dihydrofurano-[4″,5″:7,8]-4′-hydroxy-5-methoxyflavanone (**191**), and (Z)-2″-(2″-hydroxyisopropyl)-dihydrofurano-[4″,5″-6,7]-3′,4′-dihydroxy-4-methoxyaurone (**192**) were obtained by transformation of **185** in *Aspergillus ochraceus* (AM 465) culture (**Figure 19**) [106].

Incubation of xanthohumol (**185**) both with *Fusarium avenaceum* (AM11) and *F. oxysporum* (AM727) gave a single metabolite 2″-(2″-hydroxyisopropyl)-dihydrofurano-[4″,5″:3′,4′]-4′,2-dihydroxy-6′-methoxy-α,β-dihydrochalcone (**193**), which turned out to be the product of the prenyl group cyclization and α,β-double bond reduction. *F. tricinctum* reduced α,β-double bond of **185** to give 4,2′,4′-trihydroxy-6′-methoxy-3-prenyl-α,β-dihydrochalcone (**194**). *Penicillium albidum* (AM79) oxidized **185** at the double bond of prenyl group to xanthohumol H (**195**) [107]. The culture of the yeast, *Rhodotorula marina* (AM 77), converted **185** and 4-methoxychalcone (**196**) to α,β-dihydroxanthohumol (**197**) and 4-methoxydihydrochalcone (**198**) with the yields of 18% and 20%, respectively [108]. *Penicillium albidum* (AM79) dihydroxylated the $\Delta^{2''(3'')}$ double bond of xanthohumol to produce 3′-[3″-hydroxy-3″-methylbutyl]-4,2′,4′-trihydroxy-6′-methoxychalcone (199, 22.84%) (**Figure 19**).

B. bassiana AM278 and *Absidia glauca* AM177 converted isoxanthohumol (**189**) into glucoside derivatives (**200**, **201**), whereas *Fusarium equiseti* AM15 transformed it into (2R)-2-(2-hydroxyisopropyl)-dihydrofurano-[2,3:7,8]-4-hydroxy5-methoxyflavanone (**202**) (**Figure 20**) [95, 106].

C. echinulata (ATCC 9244) sulfated silybin (**203**) to silybin-7-sulfate (**204**) and 2,3-dehydrosylibin-7-sulfate (**205**). Sulfonation at the C-7 position of silybin

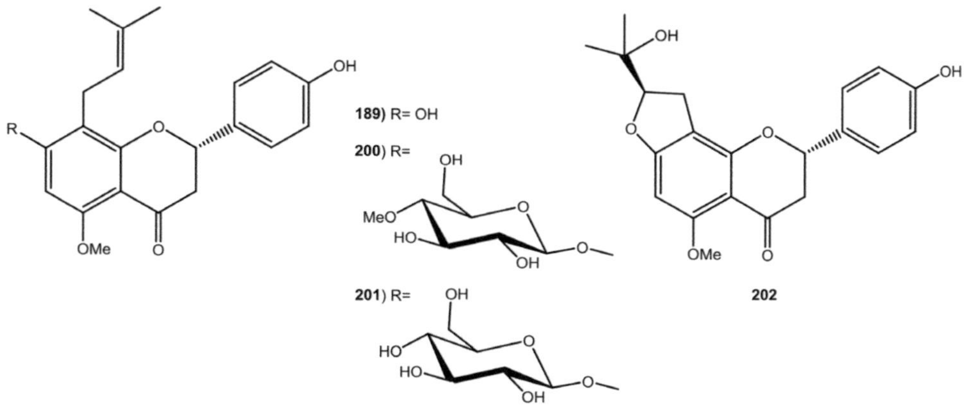

Figure 20.
Biotransformation products of isoxanthohumol (189).

significantly decreased the DPPH free radical scavenging potential; however, further dehydrogenation $\Delta^{2(3)}$ to 2,3-dehydrosilbyn-7-sulfate (**206**) drastically enhanced the DPPH free radical scavenging potential activity [109] (**Figure 21**).

2.7 Enzymes isolated from microorganisms and their application

Enzymes are the most proficient catalysts, offering much more competitive processes than chemical catalysts. A number of enzyme-based processes have been commercialized for producing several valuable products. During the 1980s and 1990s, engineering of enzymes based on structural information allowed extension of their substrate ranges, enabling the synthesis of unusual intermediates. Accordingly, the use of enzymes has been expanded to the manufacture of pharmaceutical intermediates and fine chemicals [110]. Microorganisms and enzymes (biocatalysts) are highly enantio-, chemo-, and regioselective in a wide range of reaction conditions. Selectivity is extremely desirable in the synthesis of different synthesis products, since it offers advantages such as minimizing the side reactions that do not require protection and deprotection steps, which allows for shorter synthesis. Biocatalysis provides a technology that is environmentally safer, and it effectively reduces the level of waste and even eliminates the waste generation rather than remediation and disposal of wastes at the end of the process. In addition

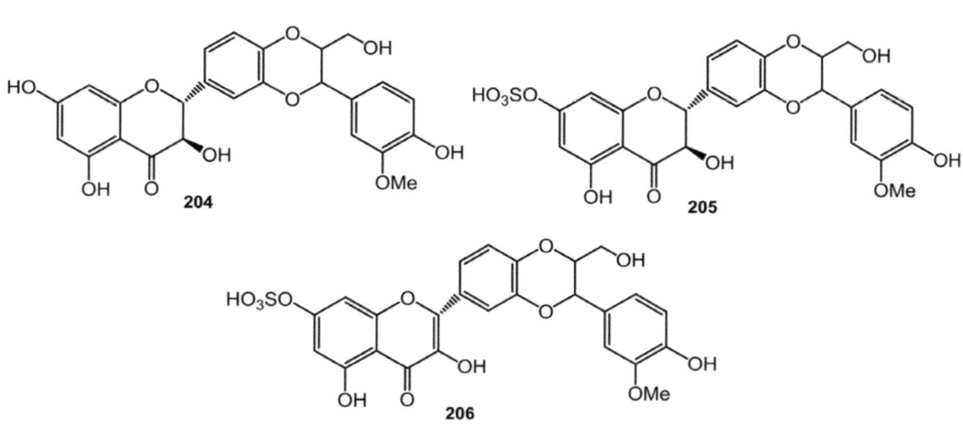

Figure 21.
Biotransformation products of sylbin (204).

to, biocatalysts have many attractive features in the context of green chemistry and sustainable development. Various enzymes used in different industrial processes have been described in the literature. **Table 6** indicates some enzymes, their source, and some applications [111–113].

2.8 Extremophiles

A very interesting research area in biology and biotechnology is the of extremophile microorganisms. Extremophiles can be divided into group according to (i) temperature tolerance, (ii) salt concentration, (iii) pH range, or (iv) pressure conditions. Enzymes from extremophilic microorganisms offer versatile tools for

Microbial enzymes	Microorganism	Application
α-Amylase	*Bacillus amyloliquefaciens* *B. stearothermophilus* *B. licheniformis*	Baking, brewing, starch liquefaction Clarification of fruit juice Textile industry Paper industry
Glucoamylase	*Aspergillus niger* *A. awamori* *Rhizopus oryzae*	Beer production High glucose and high fructose syrups
Proteases	*A. usami*	
Lactase (β-galactosidase)	*Kluyveromyces lactis* *K. fragilis*	Lactose intolerance reduction in people Prebiotic food ingredients
Lipase	*Candida antarctica* *C. cylindraceae Ay30* *Helvina lanuginosa* *Pseudomonas* sp. *Geotrichum candidum*	Cheese flavor development Textile indutry Medicinal applications Use in cosmetics Use as biosensors Use in biodegradation
Phospholipases	*Fusarium oxysporum*	Cheese flavor development
Esterases	*Bacillus licheniformis*	Enhancement of flavor and fragrance in fruit juice
Xylanases	*Streptomyces* sp. *Bacillus* sp. *Pseudomonas* sp.	Clarification of fruit juice Beer quality improvement
Glucose oxidase	*A. niger* *Penicillium glaucum* *P. adametzzi*	Food shelf life important Food flavor improvement
Laccase	*Funalia trogii* *Bacillus licheniformis* *Bacillus vallismortis*	Polyphenol removal from wine baking
Pectinases	*A. niger* *A. wentii* *Rhizopus* sp.	Clarification of fruit juice
Catalase	*A. niger* *Metarhizium anisopliae* *Psychrobacter piscatorri*	Food preservation Removal of H_2O_2 from milk prior to cheese production
Peroxidase	*Streptomyces viridosporus*	Development of flavor, color and nutritional quality of food

Table 6.
Enzymes, source, and some applications.

sustainable developments in a variety of industrial applications as they show important environmental benefits due to their biodegradability, specific stability under extreme conditions, improved use of raw materials, and decreased amount of waste products. Although major advances have been made in the last decade, our knowledge of the physiology, metabolism, enzymology, and genetics of this fascinating group of extremophilic microorganisms and their related enzymes is still limited [114–116].

The outstanding properties of thermozymes are suited to industries that employ elevated temperatures, such as the pulp and paper, food, brewing, and feed processing industries. Thermophiles are often highly resistant to harsh conditions such as chemical denaturing agents, wide pH ranges, and/or nonaqueous solvents. Examples of such enzymes are cellulases, xylanases, pectinases, chitinases, amylases, pullulanases, proteases, lipases, glucose isomerases, alcohol dehydrogenases, and esterases. Thermophilic enzymes have played important roles not only at the industrial level but also in pharmaceutical applications requiring use of specific aldolases for the synthesis of enantiopure compounds (**Table 7**) [118].

Source	Enzyme	Activity	Bioprocess/industry	Reference
Sulfolobus solfataricus *S. acidocaldarius* *Thermoproteus texas* *Hyperthermus butylicus*	Aldolase	Stereoselective C-C bond formation	Pharmaceutical industry	[117]
Pyrococcus furiosus	Hydrogenase	Final stage of glucose oxidation by oxidative pentose phosphate cycle	Enhanced production of biohydrogen	[119]
Geobacillus thermoleovorans	Carboxylesterase	Carboxyl ester hydrolysis	Agriculture, food, and pharmaceutical industries	[120]
Bacillus pumilus	Acidic thermostable lipase	Degradation of palm oil	Treatment of palm oil-containing wastewater	[121]
Geobacillus sp.	Lipase	Hydrolysis of diver's lipid substrates	Biofuel, cosmetics, or perfume production, leather and pulp industries	[122]
Microbial community from solid-state fermentation reactor	Protease	Degradation of hair waste from tannery	Leather industry	[123]
Sulfolobus tokodaii	Chitinase	Hydrolysis of β-(1, 4)-glycosidic bonds in chitin	Biomedical, pharmaceutical, food, and environmental	[124]
Acidothermus cellulolyticus	Endoxylanase	β-(1,4)-xylan cleavage	Biofuel production from lignocellulose	[125]
Thermotoga neapolitana	Pullulanase	Hydrolysis of α-(1, 6)-glucosidic linkages	Biofuel production	[126]

Table 7.
Extremophile microorganisms and some applications of their enzymes.

3. Conclusion

Due to microorganisms' abundant multienzyme systems, microbial transformation possesses advantages against chemosynthesis of environmental friendliness, mild reaction conditions, and high *stereo-*, *regio*, and *chemo*-selectivities as well as in improving conversion rates and reducing cost. Thus, microbial transformation technique is being increasingly used to structurally modify natural and synthetic compounds.

The hydrolytic and reductive capabilities of microorganisms have been known and are currently used in preparative and industrial reactions. Various classes of bioactive organic compounds have been subjected to enzymatic transformation to obtain more active and less toxic substances or to elucidate their metabolic pathways.

For example, biotransformation-derived steroids are used for a wide range of pharmacotherapeutic purposes, such as anti-inflammatory, immunosuppressive, progestational, diuretic, anabolic, as neurosteroids, and as contraceptive. Researchers continue to discover more useful steroid compounds and to isolate microorganisms that can perform the structural transformations desired. New technologies such as genomics, metanogenomics, gene shuffling, and DNA evolution provide valuable tools for improving or adapting enzyme properties to the desired requirements.

An alternative may be extremophilic microorganisms such as biocatalysts for countless future industrial applications that are more environmentally friendly.

Acknowledgements

The authors thank Carrera de Biología, FES-Zaragoza, UNAM, Al Departamento de Química Orgánica, FES-Cuautitlán, UNAM.

Conflict of interest

The authors report no conflicts of interest.

Author details

Arturo Cano-Flores[1*], Javier Gómez[1], Iker S. Escalona-Torres[1] and Benjamín Velasco-Bejarano[2]

1 Laboratorio de Biotransformaciones y Química de Productos Naturales, L-314, Facultad de Estudios Superiores Zaragoza, UNAM, CDMX, Mexico

2 Sección de Química Orgánica, Departamento de Ciencias Químicas, FES-Cuautitlán, UNAM, Edo. de México, Mexico

*Address all correspondence to: aecanomx@unam.mx; acano1750@gmail.com

References

[1] Lehman LR, Stewart JD. Filamentous fungi: Potentially useful for the biohydroxilation of non-actived carbón centres. Current Organic Chemistry. 2001;**5**:439-470. DOI: 10.2174/1385272013375490

[2] Pollard DJ, Woodley JM. Biocatalysis for pharmaceutical intermediates: The future is now. Trends in Biotechnology. 2006;**25**:66-72. DOI: 10.1016/j.tibtech.2006.12.005

[3] Straathof AJJ. The production of fine chemicals by biotransformation. Current Opinion in Biotechnology. 2002;**13**:548-556. DOI: 10.1016/S0958-1669(02)00360-9

[4] Parra A, Rivas F, Garcia-Granados A, Martínez A. Microbial transformation of triterpenoids. Mini-Reviews in Organic Chemistry. 2009;**6**:307-320. DOI: 10.2174/157019309789371569

[5] Schrewe M, Julsing MK, Buhler B, Schmid A. Whole-cell biocatalysis for selective and productive C-O functional group introduction and modification. Chemical Society Reviews. 2013;**42**: 6346. DOI: 10.1039/c3cs60011d

[6] Brenna E, Fugati C, Gatti FG, Serra S. Biocatalytic methods for the synthesis of enantioenriched odor active compounds. Chemical Reviews. 2011; **111**:4036-4072. DOI: 10.1021/cr100289r

[7] Sultana N, Saeed SZ. Enzymatic biotransformation of terpenes as bioactive agents. Journal of Enzyme Inhibition and Medicinal Chemistry. 2013;**28**:113-1128. DOI: 0.3109/14756366.2012.727411

[8] Silva EO, Gonçalves NS, Santos RA, Furtado NAJC. Microbial metabolism of atovaquone and cytotoxicity of the produced phase I metabolite. European Journal of Drug Metabolism and Pharmacokinetics. 2016;**41**:645-650

[9] Pervaiz I, Ahmad S, Madni MA, Ahmad H, Khaliq FH. Microbial biotransformation: A tool for drug designing. Applied Biochemistry and Microbiology. 2013;**49**:437-450. DOI: 10.1134/S0003683813050098

[10] Zhan Y, Liu H, Wu Y, Wei P, Chen Z, William JS. Biotransformation of artemisinin by *Aspergillus niger*. Applied Microbiology and Biotechnology. 2015;**99**:3443-3446. DOI: 10.1007/s00253-015-6464-x

[11] Ponnapalli MG, Sura MB, Sudhakar R, Govindarajalu G, Sijwali PS. Biotransformation of Artemisinin to 14-hydroxydeoxyartemisinin: C-14 hydroxylation by *Aspergillus flavus*. Journal of Agricultural and Food Chemistry. 2018;**66**:10490-10495. DOI: 10.1021/acs.jafc.8b03573

[12] Srivastava S, Luqman S, Fatima A, Dorokar MP, Negi AS, Kumar JK, et al. Biotransformation of artemisinin mediated through fungal strains for obtaining derivatives with novel activities. Scientia Pharmaceutica. 2009; **77**:87-95. DOI: 10.3797/scipharm.0803-15

[13] Parhikov IA, Muraludharan KM, Avery MA, Williamson JS. Transformation of artemisinin by *Cunninghamella elegans*. Applied Microbiology and Biotechnology. 2004; **64**:782-786. DOI: 10.1007/s00253-003-1524-z

[14] Zhan Y, Wu Y, Xu F, Bai Y, Guan Y, Williamson JS, et al. A novel dihydroxylated derivative of artemisinin from microbial transformation. Fitoterapia. 2017;**120**: 93-97. DOI: 10.1016/j.fitote.2017.05.015

[15] Goswami A, Saikia PP, Barua NC, Bordoloi M, Yadav A, Bora TC, et al.

Bio-transformation of artemisinin using soil microbe: Direct C-acetoxylation of artemisinin at C-9 by *Penicillium simplissimum*. Bioorganic & Medicinal Chemistry Letters. 2010;**20**:359-361. DOI: 10.1016/j.bmcl.2009.10.097

[16] Gaur R, Darokar MP, Ajayakumar PV, Shukla RS, Bhakuni RS. In vitro antimalarial studies of novel artemisinin biotransformed products and its derivatives. Phytochemistry. 2014;**107**:135-140. DOI: 10.1016/j.phytochem.2014.08.004

[17] Liu J-H, Chen Y-G, Yu B-Y, Chen Y-J. A novel ketone derivative of artemisinin biotransformed by *Streptomyces griseus* ATCC 13273. Bioorganic & Medicinal Chemistry Letters. 2006;**16**:1909-1912. DOI: 10.1016/j.bmcl.2005.12.076

[18] Saraswati S, Agrawal S-S, Alhaider A-A. Ursolic acid inhibits tumor angiogenesis and induces apoptosis through mitochondrial-dependent pathway in Ehrlich ascites carcinoma tumor. Chemico-Biological Interactions. 2013;**206**:153-165. DOI: 10.1016/j.phytol.2019.04.019

[19] Zhang C, Xu S-H, Ma B-L, Wang W, Yu B-Y, Zhang J. New derivatives of ursolic acid through the biotransformation by *Bacillus megaterium* CGMCC 1.1741 as inhibitors on nitric oxide production. Bioorganic & Medicinal Chemistry Letters. 2017;**27**: 2575-2578. DOI: 10.1016/j.bmcl.2017.03.076

[20] Zhang C-X, Ma W-J, Liu D-L, Jia X-J, Zhao Y-M. Biotransformation of ursolic acid by *Alternaria longipes* AS3.2875. Natural Product Research. 2018;**32**:536-543. DOI: 10.1080/14786419.2017.1327860

[21] Wang Y, Xiang L, Chen M, Zhang Z-X, He X. Substrate specify for the 2β-hydroxylation of ursolic acid by *Alternaria alternate* and the antitumor activities of those metabolites. Journal of Molecular Catalysis B: Enzymatic. 2012; **83**:51-56. DOI: 10.1016/j.molcatb.2012.07.005

[22] S-b F, Yang J-S, J-l C, Sun D-A. Biotransformation of ursolic acid by *Syncephalastrum racemosum* CGMCC 3.2500 and anti-HCV activity. Fitoterapia. 2013;**86**:123-128. DOI: 10.1016/j.fitote.2013.02.007

[23] Huang F-X, Yang W-Z, Ye F, Tian J-Y, Hu H-B, Feng L-M, et al. Microbial transformation of ursolic acid by *Syncephalastrum racemosum* (Cohn). Phytochemistry. 2012;**82**:56-60. DOI: 10.1016/j.phytochem.2012.06.020

[24] Fu S-B, Yang J-S, Cui J-L, Meng Q-F, Feng X, Sun D-A. Multihydroxylation of ursolic acid by *Pestalotiopsis microspora* isolated from the medicinal plant *Huperzia serrate*. Fitoterapia. 2011;**82**:1057-1061. DOI: 10.1016/j.fitote.2011.06.09

[25] Zhanga S-S, Hea L-S, Zhaoa Y-M, Fub S, Liua D-L, Yua Z-H, et al. Three new triterpenoids transformed from ursolic acid by *Mucor spinosus* (AS 3.3450) and their cytotoxicity. Phytochemistry Letters. 2019;**32**:33-37. DOI: 10.1016/j.phytol.2019.04.019

[26] Wang Y, Sun Y, Wang C, Huo X, Liu P, Wang C, et al. Biotransformation of 11-keto-b-boswellic acid by *Cunninghamella blakesleeana*. Phytochemistry. 2013;**96**:330-336. DOI: 10.1016/j.phytochem.2013.07.018

[27] Yu-Juan Q, Bing F, Xin-Bo S, Wen-Bin Z, He-Shui Y, Li-Li Z, et al. Biotransformation of glycyrrhetinic acid by *Cunninghamella blakesleeana*. Chinese Journal of Natural Medicines. 2010;**8**:373-381. DOI: 10.3724/SP.J.1009.2010.00373

[28] Maatooq GT, Marzouk AM, Gray AI, Rosazza JP. Bioactive microbial metabolites from glycyrrhetinic acid.

Phytochemistry. 2010;**71**:262-270. DOI: 10.1016/j.phytochem.2009.09.014

[29] Mufler K, Leipold D, Scheller M-C, Haas C, Steingroewer J, Bley T, et al. Biotransformation of triterpenes. Process Biochemistry. 2011;**46**:1-15. DOI: 10.1016/j.procbio.2010.07.015

[30] Chi H, Ji G. Transformation of Ginsenosides Rb1 and Re from *Panaxginseng* by food microorganisms. Biotechnology Letters. 2005;**27**:765-771. DOI: 10.1007/s10529-005-5632-y

[31] Han Y, Sun B, Jiang B, Hu X, Spranger MI, Zhang Y, et al. Microbial transformation of ginsenosides Rb1, Rb3 and Rc by *Fusarium sacchari*. Journal of Applied Microbiology. 2010;**109**: 792-798. DOI: 10.1111/j.1365-2672.2010.04707.x

[32] Gao J, Xu W, Fang Q, Liang F, Jin R, Wu D, et al. Efficient biotransformation for preparation of pharmaceutically active ginsenoside compound K by *Penicillium oxalicum sp*. Annales de Microbiologie. 2013;**63**:139-149. DOI: 10.1007/s13213-012-0454-3

[33] Dong A, Ye M, Guo H, Zhen J, Guo D. Microbial transformation of ginsenoside Rb1 by *Rhizopus stolonifer* and *Curvularia lunata*. Biotechnology Letters. 2003;**25**:339-344

[34] Shirane N, Hashimoto Y, Ueda K, Takbnaka H, Katoh K. Ring a cleavage 3-oxo-olean-12-en-28-oic acid by the fungus *Chaetomium longirostre*. Phytochemistry. 1996;**43**:99-104

[35] Martinez A, Perojil A, Rivas F, Parra A, Garcia-Granados A, Fernandez-Vivas A. Biotransformation of oleanolic and maslinic methyl esters by *Rhizomucor miehei* CECT 2749. Phytochemistry. 2015;**117**:500-508. DOI: 10.1016/j.phytochem.2015.07.020

[36] Kouzi SA, Chatterjee P, Pezzuto JM, Hamann MT. Microbial transformations of the antimelanoma agent betulinic acid. Journal of Natural Products. 2000; **63**:1653-1657

[37] Chatterjee P, Kouzi SA, Pezzuto JM, Hamann MT. Biotransformation of the antimelanoma agent betulinic acid by *Bacillus megaterium* ATCC 13368. Applied and Environmental Microbiology. 2000;**66**: 3850-3855

[38] Chatterjee P, Pezzuto JM, Kouzi SA. Glucosidation of betulinic acid by *Cunninghamella* species. Journal of Natural Products. 1999;**62**:761-763

[39] Akihisa T, Takamine Y, Yoshizumi K, Tokuda H, Kimura Y, Ukiya M, et al. Microbial transformations of two lupane-type triterpenes and anti-tumor promoting effects of the transformation products. Journal of Natural Products. 2002;**65**: 278-282

[40] Goswami A, Guo Z, Tullya TP, Rinaldia FA, Huangb XS, Swidorskib JJ, et al. Microbial transformations of betulinic and betulonic acids. Journal of Molecular Catalysis B: Enzymatic. 2015; **117**:45-53. DOI: 10.1016/j.molcatb.2015.04.012

[41] Wang W-W, Xu S-H, Zhao Y-Z, Zhang C, Zhang Y-Y, Yu B-Y, et al. Microbial hydroxylation and glycosylation of pentacyclic triterpenes as inhibitors on tissue factor procoagulant activity. Bioorganic & Medicinal Chemistry Letters. 2017;**27**: 1026-1030. DOI: 10.1016/j.bmcl.2016.12.066

[42] Yana S, Lina H, Huang H, Yanga M, Xua B, Chena G. Microbial hydroxylation and glycosidation of oleanolic acid by *Circinella muscae* and their anti-inflammatory activities. Natural Product Research. 2019;**33**: 1849-1855. DOI: 10.1080/14786419.2018.1477150

[43] Akihisa T, Watanabe K, Yoneima R, Suzuki T, Kimura Y. Biotransformation of cycloartane-type triterpenes by the fungus *Glomerella fusarioides*. Journal of Natural Products. 2006;**69**:604-607

[44] Wang F-Q, Li B, Wang W, Zhang C-G, Wei D-Z. Biotransformation of diosgenin to nuatigenin-type steroid by a newly isolated strain, *Streptomyces virginiae* IBL-14. Applied Microbiology and Biotechnology. 2007;**77**:771-777. DOI: 10.1007/s00253-007-1216-1

[45] Adham NZ, Zaki RA, NAim N. Microbial transformation of disogenin and its precursor furestanol glycosides. World Journal of Microbiology and Biotechnology. 2009;**25**:481-487. DOI: 10.1007/s11274-008-9913-1

[46] Wu G-W, Gao J-M, Shi X-W, Hang Q, Wei S-P, Ding K. Microbial transformations of diosgenin by the white-rot Basidiomycete *Coriolus versicolor*. Journal of Natural Products. 2011;**14**:2095-2101. DOI: 10.1021/np2003484

[47] Xu M, Huo X-K, Tian X-G, Dong P-P, Wang C, Huang S-S, et al. Microbial transformation of diosgenin by *Cunninghamella blakesleana* AS 3.970 and potential inhibitory effects on P-glycoprotein of its metabolites. RSC Advances. 2015;**78**(78):081-78089. DOI: 10.1039/c5ra122553h

[48] Zhao Y, Sun LM, Wang XN, Shen T, Ji M, Li X, et al. Microbial transformation of diosgenin by *Syncephalastrum racemosum* (Cohn) Schroeter. Chinese Chemical Letters. 2010;**21**:76-80. DOI: 10.1016/j.cclet.2009.08.015

[49] Dong T, Wua G-W, Wang X-N, Gao J-M, Chen J-G, Lee S-S. Microbiological transformation of diosgenin by resting cells of filamentous fungus, *Cunninghamella echinulata* CGMCC3.2716. Journal of Molecular Catalysis B: Enzymatic. 2010;**67**: 251-266. DOI: 10.1016/j.molcatb.2010.09.001

[50] Peterson DH, Murray HC. Microbiological oxygenation of steroids at carbon-11. Journal of the American Chemical Society. 1952;**74**:1871-1872

[51] Habibi Z, Yousefi M, Ghanian S, Mohammadi M, Ghasemi S. Biotransformation of progesterone by *Absidia griseolla* var. *igachii* and *Rhizomucor pusillus*. Steroids. 2012;**77**: 1446-1449. DOI: 10.1016/j.steroids2012.08.010

[52] Shan L, Jiao K, Yin M, Huang J, Chen Y, Qin S, et al. Biotransformation of 5-en-3β-ol steroids by *Mucor circinelloides lusitanicus*. Biocatalysis and Biotransformation. 2016;**34**:83-88. DOI: 10.3109/10242422.2015.1089865

[53] Choudhary MI, Erum S, Atif M, Malik R, Tameen K, Atta-ur-Rhaman. Biotransformation of (20S)-20-hydroxymethylpregna-1,4-dien-3-one by four filamentous fungi. Steroids. 2011;**76**:1288-1296. DOI: 10.1016/j.steroids.2011.06.007

[54] Cano-Flores A, Gómez X, Ramos R. Biotransformation of steroids using different microorganisms. In: JAR S, editor. Chemistry and Biological Activity of Steroids. Rijeka: IntechOpen; 2018. DOI: 10.5772/intechopen.85849

[55] Peart PC, McCook KP, Rusell FA, Reynolds WF, Reese PB. Hydroxylation of steroids by *Fusarium oxysporum*, *Exophiala jeanselmei* and *Ceratocystis paradoxa*. Steroids. 2011;**76**: 1317-1330. DOI: 10.1016/j.steroids.2011.06.010

[56] Kristan K, Lanisnik Rizner T. Steroid-transforming in fungi. The Journal of Steroid Biochemistry and Molecular Biology. 2012;**129**:79-91. DOI: 10.1016/j.jsbmb.2011.08.012

[57] Fernandes P, Cruz A, Angelova B, Pinheiro HM, Cabral JMS. Microbial conversion of steroid compounds: Recent developments. Enzyme and Microbial Technology. 2003;**32**:688-705. DOI: 10.1016/S0141-0229(03)00029-2

[58] Suzuki K, Sanga K, Chikaoka Y, Itagaki E. Purification and properties of cytochrome P-450 (P-450lun) catalyzing steroid 11β-hydroxylation in *Curvularia lunata*. Biochimica et Biophysica Acta. 1993;**1203**:215-223. DOI: 10.1016/0167-4838(93)90086-7

[59] Lu W, Du L, Wang M, Jia X, Wen J, Huang Y, et al. Optimisation of hydrocortisone production by *Curvularia lunata*. Applied Biochemistry and Biotechnology. 2007; **142**:17-28. DOI: 10.1007/s12010-007-0005-8

[60] Bhosale S, Saratale G, Govindwar S. Biotransformation enzymes in *Cunninghamella blakesleeana* (NCIM-687). Journal of Basic Microbiology. 2006;**46**:444-448. DOI: 10.1002/jobm.200510117

[61] Petrica S, Hakkib T, Bernhardtb R, Zigonc D, Cresna B. Discovery of a steroid 11-hydroxylase from *Rhizopus oryzae* and its biotechnological application. Journal of Biotechnology. 2010;**150**:428-437. DOI: 10.1016/j.jbiotec.2010.09.928

[62] Schmitz D, Janocha S, Kiss FM, Berhardt R. CYP106A2—A versatile biocatalyst with high potential for biotechnological production of selectively hydroxylated steroid and terpenoid compounds. BBA Proteins and Proteomics. 2018;**1866**:11-22. DOI: 10.1016/j.bbapap.2017.07.011

[63] Berrie JR, Williams RAD, Smith KE. Microbial transformation of steroids—XI. Progesterone transformation by *Streptomyces roseochromogenes*—Purification and characterisation of the cytochrome P450 progesterone 16α-hydroxylase system. The Journal of Steroid Biochemistry and Molecular Biology. 1999;**71**:153-165

[64] Berrie JR, Williams RAD, Smith KE. Microbial transformation of steroids—XII. Progesterone hydroxylation profiles are modulated by post-translational modification of an electron transfer protein in *Streptomyces roseochromogenes*. The Journal of Steroid Biochemistry and Molecular Biology. 2001;**77**:87-96

[65] Tang R, Shen Y, Wang M, Zhou H, Zhao Y. Highly efficient synthesis of boldenone from androst-4-ene-3,17-dione by *Arthrobacter simplex* and *Pichia pastoris*. Process Biochemistry. 2019;**42**:933-940. DOI: 10.1007/s00449-019-02092-y

[66] Hunter AC, Khuenl-B RH, Barret P, Dodd HT, Dedi C. Transformation of some 3-substituted steroids by *Aspergillus tamarii* KITA reveals stereochemical restriction of steroid binding orientation in the minor hydroxylation pathway A. Journal of Steroid Biochemistry and Molecular Biology. 2010;**118**:171-176. DOI: 10.1016/j.jsbmb.2009.12.003

[67] Donova MV, Ergova OV. Microbial steroid transformations: Current state and prospect. Applied Microbiology and Biotechnology. 2012;**94**:1423-1447. DOI: 10.1007/978-1-4939-7183-1

[68] Sultana N. Microbial biotransformation of bioactive and clinically useful steroids and some salient features of steroids and biotransformation. Steroids. 2018;**136**:76-92. DOI: 10.1016/j.steroids.2018.01.007

[69] Tong W-Y, Dong X. Microbial biotransformation: Recent developments on steroid drugs. Recent Patents on Biotechnology. 2009;**3**:141-153

[70] Lin Y, Song X, Fu J, Lin J, Qu Y. Microbial transformation of androst-4-ene-3,17-dione by *Bordetella* sp B4 CGMC2229. Journal of Chemical Technology & Biotechnology. 2009;**84**: 789-793. DOI: 10.1002/jctb.2113

[71] Lobastova TG, Gulevskaya SA, Sukhodolskaya GV, Donova MV. Dihydroxylation of dehydroepiandrosterone in positions 7α and 15α by mycelial fungi. Applied Biochemistry and Microbiology. 2009; **45**:617-622

[72] Li H, Liu H-M, Ge W, Huang L, Shan L. Synthesis of 7α-hydroxy-dehydroepiandrosterone and 7α-hydroxy-dehydroepiandrosterone. Steroids. 2005;**70**:970-973

[73] Lobastova TG, Khomutov SM, Vasiljeva LL, Lapitskaya MA, Pivnitsky KK, Donova MV. Synthesis of 3β-hydroxy-androsta-5,7-dien-17-one from 3β-hydroxyandrost-5-en-17-one via microbial 7α-hydroxylation. Steroid. 2009;**74**:233-237. DOI: 10.1016/j.steroids.2008.10.019

[74] Al-Awadi S, Afzal M, Oommen S. Studies of *Bacillus stearothermophilus*. Part III. Transformation of testosterone. Applied Microbiology and Biotechnology. 2003;**62**:48-52. DOI: 10.1007/s00253-003-1269-8

[75] Al-Awadi S, Afzal M, Oommen S. Studies on *Bacillus stearothermophilus*. Part IV. Influence of enhancers on biotransformation of testosterone. Steroids. 2005;**70**:323-327

[76] Janeczkoa T, Dmochowska-Gladysz J, Kostrezewa-Suslowa E, Bialonskab A, Ciunikb Z. Biotransformations of steroid compounds by *Chaetomium* sp. KCH 6651. Steroids. 2009;**74**:657-660. DOI: 10.1016/j.steroids.2009.02.006

[77] Romanoa A, Romanoa D, Ragg E, Costantinoc F, Lennac R, Gandolfid R, et al. Steroid hydroxylations with *Botryodiplodia malorum* and *Colletotrichum lini*. Steroid. 2006;**71**: 429-434. DOI: 10.1016/j.steroids.2006.01.014

[78] Lisowska K, Dlugonski J. Concurrent corticosteroid and phenanthrene transformation by filamentous fungus *Cunninghamella elegans*. The Journal of Steroid Biochemistry and Molecular Biology. 2003;**85**:63-69

[79] Manosroi J, Chisti Y, Manosroi A. Biotransformation of cortexolone to hydrocortisone by molds using a rapid color-development assay. Applied Biochemistry and Microbiology. 2006; **42**:479-483

[80] Wu Y, Li H, Zhang X-M, Gong JS, Rao Z-M, Shi J-S, et al. Efficient hydroxylation of functionalized steroids by *Colletotrichum lini* ST-1. Journal of Molecular Catalysis B: Enzymatic. 2015; **120**:111-118. DOI: 10.1016/j.molcatb.2015.07.003

[81] Agnieszka Bartmanska A, Dmochowska-Gładysz J. Transformation of steroids by *Trichoderma hamatum*. Enzyme and Microbial Technology. 2007;**40**: 1615-1621. DOI: 10.1016/j.enzmictec.2006.11.011

[82] Choudhary MI, Siddiqui ZA, Musharraf SG, Nawaz SA, Atta-ur-Rahman. Microbial transformation of prednisone. Natural Product Research. 2005;**19**: 311-317. DOI: 10.1080/14786410410001729168

[83] Cano A, Ramírez-Apan MT, Delgado G. Biotransformation of sclareolide by filamentous fungi: Cytotoxic evaluations of the derivates. Journal of the Brazilian Chemical Society. 2011;**22**:1177-1182. DOI: 10.1590/S0103-50532011000600025

[84] Atta-ur-Rahman, Farooq A, Choudhary MI. Microbial transformation of sclareolide. Journal of Natural Products. 1997;**60**:1038-1040. DOI: 10.1021/np970076h

[85] Choudhary MI, Musharraf SG, Sami A, Atta-ur-Rahman. Microbial transformation of sesquiterpenes, (−)-ambrox and (+)-sclareolide. Helvetica Chimica Acta. 2004;**87**:2685-2694

[86] Ata A, Conci LJ, Betteridge J, Orhan I, Sener B. Novel microbial transformations of sclareolide. Chemical and Pharmaceutical Bulletin. 2007;**55**: 118-123. DOI: 10.1248/cpb.55.118

[87] Cao H, Chen X, Jassbi AR, Xiao J. Microbial biotransformation of bioactive flavonoids. Biotechnology Advances. 2015;**33**:214-233. DOI: 10.1016/j.biotechadv.2014.10.012

[88] Andrae-Marobela K, Ghislain FW, Okatch H, Majinda R. Polyphenols: A diverse class of multi-target anti-HIV-1 agents. Current Drug Metabolism. 2013; **7**:392-413. DOI: 10.2174/ 13892002113149990095

[89] Liao XL, Luo JG, Kong LY. Flavonoids from *Millettia nitida* var. Hirsutissima with their anticoagulative activities and inhibitory effects on NO production. Journal of Natural Medicines. 2013;**67**:856-861. DOI: 10.1007/s11418-013-0745-4

[90] Ogungbe IV, Erwin WR, Setzer WN. Antileishmanial phytochemical phenolics: Molecular docking to potential protein targets. Journal of Molecular Graphics & Modelling. 2014; **48**:105-117. DOI: 10.1016/j.jmgm. 2013.12.010

[91] Kim H, Bartley GE, Arvik T, Lipson R, Nah SY, Seo K, et al. Dietary supplementation of chardonnay grape seed flour reduces plasma cholesterol concentration, hepatic steatosis, and abdominal fat content in high-fat diet-induced obese hamsters. Journal of Agricultural and Food Chemistry. 2014; **62**:1919-1925. DOI: 10.1021/jf404832s

[92] Asha S, Vidyavathi M. *Cunninghamella*—A microbial model for drug metabolism studies—A review. Biotechnology Advances. 2009;**27**:16-29. DOI: 10.1016/j.biotechadv.2008.07.005

[93] Alarcón J, Alderete J, Escobar C, Araya R, Cespedes CL. *Aspergillus niger* catalyzes the synthesis of flavonoids from chalcones. Biocatalysis and Biotransformation. 2013;**31**:160-167. DOI: 10.3109/10242422.2013.813489

[94] Sordon S, Madej A, Poplonski J, Bartmanska A, Tronina T, Brzezowska E, et al. Regioselective ortho-hydroxylations of flavonoids by yeast. Journal of Agricultural and Food Chemistry. 2016;**64**:5525-5530. DOI: 10.1021/acs.jafc.6b02210

[95] Bartmanska A, Huszcza E, Tronina T. Transformation of isoxanthohumol by fungfi. Journal of Molecular Catalysis B: Enzymatic. 2009; **61**:221-224. DOI: 10.1016/j. molcatb.2009.07.008

[96] Chen X, Mukwaya E, Wong MS, Zhang Y. A systematic review on biological activities of prenylated flavonoids. Pharmaceutical Biology. 2014;**52**:655-660. DOI: 10.3109/ 13880209.2013.853809

[97] Cao H, Chen X, Jassbi AR, Xiao J. Microbial transformation of bioactive flavonoids. Biotechnology Advances. 2015;**33**:214-223. DOI: 10.1016/j. biotechadv.2014.10.012

[98] Bartmanska A, Tronina T, Huszcza E. Transformation of 8-prenylnaringenin by *Absidia coerula* and *Beauveria bassiana*. Bioorganic & Medicinal Chemistry Letters. 2012;**22**: 6451-6453. DOI: 10.1016/j. bmcl.2012.08.060

[99] Tronina T, Strugala P, Poplonski J, Wloch SS, Bartmanska A, Huszcza E. The influence of glycosylation of natural and synthetic prenylated flavonoids on binding to human serum albumin and inhibition of cyclooxygenases COX-1 and COX-2. Molecules. 2017;**22**: 1230-1250. DOI: 10.3390/molecules22071230

[100] Haraguchi H, Inoue J, Tamura Y, Mizutani K. Antioxidative components of *Psoralea corylifolia* (Leguminosae). Phytotherapy Research. 2002;**16**:539e544. DOI: 10.1002/ptr.972

[101] Lim SH, Ha TY, Ahn J, Kim S. Estrogenic activities of *Psoralea corylifolia* L, seed extracts and main constituents. Phytomedicine. 2011;**18**: 425-430. DOI: 10.1016/j.phymed.2011.02.002

[102] Luo J, Liang Q, Shen Y, Chen X, Yin Z, Wang M. Biotransformation of bavachinin by three fungal cell cultures. Journal of Bioscience and Bioengineering. 2014;**117**:191-196. DOI: 10.1016/j.jbiosc.2013.08.001

[103] Tronina T, Bartmanska A, Milczarek M, Wietrzyk J, Poplonski J, Roj E, et al. Antioxidant and antiproliferative activity of glycosides obtained by biotransformation of xanthohumol. Bioorganic & Medicinal Chemistry Letters. 2013;**23**: 1957-1960. DOI: 10.1016/j.bmcl.2013.02.031

[104] Kim HJ, Lee I-S. Microbial metabolism of the prenylated chalcone xanthohumol. Journal of Natural Products. 2006;**69**:1522-1524. DOI: 10.1021/np060310g

[105] Bartmanska A, Huszcza E, Tronina T. Transformation of isoxanthohumol by fungi. Journal of Molecular Catalysis B: Enzymatic. 2009; **61**:221-224. DOI: 10.1016/j.molcatb.2009.07.008

[106] Tronina T, Bartmanska A, Poplonski J, Huszcza E. Transformation of xanthohumol by *Aspergillus ochraceus*. Journal of Basic Microbiology. 2014;**54**: 66-71. DOI: 10.1002/jobm201200320

[107] Tronina T, Bartmanska A, Filip-Psurska B, Wietrzyk J, Poplonski J, Huszcza EA. Fungal metabolites of xanthohumol with potent antiproliferative activity on human cancer cell lines in vitro. Bioorganic & Medicinal Chemistry. 2013;**21**: 2001-2006. DOI: 10.1016/j.bmc.2013.01.026

[108] Stompor M, Potaniec B, Szumny A, Zieliński P, Żołnierczyk AK, Anioł M. Microbial synthesis of dihydrochalcones using Rhodococcus and Gordonia species. Journal of Molecular Catalysis B: Enzymatic. 2013a;**97**:283-288

[109] Abourashed EA, Mikell JR, Khan IA. Bioconversion of silybin to phase I and II microbial metabolites with retained antioxidant activity. Bioorganic & Medicinal Chemistry. 2012;**20**:2784-2788. DOI: 10.1016/j.bmc.2012.03.046

[110] Choi J-M, Han S-S, Kim H-S. Industrial applications of enzyme biocatalysis: Current status and future aspects. Biotechnology Advances. 2015; **33**:1443-1454. DOI: 10.1016/j.biotechadv.2015.02.014

[111] De Carvalho CCCR. Enzymatic and whole cell catalysis: Finding new strategies for old processes. Biotechnology Advances. 2011;**29**:75-83

[112] Muñoz SD, Hoyos P, Hernáiz MJ, Alcántara AR, Sánchez-Moreno JM. Industrial biotransformations in the synthesis of building blocks leading to enantiopure drugs. Bioresource Technology. 2012;**115**:196-207. DOI: 10.1016/j.biortech.2011.11.131

[113] Patel RN. Biocatalysis for synthesis of pharmaceuticals. Bioorganic &

Medicinal Chemistry. 2018;**26**: 1252-1274. DOI: 10.1016/j. bmc.2017.05.023

[114] Van Den Burg B. Extremophiles as a source for novel enzymes. Current Opinion in Microbiology. 2003;**6**: 213-218. DOI: 10.1016/S1369-5274(03) 00060-2

[115] Antranikian G, Vorgias CE, Bertoldo C. Extreme environments as a resource for microorganisms and novel biocatalysts. Advances in Biochemical Engineering/Biotechnology. 2005;**96**: 219-262. DOI: 10.1007/b135786

[116] Elleuche S, Schroder C, Sahm K, Antranikian G. Extremozymes-biocatalysts with unique properties from extremophilic microorganisms. Current Opinion in Biotechnology. 2014;**29**:116-123. DOI: 10.1016/j. copbio.2014.04.003

[117] Falcicchio P, Wolterink-Van Loo S, Franssen MC, van der Oost J. DHAP-dependent aldolases from (hyper) thermophiles: Biochemistry and applications. Extremophiles. 2014;**18**: 1-13. DOI: 10.1007/s00792-013-0593-x

[118] Urbieta MS, Donati ER, Chan KG, Shahar S, Li Sin L, Goh KM. Thermophiles in the genomic era: Biodiversity, science, and applications. Biotechnology Advances. 2015;**33**: 633-647. DOI: 10.1016/j. biotechadv.2015.04.007

[119] Woodward J, Orr M, Cordray K, Greenbaum E. Biotechnology: Enzymatic production of biohydrogen. Nature. 2000;**405**:1014-1015

[120] Soliman N, Abdel-Fattah Y, Mostafa HE, Gaballa A. Heterologous expression of thermostable esterase gene from *Geobacillus thermoleovorans* YN under different expression promoters. International journal of Environmental Science and Technology. 2014;**11**:119-126. DOI: 10.1007/ s13762-013-0360-7

[121] Saranya P, Kumari HS, Rao BP, Sekaran G. Lipase production froma novel thermo-tolerant and extreme acidophile *Bacillus pumilus* using palmoil as the substrate and treatment of palm oil-containing wastewater. Environmental Science and Pollution Research International. 2014;**21**: 3907-3919. DOI: 10.1007/s11356-013-2354-x

[122] Gudiukaitė R, Gegeckas A, Kazlauskas D, Citavicius D. Influence of N-and/or C-terminal regions on activity, expression, characteristics and structure of lipase from *Geobacillus* sp. 95. Extremophiles. 2014;**18**:131-145. DOI: 10.1007/s00792-013-0605-x

[123] Abraham J, Gea T, Sánchez A. Substitution of chemical dehairing by proteases from solid state fermentation of hair wastes. Journal of Cleaner Production. 2014;**74**:191-198

[124] Staufenberger T, Imhoff JF, Labes A. First crenarchaeal chitinase found in *Sulfolobus tokodaii*. Microbiological Research. 2012;**167**: 262-269

[125] Barabote RD, Parales JV, Guo Y-Y, Labavitch JM, Parales RE, Berry AM. Xyn10A, a thermostable endoxylanase from *Acidothermus cellulolyticus* 11B. Applied and Environmental Microbiology. 2010;**76**:7363-7366

[126] Kang J, Park K-M, Choi K-H, Park C-S, Kim G-E, Kim D, et al. Molecular cloning and biochemical characterization of a heat-stable type I pullulanase from *Thermotoga neapolitana*. Enzyme and Microbial Technology. 2011;**48**:260-266